AF768291

LOGIC, EPISTEMOLOGY, AND THE UNITY OF SCIENCE

VOLUME 3

Logic, Epistemology, and the Unity of Science aims to reconsider the question of the unity of science in light of recent developments in logic. At present, no single logical, semantical or methodological framework dominates the philosophy of science. However, the editors of this series believe that formal techniques like, for example, independence friendly logic, dialogical logics, multimodal logics, game theoretic semantics and linear logics, have the potential to cast new light on basic issues in the discussion of the unity of science.
This series provides a venue where philosophers and logicians can apply specific technical insights to fundamental philosophical problems. While the series is open to a wide variety of perspectives, including the study and analysis of argumentation and the critical discussion of the relationship between logic and the philosophy of science, the aim is to provide an integrated picture of the scientific enterprise in all its diversity.

The Age of Alternative Logics

Assessing Philosophy of Logic and Mathematics Today

Edited by

Johan van Benthem
ILLC, University of Amsterdam, The Netherlands

Gerhard Heinzmann
Archives Henri Poincaré, Nancy University, France

Manuel Rebuschi
Archives Henri Poincaré, Nancy University, France

and

Henk Visser
ILLC, University of Amsterdam, The Netherlands

Springer

A C.I.P. Catalogue record for this book is available from the Library of Congress.

ISBN-10 1-4020-5011-9 (HB)
ISBN-13 978-1-4020-5011-4 (HB)
ISBN-10 1-4020-5012-7 (e-book)
ISBN-13 978-1-4020-5012-7 (e-book)

Published by Springer,
P.O. Box 17, 3300 AA Dordrecht, The Netherlands.

www.springer.com

Printed on acid-free paper

CONTENTS

ACKNOWLEDGMENTS

The conference on *Philosophical Insights into Logic and Mathematics* (PILM for short), of which this book is the outcome, would not have been possible without the generous financial support of the following institutions: Evert Willem Beth Foundation (Amsterdam), Centre National de la Recherche Scientifique (CNRS), Conseil Régional de Lorraine, Conseil Général de Meurthe et Moselle, Communauté Urbaine du Grand Nancy, Université Nancy 2, Université Henri Poincaré, Institut National Polytechnique de Lorraine, Laboratoire de Philosophie et d'Histoire des Sciences – Archives Henri-Poincaré (UMR 7117), Institute for Logic, Language and Computation (ILLC Amsterdam), Loria, Goethe Institut (Nancy). To all these institutions we express our warm gratitude. We are also grateful to the members of the PILM scientific committee for their invaluable help in preparing the program and reporting on so many lectures, as well as to the staff of the Poincaré Archives for their help in the preparation of the conference and their expert assistance in various ways. We particularly express our gratitude to Dr. Prosper Doh (Poincaré Archives) who has taken on the job of technical editor for this book and who has realized the index and the camera-ready copy. Finally, the editors are indebted to the editorial board of "Logic, Epistemology, and the Unity of Science" for accepting this volume in their series. They would also like to thank Springer Publishers and, in particular, Floor Oosting.

Chapter 1

INTRODUCTION: ALTERNATIVE LOGICS AND CLASSICAL CONCERNS

Johan van Benthem
University of Amsterdam/ILLC
johan@csli.stanford.edu, johan@science.uva.nl

Modern logic shows a wide variety of perspectives, application areas, and formal systems, which often go under the heading of 'alternative logics'. The lively PILM conference held in Nancy during September 2002 on which the present volume is based was intended as an encounter between modern work on alternative logics and classical issues in the foundations of mathematics and the philosophy of logic. This book contains a substantial sample of what happened in the process, but it can also be read independently as a report on the state of the art.

Actually, terms like 'alternative' or 'non-classical' logic can easily be misunderstood. Our aim with the conference and this book is not a simplistic endorsement of mass-production and 'anything goes' in logic. In particular, we do not aim for 'alternative' pop-art versions of the grand issues that initiated 20th century logic, whose praises used to be sung in the measured classical strains of Bach and Beethoven - or perhaps Wagnerian doom in times of foundational crisis. There is no need to replace these by versions in modern logics that sound of jazz, rock, or (as some critics would probably have it) punk, disco, or rap. To us, the diversity of logical systems today rather signals a natural and respectable process of growth of the discipline, not of replacement or competition. In terms of our musical metaphor, this development transforms rigid classical partitions into a more open playground for improvisation.

Some of the forces driving this process of growth are external. Over time, and throughout the last century, logic has been confronted with a growing set

J. van Benthem, G. Heinzmann, M. Rebuschi and H. Visser (eds.), The Age of Alternative Logics: Assessing Philosophy of Logic and Mathematics Today, 1–7.

of congenial developments in neighbouring disciplines, including mathematics and philosophy, but also physics, computer science, economics, or linguistics. The result has been an influx of new ideas, concerns, and logical systems reflecting a great variety of reasoning tasks. For instance, modern epistemic and dynamic logics have received their major impetus from studies of information, computation, and action beyond classical foundational concerns, paraconsistent logics arise from taking actual human argumentation seriously, while, say, non-monotonic logics reflect basic features of common sense reasoning in artificial intelligence. In tandem with these external influences, there has also been an internal dynamics of the field of logic, with its own intriguing shifts in the agenda and the way one views classical issues and results. To see such trends, one needs to step back and consider the flow of ideas over several decades. For instance, consider 'constructivism' and intuitionistic logic, the only 'alternative logic' that has been accorded somewhat of a respectable status within the classical tradition, if only as a well-encapsulated former rebel. Intuitionistic logic is about the computational content of mathematical proofs, and the epistemic status of mathematical statements. But over time, this topic has merged into the more general algorithmic study of proof and type structure, e.g. via the Curry-Howard isomorphism, and eventually into the most general 'dynamic logic' of all, viz. category theory - perhaps the most powerful alternative foundational paradigm to-day. The continued vitality of this way of thinking shows in linear logic, whose fresh look at proof structure found surprising new structure levels of computation and interaction underneath classical and intuitionistic logic. Likewise, on the epistemic side, intuitionistic logic may now be viewed as part of much more general study of arbitrary information-based assertions, which takes us again to epistemic logic, and logics for update or revision actions that add or modify information. Moreover, the two aspects: dynamic and epistemic, still come together naturally in many settings beyond the original foundations of mathematics. A prime example of this are games. Games involve a tight interplay of what agents know and how they act, and the rise of this paradigm inside logic is unmistakable. But note again that this development also involves a major extension of a classical viewpoint. Game are typically an interactive process involving several agents, and indeed, many issues in logic to-day are no longer about zero-agent notions like truth, or single-agent notions like proof, but rather about processes of verification, argumentation, communication, or general interaction.

Of course, what may - and sometimes does - happen is that the broader canvas of growth in logical issues, notions, and formal systems also changes our perspective of the traditional logical core. For instance, we understand first-order logic much better today precisely because of the emergence of variations and alternatives. Some would even argue that we are now in a position to rethink historical decisions made around 1900, and redesign predicate logic in a more powerful 'independence-friendly' style. And there are similar claims about redesigning traditional set theory as a foundation for mathematics, given the wealth of experience with category-theoretic alternatives.

Against this background of growth and reflection, we have grouped our various chapters under the following five systematic headings, with individual contributions ordered alphabetically by author. Across all these chapters, the reader will see a great variety of sources of inspiration at work, both external and internal in the above sense.

Our first group of papers falls under the heading of Proof, Knowledge and Computation. For a start, Cozic discusses the vexed problem of omniscience in epistemic logic, i.e., the unrealistic feature of the usual formal systems that agents automatically know all logical consequences of what they know. He proposes putting epistemic inference on a more discriminating base in substructural logics, with intuitionistic logic eventually serving as a reasonable compromise between classical and linear proof systems. Peregrin formally demonstrates another role of inference, viz. its potential for generating and justifying a semantics for connectives - as has been suggested more informally by modern evidentialist philosophers like Brandom. He also extends this to packages of abstract substructural rules for dynamic inference. Turning to computation, Shapiro discusses reasons that led to the historical emergence of the logical interest in computability in the 1930s - including its not wholly straightforward relationship with Hilbert's program and the incompleteness theorems. In the end, he argues against the extensionalist turn of Church's Thesis, preferring the historically prior 'intensional' conception of computation, which seems to call for a mixture of classical and constructive mathematics, as evidenced in his own work on 'epistemic mathematics'. Going back to the same classical period, Vidal-Rosset discusses the current status of Gödel's theorems, the famous limiting results on computability and provability, comparing various stances in the most recent literature on what these theorems really say. He concludes that there are only two 'stable' interpretations. One is the Platonist stance, reading the theorems as saying that truth transcends proof, the other is the Deflationist stance, which denies any transcendent realm of truth against which formal proof systems fall short: the only reality are proof systems of increasing strength. Finally, Visser looks at proof and computation in the very practical setting of mathematical problem solving, in a style which mixes contexts of discovery and justification in fresh ways. He uses a new method for solving concrete mathematical problems, such as organizing a tournament, showing the interplay of numerical and sometimes quite surprising geometrical representations in finding optimal solutions. This leads to some new issues about finite geometries, which have already inspired some new computational work. Such 'transpositions' in representations of a given problem seem essential to understanding the real workings of mathematical reasoning.

The second group of papers illustrates another perennial topic, viz. Truth Values Beyond Bivalence. Starting from a critique of possible worlds semantics for modal logic, Béziau revives many-valued models for intensional languages. In particular, he shows how these can sometimes replace possible-worlds models, while they can also co-exist quite well with them in richer settings. Da Costa and Krause give a new twist to the usual development of para-consistent logics allowing contradictions, by bringing in motivations from the foundations of physics. Their main concern here is the proper logic of reasoning about complementarity in quantum mechanics. Following a lucid discussion of major philosophers of science on the latter topic, they define complementarity more formally, and then propose para-consistent systems that capture styles of inference in line with Bohr's thinking. Indeed, many-valued logics of various sorts are making a come-back theses days across a broad range of applications, including pure mathematics, e.g., in recent work by Hajek and Mundici. Libert shows how set theory can be developed consistently in fascinating new ways based on the well-known Lukasiewicz logic with continuum many values, while keeping unrestricted comprehension axioms valid. This line of research was initiated by Skolem, and it also turns out to link up eventually with other major developments in alternative, such as Scott models and category theory.

This brings us to another pervasive theme in many of our papers, viz. Category-Theoretic Structures. A lucid introduction to the connection between category theory and higher order logic in its 'general models' guise is given by Awodey, who also explains how category theory replaces the standard notion of set by a mode dynamic notion of 'continuously variable set'. Interestingly, famous category-theoretic results like De Ligne's Theorem acquire deep logical import. The major claim is that higher-order logic indeed is the logic of continuous variation. But Hellman is critical of any foundationalist claims of category theory. He discusses two versions of these, one casting category theory in the Fregean mode of being about some genuine universe of objects (the category of categories), others more Hilbertian, casting category theory as a methodological recipe for deduction according to some useful primitive notions and axioms. Both seem to presuppose sets at some level. Hellman's own proposal is to merge both stances, providing an anchoring for mathematics in a more abstract 'theory of large domains' based on part-whole relationships and plural quantification. In this ongoing discussion of the true foundational role of category theory, our remaining two authors introduce refinements. Landry finds the usual discussions between faithful and critics like Feferman in terms of the privileged underlying mathematical objects misguided. category theory does not intend to provide a foundation in this sense, but it is rather a scheme for bringing out shared structure of families of abstract mathematical systems. More philosophically, she argues for an 'in re' view of structuralism, informed by a similar Aristotelean notion and close to Carnap's use of conceptual 'frameworks', that would be tenable against criticisms of the Hellman-Feferman type. Finally, Marquis, too, denies that categories necessarily presuppose some informal conception of set. Instead, he emphasizes a new mathematical universe based on a new hierarchy of categories, emerging in the work of Makkai, with sets just forming the base level. He claims that the usual epistemic or methodological objections to category theory evaporate on this new conception, while admitting that many technicalities remain to be explored, as this is certainly not 'business as usual', even for died-in-the-wool category theorists.

Our last two groups of papers have to do with the game-theoretic perspective on logical and mathematical activity, as involving interactions between various players with different 'logical roles'.

The part on Independence, Evaluation Games, and Imperfect Information is mainly devoted to current game-theoretical semantics for quantifiers and connectives that spring the bounds of the linear dependence format of classical systems. First, Hintikka explains the main features and guiding motivations of his 'IF-logic', showing how replacing classical first-order logic by an 'information-friendly' version leads to more delicate accounts of quantification, negation and truth, with repercussions in linguistics, computation, and indeed the foundations of mathematics. Independence in the IF sense can be explained in game-theoretic terms, using a feature different from all usual 'logic games', viz. imperfect information: players need not know all precise moves played previously by their opponents. The result is a much richer pattern of dependence and independence between variables in formal reasoning than that provided by classical logic. For a precise statement of Hintikka's current views on the status of IF-logic, and the extent of its 'gaminess', we refer to his chapter. Next, Janssen and Dechesne show that the game-theoretic account of IF logics is not as perspicuous as it might appear at first sight. In particular, the phenomenon of 'signalling' in games, i.e., passing information to oneself in indirect ways, seems to play havoc with many intuitive claims that have been made about IF logic, and it even invalidates published technical statements about it by Caicedo & Krynicki. The most startling claim in the paper is that IF logic, on Janssen & Dechesne's way of taking it, is not conservative over classical first-order logic... Continuing with issues of information flow in games, Pietarinen takes the logic-game theory interface even further than players' being ignorant of each other's precise moves. Notably, he considers logical effects of other forms of imperfect information, viz. when players do not know exactly which game they are in. This quite plausible extension considerably extends the set of shared concerns between logic and the general study of planned human behaviour in economics or artificial intelligence. Finally, Rebuschi connects IF logic with the earlier topic of epistemic logic. He discusses the connection between global IF games, where only winning strategies matters, and the fine-structure analysis of moves in extensive game trees provided by the 'epistemic action logic' EAL proposed by van Benthem as a more classical analysis of IF logic. Rebuschi's eventual proposal is to use an IF version of EAL, without incurring the potential infinite regress lurking in iterated strings of system building 'IF(EAL(IF))'...

Another, and in some ways older, use of games in logic comes through the Lorenzen program from the 1950s. In our final part on Dialogue and Pragmatics, three authors provide modern views on this enterprise and its continued potential. Heinzmann discusses how the dialogical pragmatic stance changes our conceptions of epistemological and ontological issues in the philosophy of mathematics. He provides a broad overview of various brands of nominalism versus realism through the 20th century, finding most of them wanting. He eventually endorses four guiding desiderata stated by Gonseth for any analysis of the mathematical activity, emphasizing broad features of 'duality', 'revisability', 'technicity', and 'solidarity'. Modern alternative systems, like Hintikka's game logics, or the Feferman/Hellman view of sets, have the potential of doing better in this respect by respecting the required pragmatic dynamics. Next, Lorenz, one of the pioneers of dialogical logic, looks at semiotic aspects of language, and shows how taking an speaker/interpreter game perspective helps understand the duality between the representational and communicative function of language. One of the noticeable outcomes is a rich theory of predication and 'indication'. Finally, and more technically, Rahman demonstrates the flexibility of the dialogical paradigm by providing a game-based modelling for derivations in non-normal modal logics, which go beyond standard possible worlds semantics in various ways. He also shows how to approach the expressive powers of recent 'hybrid languages' raising the expressive power of the basic modal language, thereby bringing dialogical logic in touch with state-of-the-art modal logic.

This concludes our brief summary of the contents of this book. We hope that the chapters testify to the liveliness of modern perspectives on the philosophy of logic and mathematics. Naturally, other principles of division would have been possible, e.g. by area of application. But the very fact that various groupings make sense itself speaks to the coherence of the material presented here. The field is not haphazard, but it is held together by a multi-dimensional network of concerns.

A final aspect of bringing together all this material is the better view of possible developments and desiderata for research. We conclude with a few examples of this.

At the start, we dismissed the idea of looking at the grand classical concerns with new-fangled, perhaps even light-weight, alternative tools. But of course, there is nothing wrong with asking after all whether alternative logics dealing with such issues as knowledge, non-monotonic reasoning, update and revision dynamics, or general game-theoretic interaction, have direct concrete uses in the analysis of mathematics and other areas in the logical heartland. Such analysis might serve to bring out features of mathematical proof that have been neglected so far, while also showing the unity of human intelligence, or - in Clausewitz's immortal turn of phrase - that 'mathematics is just common sense continued by other means'.

Other conspicuous questions raised by our collection have to do with possible unification of the various perspectives represented in this book. In particular, our two main headings of games and categories suggest some tantalizing analogies. Categories make the mathematical universe into a dynamic whole of primitive objects and primitive functions, emphasizing transitions from one representation structure to another. This dynamic intuition seems close in spirit to modern epistemic and dynamic logics that put logical activities at centre stage, e.g., in games. For instance, categorical arrow composition seems close to successive action composition, the engine of information update. In fact, it is known that certain sorts of games form natural categories, e.g., in the semantics of linear logic. But a unified picture still eludes us, as the various senses of dynamics (proof-theoretic, modal) do not all seem to lie at the same conceptual level.

Finally, at the strategic distance provided by this book, we can see new trends now, and natural new issues not normally raised in the standard agenda of the philosophy of logic and mathematics. For instance, the multi-agent shift in the game perspective also represents a major shift in conceptualizing the basic tasks of logic. The usual paradigm of a logical task is that of an inference or a proof, viewed either as a product, or as an activity by a single agent. But from a truly interactive perspective, an irreducibly two-agent episode like asking a question and giving an answer may be just as paradigmatic an example of a 'logical' activity! Thus, the set of defining concerns for our field may still undergo major changes, as the effects of external and internal dynamics of logic keep making themselves felt.

Of course, all these lively messages of current activity and future growth are what inspired the PILM organizers and editors in undertaking this not undemanding job. We will be amply rewarded if some of this enthusiasm transfers to our readers!

I

PROOF, KNOWLEDGE AND COMPUTATION

Chapter 2

EPISTEMIC MODELS, LOGICAL MONOTONY AND SUBSTRUCTURAL LOGICS*

Mikaël Cozic
Paris IV-Sorbonne/IHPST (CNRS)
mikael.cozic@paris4.sorbonne.fr

2.1 Introduction: logical omniscience and logical monotony

Suppose that a modeller wants to represent the cognitive state (the set of beliefs) of a given reasoner. If this modeller uses a common epistemic logic (i.e., a normal modal logic), even in its weaker form (the so-called system K), his or her model will *necessarily* ascribe to the reasoner a set of beliefs closed under the consequence relation of classical logic. In the literature, this phenomenon is called the problem of logical omniscience (PLO). It is worth noting that this is not an isolated phenomenon: beyond epistemic logic, a vast range of "epistemic models" (that is, formal models of knowledge and belief) like probability theory or belief revision exhibits an analogous form of closure:[1]

$$\frac{A \rightarrow B}{B_i A \rightarrow B_i B}$$

(Rule of Epistemic Monotony)

*This paper is based on an essay written for the DEA of Cognitive Science 2001 (Paris). I am grateful to the supervisors and referees of this essay: D. Andler, J. Dubucs, J-P. Dupuy and B. Walliser. I especially would like to thank D. Bonnay, P. Egré and P. Gochet for their comments and criticisms and C. Hill for help in translation. I am also grateful to the participants of the groups "Philosophie formelle" (IHPST, Paris), "Economie Cognitive" (CNRS, Paris) and the colloquiums "Logique et rationalité" (Paris, March 2002) and, of course, PILM (Nancy, October 2002).

[1]In the standard formalisms, $B_i A$ means that the reasoner i believes that A, $L_i^{\alpha} A$ means that the proposition expressed by A has at least probability α for i and $A \in K_i$ means that the proposition expressed by A is in the belief set of i.

J. van Benthem, G. Heinzmann, M. Rebuschi and H. Visser (eds.), The Age of Alternative Logics: Assessing Philosophy of Logic and Mathematics Today, 11–23.

$$\frac{A \rightarrow B}{L_i^\alpha A \rightarrow L_i^\alpha B}$$

(Rule of Probabilistic Monotony)

If $\vdash_{CL} A \rightarrow B$ then if $A \in K_i$, $B \in K_i$

(Monotony of Revision Property)

All these inference rules are valid in the corresponding epistemic models. All express a form of closure that might be called "logical monotony" and all are, therefore, *uneliminable* assumptions of such models. Logical omniscience is then the particular instance of logical monotony in the case of epistemic logic, and its importance comes first from the fact that it is a simple and representative instance of logical monotony. Since the seminal work of [J. Hintikka 1962], lots of solutions have been defended to solve the (PLO),[2] but there is little consensus as to which are the best; there is even more little consensus as to what would be a good solution to (PLO).

The aim of this paper is, following [J. Dubucs 1991] and [J. Dubucs 2002], to defend a family of proof-oriented solutions to the (PLO) starting from a conceptual analysis of the solutions' space, that is, the aim of the paper is to characterize what would be a good solution to (PLO) and then to propose some logics as solutions to (PLO).

The remainder of the paper proceeds as follows. Section 2 puts some constraints on the solutions' space. This results in a criterion of cognitive realism called the Principle of Epistemic Preservation (PEP). In Section 3, I shall claim that two proposals are more adequate to (PEP) than classical epistemic logic (CEL). Those proposals will be discussed in Section 4. I conclude in Section 5.

2.2 Looking for a better epistemic logic: preliminary steps

2.2.1 The core of (PLO)

There exists today a huge family of alternative epistemic logics that have been devised in order to solve the (PLO). They are characterized by the failure of closure under the classical consequence relation. Among them, the two main proposals are the logic of awareness (AEL) and the logic of impossible worlds (IWEL). In the first case, a set of formulas $\mathcal{A}(s)$ is associated to each world s of the state space S and a formula $B_i A$ is true in s iff in every world s' accessible from s A is true and $A \in \mathcal{A}(s)$. In the second case, a set of "impossible worlds" is added. In those worlds, there are no constraints on the valuation of formulas (e.g. in the same impossible world, it is possible that A and $\neg A$ are

[2] See [R. Fagin *et al.* 1995], chap. 9.

true, $A \wedge B$ is true but $B \wedge A$ false, etc.) The point is that there cannot exist more powerful solutions to (PLO) because epistemic logics where the deductive ability is weaker cannot exist. With (IWEL) or (AEL), one might represent the beliefs of reasoners who believe in a set Γ of formulas, but who do not believe any logical consequence of Γ.[3] Hence, the difficulty with (PLO) is not to find powerful enough alternative logics, but to find *good* alternative logics. In this section, my purpose is to define what is a good (or a at least a better) epistemic logic compared to the classical one. Many reasons can bring a reasoner not to believe a consequence A of a set of beliefs Γ. At one extreme, it might be that A is a trivial consequence of Γ, but that the reasoner reasons very poorly or does not pay attention; at the other extreme, it might be that there does not exist any systematic procedure to go from Γ to A. Clearly, the second is a more essential reason, whereas the first is more contingent (a Chomskyan linguist would perhaps say an "error of performance"). My first claim is that the aim of a solution to (PLO) is to capture the latter kind of reasons and to abstract from the former one. This point has several consequences.

First of all, an adequate epistemic logic should not only be deductively weaker than (CEL) but should exhibit such a weakening for essential reasons. (AEL) and (IWEL) do not fit this requirement. In (AEL), a reasoner does not believe in a consequence A of Γ only if the formula A is not a formula of which he is aware; in (IWEL), an agent does not believe in a consequence A of Γ only if A is false in some accessible impossible world. And A is false in some accessible impossible world only if the meaning of some logical connective *changes* with respect to the "true" possible worlds. Hence (AEL) and (IWEL) are not good solutions to (PLO) because neither morphological availability (a formula is morphologically available if the reasoner is aware of its existence) nor the changing nature of connectives are likely to be the essential reasons of bounded deductive ability. On the contrary, it seems to me reasonable to assume that reasoners have a minimally correct understanding of logical connectives. This last requirement, admittedly vague for the moment, can be called the Minimal Rationality Requirement (MRR). To sum up, a good solution to (PLO) has to deal with the core of (PLO), that is, it has to concern the ability to draw inferences. A second consequence is that a "good" epistemic logic will still involve a large measure of idealization with respect to the reasoners' actual deductive behaviors. This is not a bad point because the constitutive assumption of epistemic logic is arguably the fact that cognition often fit the logical standard. Hence, the idea of a *base logic*, that is a logic with respect to which an agent is omniscient, should not be rejected, but it is unreasonable to assume that this base logic is, as in (CEL), classical logic.

[3]See [H. Wansing 1990].

2.2.2 The Principle of Epistemic Preservation

How is one to go beyond this diagnosis ? The syntactic format of usual epistemic logics is the axiomatic or "Hilbert-style" format. It is probably heuristically inadequate because the focusing on information processing leads us to see logic as a set of rules of reasoning more than as a body of abstract truth. The first thing to do is then to move from this format to a "rule-based" format like Natural Deduction (ND) or Sequent Calculus (SC). However, at this stage, one might imagine two distinct approaches: a quantitative one and a qualitative one. In the quantitative approach, one keeps the classical rules, but one restricts the cognitive complexity (e.g. the size) of the possible proofs based on them. A brutal way of implementing this approach could be the following one: given a set of beliefs Γ, only the formulas deducible using proof of size smaller than k are ascribed to the agent. This is not the approach that this paper will defend. In the qualitative approach, one scrutinizes the rules themselves. It is precisely this qualitative approach that I would like to investigate here.

The rules are the basic components of the modeller's predictions concerning the reasoners' cognitive behaviour in the sense that given a rule $(r*)$: $A_1, ..., A_n \vdash B$ of the base logic, if the modeller ascribes $A_1, ..., A_n$ to a reasoner, he or she will necessarily ascribe B too. Thus one way to proceed would be to test the cognitive realism of rules separately. The cognitive realism of a rule $(r*)$ is naturally defined by the fact that if a reasoner believes $A_1, ..., A_n$, he's likely to believe B. Hence cognitive realism is defined by a form of epistemic preservation. The leading principle of the qualitative approach is then the

Principle of Epistemic Preservation (PEP). A rule $(r*) : A_1, ..., A_n \vdash B$ satisfies (PEP) iff when reasoners have justifications for $A_1, ..., A_n$, they have a justification for B.

(PEP) is very strong, but one can, at least, retain the minimal requirement that follows from (PEP), namely the

Preservability Requirement (PR). A rule $(r*)$: $A_1, ..., A_n \vdash B$ satisfies (PR) iff when reasoners have justifications for $A_1, ..., A_n$, it is possible for them to have a justification for B.

Are there base logics that would fit those principles better than (CEL)? The next sections attempts to answer this question in the affirmative.

2.3 Two proposals of weak epistemic logics epistemic logic

The aim of this section is to sketch some arguments in order to show that two proposals, intuitionistic epistemic logic (IEL) and linear epistemic logic (LEL), satisfy (PEP) and (PR) - or, at least, that they satisfy (PEP) and (PR) better than (CEL).

2.3.1 First proposal: an intuitionistic epistemic logic (IEL)

The first proposal made to satisfy (PEP) and (PR) is an intuitionistic epistemic logic (IEL), that is an epistemic logic where intuitionistic logic (IL) is the base logic. The main conceptual motivation for this proposal comes from the BHK-interpretation[4] of logical constants: following this interpretation, one may associate an elementary construction to every logical constant. For example, to the conjunction $\wedge$ is associated the operation of pairing because a justification for $A \wedge B$ is constructed by pairing a justification for A and a justification for B. Let us say that an inference rule passes the *BHK-test* if an elementary operation of this kind can be associated to it; the conceptual motivation to adopt an (IEL) comes from the fact that one can see the BHK-test as a first approximation of the Preservability Requirement (PR) since it guarantees the existence of a construction corresponding to every inference rule.

What is the result of this test? A well-known fact is that not every rule of (NK)[5] passes this BHK-test, since classical absurdity rule (ar):

$$\frac{\Gamma, \neg A \vdash \bot}{\Gamma \vdash A}(ar)$$

is rejected. In epistemic terms, the absurdity rule is rejected because no epistemic preservation is guaranteed. It is not because someone has a justification for the fact that $\neg A$ implies $\bot$ that he or she has a justification for the fact that A. Hence, if one eliminates (ar) from (NK), one obtains a logic which is cognitively more "realistic" than (NK). (NJ) is such a logic, thus (NJ) could be a first approximation of (PR). Therefore, one would obtain a more realist epistemic logic by replacing the classical base logic by an intuitionistic one. And the result would be an intuitionistic epistemic logic (IEL).

2.3.2 Second proposal: a linear epistemic logic (LEL)

It is impossible to deny that there is still lots of idealization in the first proposal, for real agents are not omniscient with respect to (IL). Can we do better? Can we find a logic which would be a better approximation than (IEL)? Following [J. Dubucs 1991] and [J. Dubucs 2002], the claim of my second proposal is that a substructural logic like linear logic would be a good candidate.

When one looks for a better approximation of logical competence, the trouble is that in the (ND)-format, it is hard to see how to weaken the base logic without changing the inference rules associated with the connectives, that is, it is hard to see how to weaken the base logic without violating the Minimal Rationality

[4] For Brouwer-Heyting-Kolmogorov, see e.g. [A. Troelstra and D. van Dalen 1988] p. 9.

[5] The usual set of inference rules for (CPL) in (ND).

Requirement of Section 2. But the move to (SC)-format provides an interesting perspective because it permits one to distinguish between different categories of inference rules. More precisely, one can distinguish, on one hand, rules that govern the behavior of logical constants or "logical rules", and, on the other hand, rules that govern the management of the sequent or "structural rules". As [K. Dosen 1993] says, "*a very important discovery in Gentzen's thesis [1935] is that in logic there are rules of inference that don't involve any logical constant.*" What is critical from our point of view is that with such a distinction, by eliminating (or controlling) the structural rules, it is in principle possible to reach a higher level of weakening while keeping the rules for connectives fixed. The main question is then to know whether there are good reasons to think that such rules conflict with (PEP) or (PR).

Among usual structural rules, the most debatable are probably the contraction rule (cr) and the weakening rule (wr). Here are their left-version in (LJ), the Sequent Calculus for (IL):

$$\frac{\Gamma, A, A \vdash B}{\Gamma, A \vdash B}(cr)$$

$$\frac{\Gamma \vdash B}{\Gamma, A \vdash B}(wr)$$

To evaluate these rules, it is necessary to give an epistemic interpretation of them.

Epistemic Interpretation of (*cr*). One may infer from the fact that the reasoners have a justification for B on the basis of several justifications for A (and other premises) that they have a justification for B on the basis of only one justification for A (and other premises).

Epistemic Interpretation of (*wr*). One may infer from the fact that reasoners have a justification for B on the basis of some premises that they still have a justification for B on the basis on these premises and a new premise A.

Historically, (wr) has been the most challenged of these two rules because it allows a conceptual gap between the premises and the conclusion of a chain of reasoning. The so-called "relevant logics" are designed to correct this point. But if one is focused on the epistemic interpretation of the structural rules, it seems to me that (cr) is the most debatable,[6] and I shall now argue against this rule.

The following argument is a conceptual argument based on the motivation often given for a well-known logic among those that challenge the (unrestricted

[6]This does not exclude the possibility that a rejection of (wr) could be relevant too for (PLO).

use of) contraction rule, namely linear logic (LL). The basic idea is to suggest an intuitive interpretation of logical constants and inference rules in terms of resources and resource-consumption. In this interpretation,

- a formula A is interpreted as a type of resource;
- an occurrence of a formula A in a sequent is a resource of type A; and
- a sequent is interpreted as a relation of resource-consumption.

For example, in [M. Okada 1999], the right rule for the tensor $\otimes$:

$$\frac{\Gamma \vdash A \quad \Delta \vdash B}{\Gamma, \Delta \vdash A \otimes B}(tr)$$

is interpreted as meaning that "if A can be generated by using resource Γ and if B can be generated by using resource Δ, then $A \otimes B$ (A and B in parallel) can be generated by using resource Γ and Δ". This interpretation was introduced by [J-Y. Girard 1987] and is systematically developed by [M. Okada 1999]. For this reason, let us call it the "GO-interpretation" to stress the parallelism with the BHK-interpretation mentioned above. And from this interpretation, one can extract a *GO-test* for inference rules similar to the BHK-test described above. The first intermediate step of this argument is that (cr) does not pass the GO-test: if one needs two resources of a given type to do some task, nothing guarantees that, with only one token of this resource, one is able to fulfill a book with two times 10 euros, one cannot infer that this person can buy the same book with only 10 euros; or to take the chemical example of [J-Y. Girard 1995], two molecules of H_2O can be generated by two molecules of H_2 (and one of O_2), but not by one molecule of H_2 (and one of O_2). Generally speaking, (cr) is not valid for a consumption relation.

An important intermediate step is still missing, however, namely: What is the relationship between this GO-test and the epistemic interpretation of (cr)? Looking at these examples from chemistry or book buying, one may note that they focus on *an objective kind of resource-sensitivity*. The question is whether logic can faithfully represent such (objective) processes as book buying or chemical reactions which imply, as noted by [J-Y. Girard 1995], that temporality and especially updating are being taken into account. It is worth noting that this objective resource-sensitivity has in itself nothing to do with a *computational resource-sensitivity*. It is only a matter of making our language and our logic more faithful to an intended interpretation. But, for this reason, this is hardly what we are looking for. We want a more accurate representation of reasoners whose deductive resources (especially their computational power) are limited. Hence we are looking for a *cognitive kind of resource-sensitivity*, as expressed in the epistemic interpretation of (cr). We did not face such a problem with the BHK-test because the BHK-interpretation is arguably intrinsically epistemic whereas the GO-interpretation is not. Consequently, the question is: Does the GO-interpretation make sense in an epistemic context?

I do not have a conclusive answer to this question, but I think the following suggestion is plausible: to fill the gap, we have to see

- a justification for a formula as resource in a reasoning process, and
- a reasoning process as a consumption-relation.

Now I shall develop an (admittedly highly speculative) argument in support of this view. Suppose a reasoner holds the belief that A. This belief has *inferential power* in the sense that a reasoner who believes that A is able to make this belief interact with other beliefs in some reasoning processes. So one can see a reason to believe A as a resource for reasoning processes. But clearly, this inferential power is bounded - otherwise, we would be logically omniscient. Therefore, the inferential power of a belief is a scarce resource. And, precisely, seeing the reasoning process as a consumption-relation permits us to represent this scarcity of inferential power. One can give a more psychological flavour to the rejection of (cr): reasons have a psychological strength and sometimes people would not hold something to be true were they to have fewer reasons for this belief than they actually have. But it seems to me that the fundamental idea is not different. One can speak of "strength" or of "resources". The main point, in both cases, is that a kind of causal power in reasoning is associated to reasons.

2.4 Discussion

This section attempts to answer the main questions and objections raised by the two proposals.

2.4.1 Question 1: Are (IEL) and (LEL) technically possible?

The answer is yes. At the syntactic level, the matter is easy: one has to substitute intuitionistic logic (or linear logic) for classical logic. At the semantic level, things are more complex. In the intuitionistic case, one might take advantage of the well-known Kripke-style semantics for intuitionistic logic. In this semantics, the accessibility relation I is reflexive and transitive and the valuation satisfies a Persistence Property: for every atomic proposition p and every state $s \in [[p]]$,[7] $\forall s'$ s.t. sIs', $s' \in [[p]]$. The basic problem is to keep the Persistence Property when one adds the epistemic modalities. It can be solved either by putting some constraints on I and R[8] (see [M. Bozic and

[7] $[[p]]$ denotes the set of states where p holds.

[8] R denotes the epistemic accessibility relation.

K. Dosen, 1983]) or by changing the satisfaction clause of the modalities. In both cases, the result is a Kripke-style semantics with two accessibility relations. Such a semantics is investigated by a growing literature on intuitionistic modal logic.[9]

In the linear case, a supplementary difficulty comes from the fact that the usual semantics of linear logic is algebraic, and not relational (in the Kripke-style). In order to design a semantics for (LEL), one can therefore either give an algebraic semantics of modal logic or give a relational semantics of linear logic. For example, [M. D'Agostino *et al.* 1997] build a linear modal logic for a simple fragment (implication and modality) by taking the latter approach. Such a semantics is based on a constrained set of states, but the constraints are considerably stronger than they can be in the intuitionistic case. Indeed, the set of spaces has to be a special complete lattice enriched by a binary operator, usually called *quantale*.[10]

2.4.2 Question 2: Are there concrete failures of logical omniscience that could be modelled by (IEL) or (LEL)?

In the previous section, I have defended (IEL) and (LEL) from an abstract point of view. The underlying claim was that choosing a set of beliefs closed by intuitionistic or linear logic is more realistic than choosing a set of beliefs closed by classical logic. But it would be nice to exhibit concrete types of failures of logical omniscience that could be modelled by (IEL) or (LEL). In the intuitionistic case at least, the answer is, not surprisingly, that we can.

Suppose that a reasoner i has a proof that $\neg A$ implies a contradiction, e.g. he or she has a proof that if a continuous function $f : C \times C$ on a n-dimensional simplex C has a fixed point, a contradiction follows. One can then ascribe to him or her the belief $B_i\neg\neg A$. By (ar), it follows that in (CEL), B_iA holds. This will not necessarily be the case in (IEL) since (ar) is not a valid rule. Hence, we will be able to model the common situation where the value x^* s.t. $f(x^*) = x^*$ is not available to the reasoner. That is, the situation where B_iA does not hold. This power of (IEL) can find several applications: in general, it permits one to represent a kind of mathematical ignorance (which is forbidden to (CEL)); in particular, it can be used to model boundedly rational *agents* (reasoners who have to act on the basis of their beliefs) who know that there is a best choice among a given set of possible actions but who do not know how to determine its value.

[9] See e.g. [F. Wolter *et al.* 1999].

[10] On those general semantics, cf. [H. Ono 1993].

2.4.3 Question 3: In shifting from (CEL) to (IEL) or (LEL), are we not shifting from one kind of omniscience to another one?

This question raises what can be called "the Big Objection". It is indeed the most common objection made to the kind of approach advocated in this paper, that of proposing a replacement for the base logic. Some remarks have already been made in Section 2. Nevertheless, I now should like to give an extended answer to the Big Objection.

1. Strictly speaking, it is of course correct to say that one is shifting from one sort of omniscience (with respect to classical logic) to another sort of omniscience, but, I repeat, this alone cannot be considered as a sound objection. Why? Because all depends on the consequence relation with respect to which the reasoner will be supposed to be omniscient. If the consequence relation of the new base logic is more realistic than the consequence relation of classical logic, progress has been made. Furthermore, as noted above, with (AEL) or (IWEL), one already knows how to weaken epistemic logic as much as possible; what is important is to find deductively significant weakening of (CEL). Hence, I agree with [R. Fagin *et al.* 1995]: "*It may not be so unreasonable for an agent's knowledge to be closed under logical implication if we have a weaker notion of logical implication.*"

2. However, the previous point is not the core of the Big Objection. Its very core is the following point: is intuitionistic (or linear) omniscience more realistic than classical omniscience? My arguments for an affirmative answer were given in Section 3. But one has to recognize that this is debatable. For example, from a computational point of view, the consequence problem is co-NP-complete in the (propositional) classical case, but PSPACE-complete in the intuitionistic one.[11] Therefore, it seems that (IL) or (LL) are not necessarily more realistic as base logics.

Suppose first that this computational point of view *with respect to the base logics* is relevant. It is not an obvious assumption because while it is clearly relevant concerning the deductive problems that the reasoner faces (e.g., as above, the search for a fixed point of a function), the base logic is above all a model of the reasoner's competence when facing these problems. Then, it is worth noting that the computational point of view is not "univocal" concerning our question: e.g., at the first-order level, the fragment (MALL)[12] of linear logic is decidable and at most NEXPTIME-hard. Hence, proof-oriented weakening does not necessarily increase computational difficulty. What we can conclude is that there is not necessarily a convergence between different criteria of cognitive realism.

[11] Cf. [R. Statman 1979].

[12] Multiplicative Additive Linear Logic: the linear modalities ! and ? do not appear in this fragment.

This is a phenomenon that one can meet elsewhere in the modelling of bounded deductive ability, and even between different kinds measurements of computational complexity, e.g. in the theory of Repeated Games, [C. Papadimitriou 1992] has proved that in the Repeated Prisoner's dilemma, when one limits the space of possible strategies using an upper bound on the size of automata implementing them, the computational complexity of finding a best-response becomes NP-complete.[13]

But is the computational point of view *with respect to the base logics* really relevant? My answer would be less conclusive on this point, but my claim is that it is not really relevant. The reason is this: The main proposal of the paper is not that the agents are reasoning *in* the base logic, but that base logics like (IL) or (LL) promise to fit the reasoners' deductive competence better because they eliminate rules that were unrealistic when interpreted epistemically (that is, interpreted as predictions about the reasoners' justifications, cf. Section 2), e.g. it is not reasonable to suppose that if a reasoner has a justification for $\neg\neg A$, he or she has a justification for A. The computational point of view with respect to the base logic seems, therefore, to confuse the reasoners' level and that of the modeller. Moreover, only an epistemic logic like (IEL) can model computational difficulty, e.g. the fact that reasoners may not be able to find a solution to an instance of the Travelling Salesman Problem whereas they know that such a solution does exist.

To sum up, my answer to the core of the Big Objection is twofold: first, in general, there is no guarantee that the different criteria of cognitive realism are convergent, and it is a difficult challenge to satisfy several of them; second, concerning the computational complexity of the base logic, it is not clear that it is itself a relevant criterion of cognitive realism.

2.5 Conclusion

There are of course many more questions raised by the two proposals made in this paper than those discussed in the previous section. For example, it is well-known that, in the absence of certain structural rules, a phenomenon of *splitting* appears among logical constants. It is important to note that this phenomenon does not in itself violate the Minimal Rationality Requirement since in a (LEL) the logical rules are fixed and well-defined. But, if one uses the expressive power of linear logic (even with additive and multiplicative constants only), one introduces a gap between our ordinary and intuitive grasp of the meaning of logical constants and the logical constants of epistemic logic. On topics like the previous one, the discussion isn't closed. But I would like to conclude by making a more general point.

[13] It is polynomial when the space of strategies is unbounded.

The central Principle of Epistemic Preservation opens the way to a wide spectrum of weak epistemic logics, that is epistemic logics where the consequence relation is weakened.

First, concerning the whole spectrum of weak epistemic logics, this "qualitative approach" has its limits because it does not permit a step-by-step control of inferences processes like, e.g. the proposal of [H. N. Duc 2001], more akin to what I labelled earlier the "quantitative approach", but it has comparative advantages too, e.g. the fact that a true semantics for belief is still possible.

Second, concerning the different logics in the spectrum, it is worth noting that the weaker that one makes the base logic, the less the formal implementation of the corresponding epistemic logic is manageable. The semantics of (CEL) is simpler than the semantics of (IEL), which in turn is simpler than the semantics of (LEL). I do not think that there is any paradox to be found in this fact. One can observe a quite similar phenomenon in decision theory in case of uncertainty where, for reasons of descriptive realism, the (simple) model of Subjective Expected Utility is generalized by non-additive probabilities (in general, much less simple ones), but a loss of simplicity is often considered as the price to be paid for this descriptive gain. From this point of view, (IEL) could, in the short term at least, be a good trade-off between the simplicity of (CEL) and the accuracy of (LEL).

References

Bozic, M. and K. Dosen (1983). Models for normal intuitionistic modal logics. *Studia Logica*, XLIII(3): 217–245, 1983.

D'Agostino, M., D. Gabbay and A. Russo (1997). Grafting Modalities onto Substructural Implication Systems. *Studia Logica*, VII, 1997.

Dosen, K. (1993). A historical introduction to substructural logics. In P. Schroeder-Heister and K. Dosen, editors, *Substructural Logics*, pages 1–30. Oxford UP, 1993.

Dubucs, J. (1991) On logical omniscience. *Logique et analyse*, 133-134: 41–55, 1991.

Dubucs J. (2002). Feasibility in Logic. *Synthese*, 132: 213–237, 2002.

Duc, H. N. (2001). *Resource Bounded Reasoning about Knowledge*. PhD thesis, University of Leipzig, 2001.

Fagin, R., J. Halpern, Y. Moses, and M. Vardi (1995). *Reasoning about Knowledge*. MIT Press, 1995.

Girard, J-Y. (1987). Linear logic. *Theoretical Computer Science*, 50: 1–102, 1987.

Girard, J-Y. (1995). Linear logic: its syntax and semantics. In *Advances in Linear Logic*. Cambridge UP, 1995.

Hintikka, J. (1962). *Knowledge and Belief*. Cornell UP, 1962.

Okada, M. (1999). An introduction to linear logic: Expressiveness and phase semantics. In M.Takahashi, editor, *Theories of Types and Proofs*, pages 376–420. Mathematical Society of Japan, 1999.

Ono, H. (1993). Semantics for substructural logics. In P. Schroeder-Heister and K. Dosen, editors, *Substructural Logics*, pages 259–2921. Oxford UP, 1993.

Papadimitriou, C. (1992). On Players with a Bounded Number of States. *Games and Economic Behavior*, 4: 122–131, 1992.

Statman, R. (1997). Intuitionnistic logic is polynomial-space complete. *Theoretical Computer Science*, 9, 1979.

Troelstra, A. and D. van Dalen (1988). *Constructivism in Mathematics. An Introduction*. North-Holland, 1988.

Wansing, H. (1990). A general possible worlds framework for reasoning about knowledge and belief. *Studia Logica*, 49(4): 523–539, 1990.

Wolter, F. and M. Zakharyashev (1999). Intuitionnistic Modal Logic. In A. Cantini, E. Casari, and P. Minari, (eds.), *Logic and Foundations of Mathematics*, pages 227–238. Kluwer, 1999.

Chapter 3

SEMANTICS AS BASED ON INFERENCE

Jaroslav Peregrin*
Dept. of Logic, Institute of Philosophy
Academy of Sciences of the Czech Republic, Prague
peregrin@ff.cuni.cz

3.1 There is more to semantics than inference...

We may say that logic is the study of consequence; and the pioneers of modern formal logic (especially Hilbert, but also, e.g., the early Carnap) hoped to be able to theoretically reconstruct consequence in terms of the relation of *derivability* (and, consequently, necessary truth in terms of *provability* or *theoremhood* – derivability from an empty set of premises). The idea was that the general logical machinery will yield us derivability as the facsimile of the relation of consequence, and once we are able to formulate appropriate axioms of a scientific discipline, the class of resulting theorems will be the facsimile of the class of truths of the discipline.

These hopes were largely shattered by the incompleteness proof of [Gödel 1931]: this result appeared to indicate that there was no hope for fine-tuning our axiom systems so that theoremhood would come to align with truth. [Tarski 1936] then indicated that there are also relatively independent reasons to doubt that we might ever be able to align derivability with consequence: he argued that whereas intuitively *every natural number has the property P* follows from the set:

$$\{n \text{ } \textit{has the property P for all } n = 1, \ldots, \infty\},$$

it can never be made derivable from it (unless, of course, we stretch the concept of derivability as to allow for *infinite* derivations).

*Work on this paper has been supported by grant No. A0009001/00 of the Grant Agency of the Academy of Sciences of the Czech Republic. I am grateful to Johan van Benthem for helpful comments.

J. van Benthem, G. Heinzmann, M. Rebuschi and H. Visser (eds.), The Age of Alternative Logics: Assessing Philosophy of Logic and Mathematics Today, 25–36.

These results reinforced the picture, present in the back of many logicians' minds anyway, of logic as trying to capture, using our parochial and fatally imperfect means, truth and consequence that are somewhere 'out there', wholly independent of us. And as whether a sentence is (necessarily) true[2] and what follows from it is a matter of its *meaning*, it also appeared to indicate that there must be much more to meaning than can be captured by inference rules. In particular, there must be more to the meanings of logical and mathematical constants than is captured by the inference rules we are able to construct as governing them.

Symptomatic of this state of mind is Arthur Prior's famous denial (see [Prior 1960] and [Prior 1964]) of the possibility of assigning a logical constant its meaning by means of inferential rules:

> It is one thing to define 'conjunction-forming sign', and quite another to define 'and'. We may say, for example, that a conjunction-forming sign is any sign which, when placed between any pair of sentences P and Q, forms a sentence which may be inferred from P and Q together, and from which we may infer P and infer Q. Or we may say that it is a sign which is true when both P and Q are true, and otherwise false. Each of these tells us something that could be meant by saying that 'and', for instance, or '&', is a conjunction–forming sign. But neither of them tells us what is meant by 'and' or by '&' itself. Moreover, each of the above definitions implies that the sentence formed by placing a conjunction-forming sign between two other sentences already *has* a meaning. For only what already has a meaning can be true or false (...), and only what already has a meaning can be inferred from anything, or have anything inferred from it (Prior 1964, p.191).

3.2 ... but there cannot be more!

Some of the most outstanding philosophers of language of the XX. century, on the other hand, arrived at the conclusion that there could be hardly any way of furnishing our words with meanings save by subordinating them to certain rules – the rules, as [Wittgenstein 1953] famously put it, of our *language games*.

The point of departure of Wittgenstein's later philosophy was the recognition that seeing language, as he himself did earlier in the *Tractatus*, as a complex set of *names* is plainly unwarranted. Most of our words are not names in any reasonable sense of the word *name*, and hence if they have meaning, they must have acquired it in a way other than by having been used to christen an object. And Wittgenstein concluded that the only possible way this could have happened is that the words have come to be governed by various kinds of rules. Thus, in his conversation with the members of the Vienna Circle he claims in [Waisman 1984], p. 105:

[2]In this paper we will have nothing to say about empirical statements and hence about other than necessary truths.

> For Frege, the choice was as follows: either we are dealing with ink marks on paper or else these marks are signs of *something,* and what they represent is their meaning. That these alternatives are wrongly conceived is shown by the game of chess: here we are not dealing with the wooden pieces, and yet these pieces do not represent anything – in Frege's sense they have no meaning. There is still a third possibility; the signs can be used as in a game.

This indicates that for Wittgenstein, Prior's claim "only what already has a meaning can be inferred from anything, or have anything inferred from it" would be no more plausible than the claim that only what is already a pawn, a knight etc. can be subordinated to the rules of chess: just like we *make* a piece of wood (or of something else) into a pawn or a knight by choosing to treat it according to the rules of chess,[3] we make a sound- or an inscription-type into a meaningful word by subordinating it to the rules of language.

As another famous proponent of the 'rule-basedness' of semantics, Wilfrid Sellars puts it, there are essentially three kinds of rules governing our words (see [Sellars 1974]):

> *language entry transitions*, or rules of the *world-language type*
> *intralinguistic moves*, or rules of the *language-language type*
> *language departure transitions*, or rules of the *language-world type*

Whereas the first and the last type is restricted to *empirical* words, nonempirical words are left with being furnished with meaning by means of the middle one, which are essentially *inferential* rules. Meaning of such a word thus comes to be identified with its *inferential role*.[4]

In some cases, this view appears to be markedly plausible (*pace* Prior[5]). How could "and" come to mean what it does? By being attached, as a label, to the standard truth function? But surely we were in possession of "and" long before we came to possess an explicit concept of function – hence how could we have agreed on calling it "and"? The Sellarsian answer is that by accepting the inference pattern

$$A \textit{ and } B \Rightarrow A$$
$$A \textit{ and } B \Rightarrow B$$
$$A, B \Rightarrow A \textit{ and } B$$

---

[3] Note that though the shape of the piece is usually indicative of its role, having a certain shape is neither necessary, nor sufficient to be, say, a pawn.

[4] Inferentialism in Wittgenstein's later philosophy is discussed by (Medina 2001); for an account of Sellars' semantics see [Marras 1978].

[5] What Prior did show was that not every kind of inferential pattern can be reasonably taken as furnishing a sign with a meaning. (His famous example is the 'vicious' pattern $A \Rightarrow (A \text{ tonk } B)$; $(A \text{ tonk } B) \Rightarrow B$). But it is hard to see why it should follow that the meaning of 'and' is not determined by the obvious pattern: is not what one learns, when one learns the meaning of 'and', precisely that *A* and *B* is true (or correctly assertible) just in case both *A* and *B* are? See (Peregrin, 2001, Chapter 8).

(which need not have been, and surely was not, a matter of accepting an explicit convention, but rather of handling the signs involved in a certain way).

In other cases it is perhaps less straightforwardly plausible, but still urged by many theoreticians. How could numerals come to mean what they do? By being attached, as labels, to numbers? But how could we achieve this? Even if we submit that numbers quite unproblematically exist (within a Platonist heaven), we surely cannot point at them, so how could we have agreed on which particular number would be denoted by, say, the numeral "87634"? The inferentialist has an answer: numbers are secondary to the rules of arithmetic, such as those articulated by means of Peano axioms; and hence 87634 is simply a 'node' within the structure articulated by the axioms, namely the node which is at a particular distance from zero. Its very *identity* is a matter of this distance; hence there is no need to identify it in any other way. As [Quine 1969], p. 45 puts it: "There is no saying absolutely what the numbers are, there is only arithmetic."

All of this appears to suggest that there cannot be more to the meanings of logical & mathematical constants than is captured by the inference rules governing them. Hence we appear to face the following question: *Can the standard meanings of logical and mathematical constants be seen (pace Tarski & comp.) as creatures of entirely inferential rules?*

3.3 Disjunction

An inferentialist has an easy time while grappling with "and"; but troubles begin as soon as he turns his attention to the (classical) "or". There seems to be no set of inferential rules pinning down the meaning of "or" to the standard truth-function. Indeed, "or" can be plausibly seen as governed by

$$A \Rightarrow A \text{ } or \text{ } B;$$
$$B \Rightarrow A \text{ } or \text{ } B,$$

but then we would need to stipulate that *A or B* is not true unless either *A* or *B* is. Of course we might have

$$not \text{ } A, not \text{ } B \Rightarrow not \text{ } (A \text{ } or \text{ } B),$$

but this presupposes the (classical) *not*, and hence only shifts the burden of inferential delimitation from *or* to *not*, which is surely no easier.

In fact as long as we construe inference as amounting to truth preservation, there can be no way to inferentially express that a sentence is, under some conditions, *not* true. (And it is well-known that the axioms of the classical propositional calculus admit theories with true disjunctions of false disjuncts[6]).

[6]It is usually assumed that the proofs of soundness and completeness of the propositional calculus establish that its axiomatics and its truth-functional semantics are two sides of the same coin. But this is not true in

Perhaps the way out of this could be to part company with classical logic (and subscribe to, presumably, intuitionism) – but how, then, could the classical "or" have come into being?

I envisage two kinds of answers to this question:

(1) It did not exist prior to our having an explicit idea of function and was only later procured from the pre-classical "or" by means of an explicit rectification.

(2) There is a stronger (and still reasonable) concept of inferential pattern such that there *is* an inferential pattern able to grant "or" its *classical* semantics.

In this paper I aim to explore the second alternative. The proposal I will make is that an inferential pattern should be read not simply as giving a list of (schematic) instances of inference, but rather as giving a list which is purported to be *exhaustive*. Why should it be read in this way? Because that is what we standardly mean when we make lists or enumerate. If I say "My children are Tom and Jerry", then what is normally taken for granted is that these are *all* my children. This has been noted by [McCarthy 1980], whom it led to the model-theoretic concept of *circumscription*; and indeed our proposal is parallel to McCarthy's (see also [Hintikka 1988] for an elaboration).

To say that we should construe inferential patterns in this way is to say that we should read them as containing, as it were, an implicit ... *and nothing else*: hence

$$A \Rightarrow A \text{ } or \text{ } B$$
$$B \Rightarrow A \text{ } or \text{ } B$$

should be read as "*A or B* is true if either *A* is true, or *B* is true – *and in no other case*". It is clear that by this reading, the correct classical semantics for "or" is secured.

Conjunction can be treated in an analogous way (though in its case this is not necessary). The point is that we can say "*A* is true and *B* is true if *A and B* is true – *and in no other case*". Thus, while the disjunction of *A* and *B* is the *maximal* statement which satisfies the above pattern, conjunction is the *minimal* one which satisfies

$$A \text{ } and \text{ } B \Rightarrow A$$
$$A \text{ } and \text{ } B \Rightarrow B$$

Also negation can be approached analogously; it can be seen as the minimal statement fulfilling

the sense that the axiomatics would pin down the meanings of the connectives to the usual truth-functions. It fixes their meanings in the sense that if the meanings are truth-functions, then they are the usual ones, but it is compatible also with certain non-truth functional interpretations.

$$A, \textit{not}\, A \Rightarrow B.$$

However, it is easy to see that this does not yield classical, but rather intuitionistic negation. To make it classical we have to require also

$$\textit{not}\;\textit{not}\, A \Rightarrow A,$$

which partly spoils the picture of logical operators as extremalities of inferential patterns and which also indicates why it is intuitionistic logic which should be seen as *the* logic of inference. It is nevertheless the case that even the classical operators can be seen as definable in terms of inferences and 'extremality conditions'.[7]

3.4 Standard properties of inference

Let us make our conceptual framework a bit more explicit. What we call *inference* is a relation between sequences of sentences and sentences – we assume that languages come with their relations of inference (which is constitutive of their semantics). An inference is called *standard* if it has the following properties (where A, B, C stand for sentences and X, Y, Z for sequences thereof):

Ref ['reflexivity']: $A \Rightarrow A$
Cut ['transitivity']: if $X \Rightarrow A$ and $YAZ \Rightarrow B$, then $YXZ \Rightarrow B$
Con ['contractibility']: if $XAYAZ \Rightarrow B$, then $XYAZ \Rightarrow B$ and $XAYZ \Rightarrow B$
Ext ['extendability']: if $XY \Rightarrow B$, then $XAY \Rightarrow B$

Note that **Con** and **Ext** together entail

Perm ['permutability']: if $XABY \Rightarrow C$, then $XBAY \Rightarrow C$

Indeed, if $XABY \Rightarrow C$, then, by **Ext**, $XABAY \Rightarrow C$, and hence, by **Con**, $XBAY \Rightarrow C$.

An *inferential structure* is a set of sentences with an inference relation. A *standard inferential structure* is an inferential structure whose inference obeys **Ref**, **Cut**, **Con** and **Ext**. It is clear that within a standard inferential structure, inference can be construed as a relation between *sets* of sentences and sentences.

We will also consider 'more global' properties of inferential structures. In an inferential structure each sentence can have a negation, each pair of sentences can have a conjunction, disjunction etc. Using 'extremality conditions' discussed in the previous section, we can characterize (the basic, intuitionistic versions of) the logical operators as follows, cf. [Koslow 1992]:[8]

---

[7] I gave a detailed discussion of inferential specifiability of classical operators elsewhere, see [Peregrin 2003].
[8] Note that here 'conjunction' does not refer to a specific sign (and similarly for the other connectives). 'Conjunction of A and B' is a sentence with certain inferential properties, and not necessarily of a certain syntactic structure (such as A and B joined by a conjunction-sign). 'Conjunction' can then be seen as a

Conj: $A \wedge B \Rightarrow A$; $A \wedge B \Rightarrow B$; if $C \Rightarrow A$ and $C \Rightarrow B$, then $C \Rightarrow A \wedge B$
Disj: $A \Rightarrow A \vee B$; $B \Rightarrow A \vee B$; if $A \Rightarrow C$ and $B \Rightarrow C$, then $A \vee B \Rightarrow C$
Neg: $A\ \neg A \Rightarrow B$; if $A\ C \Rightarrow B$ (for every B), then $C \Rightarrow \neg A$

If we assume exhaustivity in the sense of the previous section, there is no need to spell out the extremality conditions explicitly: $A \Rightarrow A \vee B$ and $B \Rightarrow A \vee B$ together come to mean that the disjunction is true if one of the disjuncts is, and the exhaustivity assumption yields that it is true *in no other case* – hence that it is false for both disjuncts being false. Hence, given this, we can abbreviate the definitions to

Conj: $A \wedge B \Rightarrow A$; $A \wedge B \Rightarrow B$
Disj: $A \Rightarrow A \vee B$; $B \Rightarrow A \vee B$
Neg: $A\ \neg A \Rightarrow B$

A standard inferential structure will be called *explicit* if it has conjunctions, disjunctions and negations. It will be called *classical* if, moreover, $\neg\neg A \Rightarrow A$ for every A.

3.5 From inferential roles to possible worlds

By *the (inferential) role of* A we will understand the specification of what A is inferable from and what can be inferred from it together with other sentences. Hence the role of A can be represented as $\langle A^+, A^- \rangle$, where

$$A^+ = \{X \mid X \Rightarrow A\}\ A^- = \{\langle X_1, X_2, Y\rangle \mid \langle X_1 A X_2 \Rightarrow Y\rangle\}.$$

If the inference obeys **Cut** and **Ref**, then $A^+ = B^+$ iff $A \Leftrightarrow B$ iff $A^- = B^-$. Indeed: (1) If $A^+ = B^+$, then as $A \in A^+$ (in force of **Ref**), $A \in B^+$, and so $A \Rightarrow B$. By parity of reasoning, $B \Rightarrow A$, and hence $A \Leftrightarrow B$. (2) If $A \Leftrightarrow B$, then if $X \in A^+$ and hence $X \Rightarrow A$, it follows (by **Cut**) that $X \Rightarrow B$ and hence $X \in B^+$. This means that $A^+ \subseteq B^+$. Conversely, $B^+ \subseteq A^+$ and hence $A^+ = B^+$. (3) If $A^- = B^-$, then as $\langle\langle\rangle, \langle\rangle, A\rangle \in A^-$ (in force of **Ref**), $\langle\langle\rangle, \langle\rangle, A\rangle \in B^-$ (where $\langle\rangle$ denotes the empty sequence) and so $B \Rightarrow A$. By parity of reasoning, $A \Rightarrow B$, and hence $A \Leftrightarrow B$. (4) If $A \Leftrightarrow B$, then if $\langle X_1, X_2, Y\rangle \in A^-$, and hence $X_1 A X_2 \Rightarrow Y$, it follows (by **Cut**) that $X_1 B X_2 \Rightarrow Y$ and hence $\langle X_1, X_2, Y\rangle \in B^-$. This means that $A^- \subseteq B^-$. Conversely, $B^- \subseteq A^-$, and hence $A^- = B^-$.

It follows that in this sense we can reduce the inferential role of A to whichever of its halves, in particular to A^+. Moreover, if we write $Y_1 \oplus \ldots \oplus Y_n$ for the set of n-tuples of strings of formulas $X_1 \ldots X_n$ such that $X_1 \in Y_1, \ldots, X_n \in Y_n$, it is the case that

$$A_1 \ldots A_n \Rightarrow A \text{ iff } A_1{}^+ \oplus \ldots \oplus A_n{}^+ \subseteq A^+,$$

relation between pairs of sentences and sentences (not generally a function, for we may have more than one different – though logically equivalent – conjunction of A and B).

and if inference obeys also **Con** and **Ext** (hence if the inferential structure is standard), then

$$A_1...A_n \Rightarrow A \text{ iff } A_1{}^+ \cap ... \cap A_n{}^+ \subseteq A^+.$$

Indeed: (1) If $A_1...A_n \Rightarrow A$ and $X \in A_1{}^+ \cap ... \cap A_n{}^+$, then, due to **Cut**, $X...X \Rightarrow A$, which, in force of **Con**, reduces to $X \Rightarrow A$ and hence to $X \in A^+$. (2) On the other hand, it follows by **Ref** that $A_i \in A_i{}^+$ for i=1,...,n, and it then follows by **Ext** that $A_1...A_n \in A_i{}^+$ for i=1,...,n; hence if $A_1{}^+ \cap ... \cap A_n{}^+ \subseteq A^+$, then $A_1...A_n \in A^+$ and hence $A_1...A_n \Rightarrow A$. (See [Van Benthem 1977, Chapter 7] for a more extensive exposition.)

Logical equivalence, $\Leftrightarrow$, is a congruence w.r.t. conjunctions, disjunctions, negations and implications. This means that if $A \Leftrightarrow A'$, $B \Leftrightarrow B'$, C is a conjunction of A and B, and C' is a conjunction of A' and B', then $C \Leftrightarrow C'$ (and similarly for the other connectives). Indeed: If C is a conjunction of A and B, then $C \Rightarrow A$ and $C \Rightarrow B$, and hence, in force of the fact that $A \Rightarrow A'$ and $B \Rightarrow B'$, $C \Rightarrow A'$ and $C \Rightarrow B'$. But as for every D such that $D \Rightarrow A'$ and $D \Rightarrow B'$ it is the case that then $C' \Rightarrow D$, it follows that $C' \Rightarrow C$. By parity of reasoning, $C \Rightarrow C'$; and hence $C \Leftrightarrow C'$.

This means that we can pass from an explicit standard inferential structure to what in algebra is called its *quotient*, i.e. a structure consisting of the equivalence classes (modulo $\Leftrightarrow$) of sentences with conjunctions etc. adjusted to act on them (which we know can be done precisely because $\Leftrightarrow$ is a congruence). This can be observed as passing from sentences to *propositions*, and from sentential operators to *propositional* operators.[9] The quotient structure is obviously what is known as the *Lindenbaum algebra*; and if the language is classical, this algebra is clearly Boolean (with the adjusted operations of conjunction, disjunction and negation playing the role of join, meet and complement, respectively). Then, in force of Stone's representation theorem,[10] it can be represented as an algebra of sets of its own ultrafilters. And as its ultrafilters correspond to just the maximal consistent theories, each sentence A can be, from the viewpoint of its 'inferential potential', characterized in terms of the set of those maximal consistent theories to which it belongs.

Now these theories can be seen as descriptive of 'possible worlds' representable by the language. Moreover, if the language in question has the usual structure of that of the predicate calculus, then the theories can be used to

[9]In fact, the inferential roles as defined here (which I called the *primary roles* elsewhere – see [Peregrin 2003]) are reasonably taken to explicate propositions only in case of inferential structures which are at most intensional (*i.e.*, for which logical equivalence entails intersubstitutivity w.r.t. logical equivalence, in the sense that for every A, B and C, $A \Leftrightarrow B$ entails $C \Leftrightarrow C[A/B]$). For hyperintensional languages we should consider *secondary inferential roles*, which are not only a matter of what A itself is inferable from and what can be inferred from it, but also of what the sentences containing A are inferable from and what can be inferred from them.

[10]See, *e.g.*, [Bell & Machover 1977, Chapter 4].

directly produce the 'worlds' – i.e. models – by means of the well-known construction of [Henkin 1950]. This means that the usual possible-worlds-variety of semantics can be seen as a means of representing a certain (usual) kind of inferential structure.

3.6 Representing non-standard inferential structures

If we have a negation, and consequently a structure that is not classical, the quotient algebra is no longer Boolean (it is rather a Heyting algebra). Hence here the Stone's theorem cannot be applied so straightforwardly. But as Kripke showed, there is still a way to go over to a kind of possible-world semantics. The same is the case with the modal algebras which result from modal logics.

Non-standard inferential structures yield us, in this way, non-standard varieties of semantic representation. If we withdraw **Con** and **Ext** (which is suggested, e.g., by considering the anaphoric structure of natural language, which appears to violate **Perm**), then the inferential structure ceases to be Boolean and does not yield the standard possible world semantics.

The closest analogue of conjunction within such a setting is what is usually called *fusion* (Restall, 2000):

Fusion: if $X \Rightarrow A$ and $Y \Rightarrow B$, then $XY \Rightarrow A{\circ}B$;
if $X \Rightarrow A{\circ}B$ and $YABZ \Rightarrow C$, then $YXZ \Rightarrow C$

Assuming **Cut** and **Ref**, we can prove the associativity of $\circ$: The definition yields us, via **Ref**,

(i) $AB \Rightarrow A{\circ}B$; and

(ii) if $YABZ \Rightarrow C$, then $Y(A{\circ}B)Z \Rightarrow C$.

These can then be used to prove, with the employment of **Cut**, that $A{\circ}(B{\circ}C) \Leftrightarrow ABC \Leftrightarrow (A{\circ}B){\circ}C$. Moreover, under such assumptions we can show that if C is such that $<> \Rightarrow C$ (i.e. it is a theorem), then $B{\circ}C \Leftrightarrow B \Leftrightarrow C{\circ}B$: for (i) yields $A \Rightarrow A{\circ}C$ and $A \Rightarrow C{\circ}A$, whereas (ii) yields the converse.

This means that if the structure has fusions and there exists a C of this kind (which is certainly the case if we assume some suitable 'proto-classical' versions of disjunction, implication and negation), the corresponding propositional structure (i.e. the quotient structure modulo $\Leftrightarrow$) is a *monoid*.[11] In this case, the most natural thing appears to be to represent the propositions as some kinds of functions. And indeed it turns out that the inferential potentials of the sentences A can be now represented as

$$A^* = \{<X, XY> \mid Y \Rightarrow A\}.$$

[11]That dynamic semantic is "monoidal semantics" is urged by [Visser 1997].

In this case, it follows from the results of [Van Benthem 1977, Chapter 7] that (where • represents functional composition)

$$A_1...A_n \Rightarrow A \text{ iff } A_1{}^*\bullet...\bullet A_n{}^* \subseteq A^*$$

Indeed: (1) If $A_1...A_n \Rightarrow A$, then if $<X, XY>\in A_1{}^*\bullet...\bullet A_n{}^*$, then $Y = Y_1...Y_n$, where $Y_i \Rightarrow A_i$, and hence (due to **Cut**) $Y_1...Y_n \Rightarrow A$; hence $<X, XY_1...Y_n>\in A^*$. (2) If, on the other hand $A_1{}^*\bullet...\bullet A_n{}^* \subseteq A^*$, then, due to **Ref**, $<X, XA_1...A_n>\in A^*$ for every X, which means that $A_1...A_n \Rightarrow A$.

Thus, in this way inferential roles yield one of the common varieties of dynamic semantics based on the so-called *updates*.[12]

3.7 Consequence via inference

All of this apparently suggests that we can construe the common creatures of formal semantics, such as intensions or updates, as 'encapsulated inferential roles'. However it seems that this yields us straightforwardly always only the Henkin semantics, not the standard one – and hence also never the Tarskian 'second-order' consequence. (Thus, the inferential structure of Peano arithmetic yields us more than one 'possible world' [= model], which blocks *every natural number has the property P* being the consequence of $\{n\ has\ P\}_{n=1,...,\infty}$). However, if we admit that 'enumerative' inferential patterns, such as those governing the expressions of Peano arithmetic, incorporate implicit exhaustivity assumptions (in the very way inferential patterns characterizing logical operators do) and hence involve extremality (in the sense of [Hintikka 1989]), we can see inferential roles as yielding even the standard semantics and 'second-order' consequence.

Indeed: look at the Peano axioms as a means of enumeration of natural numbers (and there is little doubt that this was their original aim). What they say is that zero (or one, which was their original starting point) is a number, and the successor of a number is always again a number. This yields us the standard natural numbers, but cannot block the occurrence of the non-standard ones after them. However, if we add *and nothing else is a number*, we cut the number sequence down to size: only those numbers which are *needed* to do justice to the Peano axioms are admitted; the rest are discharged.

To avoid misunderstanding, I do not think that the exhaustivity assumption can be somehow directly incorporated into logic to yield us a miraculous system which would be both complete and have the standard semantics – this, of course, would be a sheer daydream. If we admit that our inferential patterns do contain the implicit exhaustivity assumption, we must condone the fact that therefore

---

[12] [Van Benthem 1977, Chapter 7] also discusses other varieties of dynamic semantics corresponding to other 'subclassical' sets of assumptions about inference.

the patterns cease to be directly turnable into proof-procedures. My point here was that we *can* get semantics, even the 'most semantical one', out of something which can still reasonably be seen as inferential patterns; and thus we vindicate the Wittgensteino-Sellarsian claim that what our words mean cannot ultimately rest solely on the rules we subordinate them to.

References

Bell, J. L. and Machover, M. (1977): *A Course in Mathematical Logic*, North-Holland, Amsterdam.

Benthem, J. van (1997): *Exploring Logical Dynamics*, CSLI, Stanford.

Gödel, K. (1931): 'Über formal unentscheidbare Sätze der Principia Mathematica und verwandter Systeme I', *Monatshefte für Mathematik und Physik* 38, 173-198.

Henkin, L. (1950): 'Completeness in the Theory of Types', *Journal of Symbolic Logic* 15, 81-91.

Hintikka, J. (1988): 'Model Minimization – an Alternative to Circumscription', *Journal of Automated Reasoning* 4, 1-13; reprinted in Hintikka (1998), 212-224.

Hintikka, J. (1989): 'Is there completeness in mathematics after Gödel?', *Philosophical Topics* 17, 69-90; reprinted in Hintikka (1998), 62-83.

Hintikka, J. (1998): *Language, Truth and Logic in Mathematics* (*Selected Papers*, vol. 3), Kluwer, Dordrecht.

Koslow, A. (1992): *A Structuralist Theory of Logic*, Cambridge University Press, Cambridge.

Marras, A. (1978): 'Rules, Meaning and Behavior: Reflections on Sellars' Philosophy of Language', in *The Philosophy of Willfrid Sellars: Queries and Extensions* (ed. Pitt, J.C.), Dordrecht, Reidel, 163-187.

McCarthy, J. (1980): 'Circumscription – a form of non-monotonic reasoning', *Artificial Intelligence* 13, 27-39.

Medina, J. (2001): 'Verificationism and Inferentialism in Wittgenstein's Philosophy', *Philosophical Investigations* 24, 304-313.

Peregrin, J. (2000): 'The "Natural" and the "Formal" ', *Journal of Philosophical Logic* 29, 75-101.

Peregrin, J. (2001): *Meaning and Structure*, Ashgate, Aldershot.

Peregrin, J. (2003): 'Meaning and Inference', in T. Childers and O. Majer (eds.): *The Logica Yearbook 2002*, Filosofia, Prague, pp. 193-205.

Prior, A. N. (1960): 'The Roundabout Inference Ticket', *Analysis*, vol. 21, pp. 38-9.

Prior, A. N. (1964): 'Conjunction and Contonktion Revisited', *Analysis*, vol. 24, pp. 191-95.

Quine, W.V.O. (1969), *Ontological Relativity and Other Essays*, Columbia University Press, New York.

Restall, G. (2000): *Introduction to Substructural Logics*, Routledge, London.

Sellars, W. (1974): 'Meaning as Functional Classification', *Synthese* 27, 417-437.

Sellars, W. (1992): *Science and Metaphysics*, Ridgeview, Atascadero.

Tarski, A. (1936): 'Über den Begriff der logischen Folgerung', *Actes du Congrès International de Philosophique Scientifique* 7, 1-11; English translation 'On the Concept of Logical Consequence' in Tarski: *Logic, Semantics, Metamathematics*, Clarendon Press, Oxford, 1956, 409-420.

Visser, A. (1997): 'Prolegomena to the Definition of Dynamic Predicate Logic with Local Assignments', *Logic Group Preprint Series*, Department of Philosophy – Utrecht University.

Waisman, F., ed. (1984): *Wittgenstein und der Wiener Kreis: Gespräche*, Suhrkamp, Frankfurt.

Wittgenstein, L. (1953): *Philosophische Untersuchungen*, Blackwell, Oxford; English translation *Philosophical Investigations*, Blackwell, Oxford, 1953.

Chapter 4

EFFECTIVENESS

Stewart Shapiro
The Ohio State University
University of St. Andrews
shapiro.4@osu.edu

A function f is *computable* if there is an algorithm, or mechanical procedure, A that computes f. For every m in the domain of f, if A were given m as input, it would produce fm as output. Put this way, computability applies to functions on things that can be processed by algorithms or machines: marks on paper, electronic charges, and the like. In mathematics, of course, we do not usually deal with things that are that concrete. Idealizing a little, the field of computability consists of functions on strings, finite sequences of characters on a fixed alphabet. Since strings are structurally equivalent to natural numbers, it is common to think of computability as applying to number-theoretic functions, via some standard notation (see Corcoran, Frank, and Maloney [1974], and Shapiro [1982]).

In [1936], Alonzo Church proposed that computability be "defined" as recursiveness. This equation became known as *Church's thesis*. In the same paper, it was shown that recursiveness is coextensive with λ-definability. The same year, Alan Turing [1936] published his own characterization of computability, presenting the celebrated notion of Turing machine. Turing's work was independent of that of Church, but on learning of the latter, Turing showed that his own brand of computability is also coextensive with recursiveness and λ-definability. Emil Post [1936] also published a characterization of computability remarkably similar to that of Turing. Post's work was independent of Turing's, but not of the activity of Church and his students at Princeton. It was a very good year.

There are three connected themes that I would like to raise in this article. First, what led to the development of computability in the 1930's? Algorithms

J. van Benthem, G. Heinzmann, M. Rebuschi and H. Visser (eds.), The Age of Alternative Logics: Assessing Philosophy of Logic and Mathematics Today, 37–49.

have been known since antiquity, and computation was occasionally discussed in the history of mathematics and philosophy, but as far as I know, the period under study here produced the first attempts at a characterization of computability *per se*. Second, what convinced the major players – Church, Church's students, Turing, Kurt Gödel, and Post – of Church's thesis (or its equivalent)? What led each of them to believe that the characterizations at hand are successful? Finally, the pre-formal notion of effectiveness is pragmatic, or epistemic. It relates to the ability of humans to accomplish tasks in a certain manner. As such, effectiveness is intensional. Whether a given task is effective may depend on how it is described. By contrast, computability, and its formal counterparts of recursiveness, λ-definability, and Turing computability, are extensional properties of functions. What is the relation between the extensional and the pragmatic, intensional notions of effectiveness?

These are large topics, but my presentation will be brief. I will include a few questions, thus reversing the usual relationship between author and reader.

My first question is whether the work in the 1930's was in fact the first attempt at characterizing computability. It seems safe to say that we have before us the first *successful* characterization of computability, but perhaps there were previous, aborted attempts.

Q1. Were there any attempts to capture a general notion of computability before, say, 1930, perhaps less comprehensive and less precise than recursiveness?

Robin Gandy's excellent article "The confluence of ideas in 1936" ([1988]) contains a description of Charles Babbage's design of a computing device. Gandy shows that suitably idealized, and suitably programmed, a Babbage machine can compute any recursive function. But it does not seem that Babbage was interested in the limits of computability, nor did he make a claim like Church's thesis. He did assert that his work shows that "the whole of the development and operations of analysis are ... capable of being executed by machinery". But this is not the issue here. A statement that a certain area is capable of mechanization is not a statement on the overall limits of computation. Gandy suggests that had Babbage come to speculate on the *limits* of mechanization, he would surely have proposed a version of Church's thesis. But presumably, he did not come to speculate. Who did, and why?

Another possible precursor is Leibniz' aborted Universal Character, an attempt to reduce all of mathematics, science, and philosophy to algorithms. Leibniz is quite explicit concerning the goals of his program of Universal Characteristic, a forerunner of mathematical logic:

> What must be achieved is in fact this: That every paralogism be recognized as an *error of calculation*, and that every *sophism* when expressed in this new kind of notation ... be corrected easily by the laws of this philosophical grammar ... Once this is done, then when a controversy arises, disputation will no more be needed between two philosophers than between two computers. It will suffice

> that, pen in hand, they sit down and say to each other "let us calculate". (Leibniz [1686, XIV])

Here, Leibniz expresses interest in the nature and extent of all of philosophy, but not with the extent of computation (see Schrecker [1947]).

Nowadays, there is a consensus that a function is an arbitrary correspondence between collections of mathematical objects. For a given function f, there need be no rule, or even an independent description that specifies for each x in the domain of f, its value fx. In the set-theoretic foundation, a function is defined to be a set of ordered pairs, no two of which have the same first element. And a set is an arbitrary collection of objects (that is not too big). From this perspective, it is natural to speculate that some functions may not be computable. Indeed, the existence of non-computable functions follows from simple cardinality considerations, and a premise that there is a single language, based on countable alphabet, in which every algorithm can be expressed. There are uncountably many number-theoretic functions, but a language on a finite or countable alphabet has only countably many strings, each of which (presumably) expresses at most one algorithm in the given language, and each algorithm computes at most one function. Of course, prior to the work in the 1930's, it was not obvious that there is a single language capable of expressing every algorithm. In a sense, the discovery of such a language was the major achievement of the pioneering work on computability. Gödel [1946] once remarked that with computability,

> . . . one has for the first time succeeded in giving an absolute definition of an interesting epistemological notion, i.e., one not depending on the formalism chosen. In all other cases treated previously, . . . one has been able to define them only relative to a given language, and for each individual language it is clear that the one thus obtained is not the one looked for.

The cardinality argument also involves Cantor's theorem theorem and an ability to treat a language as a formal object, as a collection of strings.

Q2. Was the cardinality argument for the existence of non-computable functions offered by anyone before Church's thesis was formulated?

One reason for the relatively late development of computability is that the "classical" notion of function as an arbitrary correspondence, and, with this, the very possibility of a non-computable function, had only recently emerged. In a perceptive article, Howard Stein [1988] shows that what is today called "classical mathematics" is scarcely a century old. Until recently, there was no clear consensus of such notions as function and set. This point is well illustrated in the early debates over the axiom of choice (for which, see Moore, G. H., [1982]). Many opponents of the axiom, notably Baire, Borel, and Lebesgue, raised doubts about the very intelligibility of functions and sets that are not uniquely specifiable. It is not a big step from this to question the very intelligibility of non-computable functions of natural numbers. To this day, intuitionists and constructivists make this step.

From the classical perspective, once the notion of an arbitrary function or set is available, then perhaps it becomes natural to ask for an algorithm to compute a particular function or to decide membership in a particular set. That is, once the notion of a potentially non-computable function or undecidable set is on the table, then one can ask about the computation-status of particular functions and sets. And such problems were formulated and pursued early in the twentieth century. For example, some of Hilbert's "Mathematical Problems", proposed in [1900], are decision problems. Consider the once elusive tenth:

> Determination of the Solvability of Diophantine Equations. Given any diophantine equation with any number of unknown quantities and with rational integer coefficients: to devise a process according to which it can be determined by a finite number of operations whether the equation is solvable in rational integers.

Clearly, the best way to provide a positive solution to a decision problem is to give an algorithm and show that it does the required task. There rarely is a question concerning whether a purported algorithm really is an algorithm (although the subsequent debate over Church's thesis produced a number of exceptions to this seeming truism). On the other hand, a *negative* solution to a decision problem amounts to a theorem that *no* algorithm accomplishes the task at hand. To prove such a theorem, one must first identify a property shared by *all* algorithms, or all computable functions. The straightforward way to do this would be a precise characterization of computability. Thus, I propose that unsolved algorithm problems were a major motivation behind the development of computability. It became important to make coherent and useful statements about all algorithms, or all computable functions, in order to show that a certain function is not computable or that a certain set is not algorithmically decidable.

If this is correct, then in a sense, Hilbert himself foreshadowed the development of computability. In the aforementioned "Mathematical Problems" lecture, he recognized the possibility of negative solutions to some of his problems: "Occasionally it happens that we seek the solution... and do not succeed. The problem then arises: to show the impossibility of the solution..." Hilbert illustrates this with historical instances of problems which eventually found "fully satisfactory and rigorous solutions, although in another sense than originally intended". He mentions the problems of squaring the circle and proving the axiom of parallels. The solutions of these problems involved the development of new *theory* to provide mathematical characterizations of pre-formal notions. It is curious that although many unsolved algorithm problems originated from Hilbert's thinking, the development of computability did not come from that school.

A central goal of the Hilbert program is to justify at least apparent "reference" to infinite sets while sidestepping thorny metaphysical and epistemic issues. Hilbert proposed that proofs themselves be studied as formal, linguistic entities. That is, the program called for the discourse of each branch of mathematics to be cast in axiomatic deductive systems which, in turn, are to be studied syntactically. For this "meta-mathematics", only "finitary" methods are to be employed. The usefulness of recursively defined functions for this purpose had been established in Thoralf Skolem's [1923] paper developing the "recursive mode of thought" in an attempt to avoid reference to infinite sets in arithmetic. The subsequent research on recursion and meta-mathematics culminated in Gödel's [1931] paper on the incompleteness of arithmetic. The main result of this, of course, is that the axiomatization of *Principia Mathematica* (Whitehead and Russell [1910]) does not meet Hilbert's criterion of completeness. There is a formula in the relevant language which is neither provable nor refutable in the deductive system. Gödel's methods at least appear to be general. They do not depend on any idiosyncratic feature of *Principia Mathematica.* At least with hindsight, the obvious generalization is that no suitable axiomatization of arithmetic is complete. In order to clarify and, perhaps, prove this generalization, the concept of "suitable axiomatization", vis-à-vis the Hilbert's program, had to be developed, explicated, and studied.

If we deploy the contemporary notion of a set as an arbitrary collection, then it is clear that not every set of sentences of a given language is a suitable axiomatization. The resources to establish the existence of complete sets of sentences of *Principia Mathematica* are at hand. Notice that the Henkin-Lindenbaum version of the completeness theorem produces such sets, now routinely. Why not just take the set of arithmetic truths to be the axioms? Clearly, this is not a suitable axiomatization, in the sense of the Hilbert's program.

In the clarification of "suitable axiomatization", considerations of effectiveness, and computability arise. Axiomatic deductive systems are to represent or codify actual mathematical discourse, one purpose of which is to communicate proofs. This suggests that in a suitable axiomatization, one should be able to determine, "using only finite means", whether an arbitrary string of characters is a formula and whether an arbitrary sequence of formulas is a proof. In short, the syntax of a suitable axiomatization should be effective. This fails for the aforementioned system in which each truth is taken to be an axiom. The proposed generalization of Gödel's 1931 result is this: There is no ω-consistent, effective axiomatization of arithmetic in which every sentence is either provable or refutable. Here we have a conjecture of a negative theorem about algorithms: No algorithm decides the syntax of and consistent, complete axiomatization of arithmetic.

Gödel's [1934] lecture notes contain an elaboration and extension of the incompleteness result. The paper opens with an explicit statement that the syntax of an axiomatic deductive system must be effective. The notes also contain a theorem, similar to one in [1931], that every primitive recursive axiomatization of arithmetic is either ω-inconsistent or incomplete. Gödel states that the condition of primitive recursiveness "in practice suffices as a substitute for the imprecise requirement" of effectiveness. It does not follow, however, that no suitable axiomatization is complete because, as was known at the time, there are computable functions that are not primitive recursive. After mentioning that the primitive recursive functions are all computable, Gödel adds a footnote that "the converse seems to be true, if ... recursions of other forms ... are admitted". The lecture notes close with the formulation of a generalization of primitive recursion, attributed to Herbrand. The defined property of functions, now known as *Herbrand-Gödel* computability, is coextensive with Turing computability and λ-definability. All told, then, it appears that the lecture notes contain a precursor of Church's thesis. I shall return to this presently. In any case, this work motivated Kleene's [1935] detailed study of Herbrand-Gödel computability and this was followed, in short order, with Church's and Post's characterizations of computability. As noted, Turing's work appeared almost simultaneously.

I do not claim, of course, that this line of thought is the whole story behind the development of computability. Of the 1936 papers, only Post explicitly acknowledges that it is a contribution to the discussion of the completeness theorem. Turing merely mentions that his results are superficially similar to Gödel's, and this was almost certainly an afterthought. Church's paper contains a treatment of Herbrand-Gödel computability, with reference to the lecture notes and Kleene's 1935 study. Church also states that the possibility of a relationship between computability and recursiveness was raised by Gödel in conversation, but the paper does not directly indicate that the treatment of computability was motivated by the incompleteness theorem.

In a recent note, "Reflections on Church's thesis," Kleene [1987] argues that the connection between Church's work and Gödel's is overrated. The development and study of λ-definability was independent of the incompleteness theorem, and would have continued under its own momentum. This is surely correct, but one can still wonder whether the connection with computability would have been made and, if it was, what would have motivated it. Kleene's "Reflections" include the discussion of a number of fascinating counterfactuals in the form: What would X be like had Y not occurred first.

Before leaving this topic, let me note that Turing and Church seem to have been interested in another aspect of the Hilbert's program. One of the desiderata was that for each axiomatization, there should be an algorithm that enables one to decide, of any formula, whether it is derivable in the deductive system thereof. This is a decision problem, not unlike Hilbert's tenth problem. Also, the Entscheidungsproblem is the task of giving an algorithm to decide whether a given sentence is a logical truth. If attention is restricted to first-order languages, the Gödel *completeness* theorem shows that the *Entscheidungsproblem* is, in fact, the decision problem for the first-order predicate calculus. Hilbert characterized it as the fundamental problem of mathematical logic. The *Entscheidungsproblem* remained open despite the best efforts of many great mathematical minds. Some may have conjectured that it does not have a positive solution. Again, such a conjecture could lead to a characterization of computability.

Turing and Church both show how their characterizations result in a negative solution to the *Entscheidungsproblem* for first-order languages. This is now known as "Church's theorem". Turing calls in an "application" of the studies of computability.

Q3. Is there any further evidence that Turing and Church felt that they were contributing to a discussion of various aspects of the Hilbert's program or the generalization of Gödel's incompleteness theorem?

Q4. Prior to 1936, did anyone conjecture in print that a decision problem, or an instance of the Entscheidungsproblem, might have a negative solution?

I turn now to the acceptance of Church's thesis. It is clear that every λ-definable function is computable, since a λ-definition suggests an algorithm. The same goes for recursive functions and, especially Turing computable functions. The converses of these statements are another story. In a number of places, Kleene provided first-hand recollections of the events that led him and Church to hold that every computable function is λ-definable (e.g., Kleene [1979], [1981]). Church and his students began investigating individual functions, to determine whether they are λ-definable. The goal of this activity was to explore the extent of the newly defined property of λ-definability, not the extent of computability. Every computable function that was "tested" in this way was shown to be λ-definable, some more easily than others. A major breakthrough was Kleene's proof that the predecessor function is λ-definable (accomplished "one day late in January or early in February 1932, while in the dentist's office"). This result surprised Church, who had come to speculate that the predecessor function might not be λ-definable. When Church finally proposed the connection between computability and λ-definability, Kleene

> . . . sat down to disprove it by diagonalizing out of the class of λ-definable functions. But quickly realizing that the diagonalization cannot be done effectively, I [Kleene] became overnight a supporter of the thesis.

In Church [1936], the identification of computability with recursiveness is proposed as a "definition". Of course, he did not intend to introduce a new word with the equation. Computability is a vague concept from ordinary language. Starting with the word "compute", we modalize to obtain "computable", and then nominalize to "computability". Church's proposal was that recursiveness (or λ-definability) be substituted for this imprecise notion. Notice that if Church's thesis is construed this way, there is no question of an attempted "proof" of it. How does one establish that a vague notion is coextensive with a precise mathematical one? Indeed, from this perspective, Church's thesis does not have a truth value. It is a pragmatic matter, depending on how useful the identification is, for whatever purpose is at hand. One can, of course, demand that the precise notion correspond more-or-less with the vague, pre-formal counterpart; and this much was amply confirmed by the extensive study of examples. The usefulness of the definition was also bolstered with the discovery that different formulations of computability, different "definitions" come to the same thing. Such results show that independently motivated attempts to characterize the same notion converge on a single class; a good indication that all of them are on target.

It seems that some of the other main players did not share this attitude toward Church's thesis. We have seen that Gödel suggested that there may be a connection between recursiveness and computability. In fact, he proposed the identification of the two as a "heuristic". But at the time, Gödel remained skeptical, to say the least. Kleene reports that in a letter of November 1935, Gödel regarded Church's proposed "definition" as "thoroughly unsatisfactory". Church "... replied that if [Gödel] would propose any definition of effective calculability [Church] would undertake to prove that it was included in λ-definability." This, of course, would be more of the same kind of evidence that Church relied on. But there was already plenty of that, and Gödel was not convinced by it.

I do not think that this is a matter of Gödel being more stubborn than Church. Rather, Gödel saw the thesis differently. He preferred the rigor of conceptual analysis to the wealth of examples and the impressive convergence of various efforts. Presumably, one would attempt to formulate basic premises, or axioms, about algorithms or computation, and derive Church's thesis from those. This would be an instance of what Kreisel [1967] calls "informal rigor", the cutting edge of the interaction between mathematics and science, or philosophy. Perhaps, for Church, the informal rigor is not needed or not possible. One cannot precisely analyze a vague concept from ordinary language. One can only propose that a precise one be substituted for it. For more detail on this see Davis' "Why Gödel didn't have Church's thesis"[1982].

It was not long before Gödel was convinced of Church's thesis, and it was Turing's work that did it. Turing [1936] carefully considered the possible actions of a human following a previously specified algorithm, and he shows that every such action can be carried out by a Turing machine. This section of Turing's paper is an impressive piece of informal rigor. Gandy [1988] calls it a *proof* of Church's thesis (see also Sieg [1994]).

It seems that Post also rejected the idea that Church's thesis be accepted as a "definition". Moreover, in an early article not published until Davis [1965], he wrote, "Establishing [the thesis] is not a matter of mathematical proof but of psychological analysis of the mental processes involved in combinatory mathematical processes" (Post [1941]). I presume that Turing's work would satisfy this demand as well, or at least contribute to it. Later writings indicate that Post did accept Church's thesis. In fact, Post [1944] proposed that the identification of computability with recursiveness go beyond heuristic, and be used to develop a full fledged theory of computability, perhaps along the lines of Hartly Rogers' *Theory of recursive functions and effective computability* [1967]. The idea that

Church's thesis is somehow connected with psychology was also suggested by John Myhill [1952].

This leads to my third theme, intensionality. Notice first that the pre-formal notion of effectiveness is pragmatic, or epistemic. It is not a property of sets or functions themselves, independent of the way they are presented. To focus on an example, let HT be a function on strings whose value at any solvable diophantine equation is "yes" and whose value at any unsolvable diophantine equation is "no". The name "HT" abbreviates "Hilbert's tenth". To engage in counter-logical fantasy, suppose that there was a non-constructive proof that HT is recursive, one which did not specify a recursive derivation for it. It would then be known that there is such a derivation (and a Turing machine, etc.), but no one would know of one. This would not count as a solution to Hilbert's tenth problem, as originally stated. It would still be the case that no one has devised "a process according to which it can be determined by a finite number of operations whether [a given diophantine] equation is solvable in rational integers". The non-constructive proof would assure us that there is, in fact, such a process, but the envisioned development would not even assure us that a positive solution could be found. For all we would know, there might be some sort of epistemic barrier preventing us from knowing *of* a particular Turing machine that it computes HT.

Consider also the aforementioned requirement that the syntax of a deductive system be effective. As articulated in Church's *Introduction to mathematical logic* [1956], the reason is that for a system to be useful for communication, one needs a method to determine whether a given string is a well-formed formula and whether a given sequence of formulas is a proof. Suppose that someone proposed a system in which it is known that, say, the set of proofs is recursive, but no decision procedure for this set is known. The recursiveness of the set of proofs might be established with a non-constructive argument. Actually, systems much like this have been studied by Solomon Feferman [1960] and others in connection with Gödel's second incompleteness theorem (see also Detlefsen [1980]). Again, it seems to me that in these circumstances, the requirement of effectiveness has not been met. Even though the syntax is recursive, there is no method that is *known* to determine whether a sequence is a proof and, because of this, the system cannot be used to communicate mathematics.

The upshot of these considerations is that the pre-formal notion of effectiveness is an intensional concept. It is not a property of sets or functions themselves. Perhaps effectiveness can be thought of as a property of *presentations*, interpreted linguistic entities that denote sets and functions. A presentation Φ is effective if there is an algorithm A such that it is known (or at least knowable) that A computes Φ. The above examples indicate that if Φ is effective and Γ denotes the same function as Φ, it does not follow that Γ is effective.

On the other hand, the notion of computability, as defined above, is an extensional property of functions. If f is computable and g is the same function as f, then g is computable. Clearly, effectiveness and computability are closely connected. A function is computable if and only if there is an effective presentation that denotes it. But functions are not the same as presentations of functions. Some of the later, philosophical literature on Church's thesis suffers from confusing these concepts.

I suggest that the intensional notion is the more basic of the two, in an epistemic sense. In particular, judgements that a function is computable typically involve judgments concerning the effectiveness of a presentation. Ideally, to show that a function f is computable, one gives an effective presentation Φ and shows that Φ denotes f. In more non-constructive cases, one shows that f is denoted by one of a fixed set of effective presentations.

Of course, the extensional notion of computability has received the vast majority of attention. This is an instance of a more general phenomenon in which pre-formal, pragmatic notions lead to the formulation and study of precise, extensional counterparts. Consider such adjectives as measurable, countable, coverable, separable, and provable. The preference for extensional notions is a major theme in the writings of W. V. O. Quine. This dates back at least to his [1941] criticism of the higher-order logic of *Principia Mathematica*. Quine argues that Russell and Whitehead's "attributes" are unacceptable, because they are intensional. Sets should be used instead. Notice, incidentally, that the definition of effectiveness that I gave just above violates a closely related Quinean dictum. To indulge in jargon, the definition "quantifies in". There is a variable A (ranging over algorithms) within the scope of an epistemic operator "it is known that" which is itself in the scope of a quantifier binding A.

As above, a primary motivation of the development of computability was to establish negative theorems about computation. The extensional notion serves this purpose well. Given Church's thesis, the fact that the function HT is not recursive settles Hilbert's tenth problem in a decisive manner. It entails that there is no effective presentation of the function, nor can there be. One cannot devise a means to determine, in a finite number of steps, whether a given diophantine equation is solvable.

Nevertheless, it seems to me that the pragmatic, epistemic notion of effectiveness is useful and ought to be developed. And it will not do to simply cast recursive function theory within intuitionistic, or constructive mathematics. That would be too restrictive. Suppose, for example, that one gives a particular algorithm A and shows non-constructively that A computes the function denoted by Φ. This would establish that Φ is effective, at least to a classical mathematician. Many of results involving priority arguments, such as the solution to Post's problem, fit this mold. One gives a particular algorithm and shows, non-constructively, that the function computed by this algorithm has a particular property. Thus, the development of effectiveness seems to involve a mixture of classical and constructive notions in one and the same context. A good foundation for this is the work on intensional mathematics, some of which is in the Shapiro [1985] anthology.

References

Church, A. [1936], "An unsolvable problem of elementary number theory", *American Journal of Mathematics 58*, 345-363; reprinted in Davis [1965], 89-107.

Church, A. [1956], *Introduction to mathematical logic*, Princeton, Princeton University Press.

Corcoran, J., W. Frank, and M. Maloney [1974], "String theory", *Journal of Symbolic Logic 39*, 625-637.

Davis, M. [1965], *The undecidable*, Hewlett, New York, The Raven Press.

Davis, M. [1982], "Why Gödel didn't have Church's thesis", *Information and control 54*, 3-24.

Detlefsen, M. [1980], "On a theorem of Feferman", *Philosophical Studies 38*, 129-140.

Feferman, S. [1960], "Arithmetization of mathematics in a general setting", *Fundamenta Mathematicae 49*, 35-92.

Gandy, R. [1988], "The confluence of ideas in 1936", in *The universal Turing machine*, edited by R. Herken, New York, Oxford University Press, 55-111.

Gödel, K. [1931], "Über formal unentscheidbare Sätze der *Principia Mathematica* und verwandter Systeme I", *Montatshefte für Mathematik und Physik 38*, 173-198; translated as "On formally undecidable propositions of the *Principia Mathematica*", in Davis [1965], 4-35, and in van van Heijenoort [1967], 596-616.

Gödel, K. [1934], "On undecidable propositions of formal mathematical systems", in Davis [1965], 39-74.

Gödel, K. [1946], "Remarks before the Princeton Bicentennial Conference on Problems in Mathematics", in Davis [1965], 84-88.

Hilbert, D. [1900], "Mathematische Probleme", *Bulletin of the American Mathematical Society 8* [1902], 437-479.

Kleene, S. [1935], "General recursive functions of natural numbers", *Mathematische Annalen 112*, 727-742; reprinted in Davis [1965], 237-253.

Kleene, S. [1979], "Origins of recursive function theory", *Twentieth Annual Symposium on Foundations of Computer Science*, New York, IEEE, 371-382.

Kleene, S. [1981], "Origins of recursive function theory", *Annals of the History of Computing 3* (no. 1), 52-67.

Kleene, S. [1987], "Reflections on Church's thesis", *Notre Dame Journal of Formal Logic 28*, 490-498.

Kreisel, G. [1967], "Informal rigour and completeness proofs", *Problems in the philosophy of mathematics*, edited by I. Lakatos, Amsterdam, North Holland, 138-186.

Leibniz, G. [1686], "Universal science: Characteristic XIV, XV", *Monadology and other philosophical essays*, translated by P. Schrecker, Indianapolis, Bobbs-Merill, 1965, 11-21.

Moore, G. H. [1982], *Zermelo's axiom of choice: Its origins, development, and influence*, New York, Springer-Verlag.

Myhill, J. [1952], "Some philosophical implications of mathematical logic", *Review of Metaphysics 6*, 165-198.

Post, E. [1936], "Finite combinatory processes. Formulation I", *Journal of Symbolic Logic 1*, 103-105; reprinted in Davis [1965], 289-291.

Post, E. [1941], "Absolutely unsolvable problems and relatively undecidable propositions", in Davis [1965], 338-433.

Post, E. [1944], "Recursive sets of positive integers and their decision problems", *Bulletin of the American Mathematical Society 50*, 284-316; reprinted in Davis [1965], 305-337.

Quine, W. V. O. [1941], "Whitehead and the rise of modern logic", in P. A. Schilpp, *The philosophy of Alfred North Whitehead*, New York, Tudor Publishing Company, 127-163.

Rogers, H. [1967], *Theory of recursive functions and effective computability*, New York, McGraw-Hill.

Schrecker, P. [1947], "Leibniz and the art of inventing algorithms", *Journal of the History of Ideas 8*, 107-116.

Shapiro, S. [1982], "Acceptable notation", *Notre Dame Journal of Formal Logic 23*, 14-20.

Shapiro, S. [1985], *Intensional mathematics*, Amsterdam, North Holland Publishing Company.

Sieg, W. [1994], "Mechanical procedures and mathematical experience", in *Mathematics and Mind*, edited by Alexander George, Oxford, Oxford University Press, 71-140.

Skolem, T. [1923], "Begründung der elementaren Arithmetik durch die rekurrierende Denkweise", *Videnskapsselskapets skrifter I. Matematisk-naturvidenskabelig klasse*, no. 6; translated as "The foundations of arithmetic established by the recursive mode of thought" in van van Heijenoort [1967], 303-333.

Stein, H. [1988], "*Logos*, logic, and *Logistiké*: Some philosophical remarks on the Nineteenth Century transformation of mathematics", in *History and philosophy of modern mathematics*, edited by W. Aspray and P. Kitcher, Minneapolis, Minnesota Studies in the Philosophy of Science, Volume 11, University of Minnesota Press, 238-259.

Turing, A. [1936], "On computable numbers, with an application to the *Entscheidungsproblem*", *Proceedings of the London Mathematical Society 42*, 230-265; reprinted in Davis [1965], 116-153.

Van Heijenoort, J. [1967], *From Frege to Gödel*, Cambridge, Massachusetts, Harvard University Press.

Whitehead, A. N., and B. Russell [1910], *Principia Mathematica 1*, Cambridge, Cambridge University Press.

Chapter 5

DOES GÖDEL'S INCOMPLETENESS THEOREM PROVE THAT TRUTH TRANSCENDS PROOF?

Joseph Vidal-Rosset*
University of Nancy 2/Archives Poincaré (CNRS)
joseph.vidal-rosset@laposte.net

5.1 Introduction

Since their appearance in 1931, Gödel's incompleteness theorems have been the subject of intense philosophical examination. Though the demonstrations of the famous theorems are rather complex, but nevertheless clear, their philosophical implications are far from transparent. Contemporary philosophical logicians disagree on the philosophical significance of the incompleteness theorems, as did Carnap and Gödel themselves, and I believe that the state of the discussion has not changed a lot since the original Carnap-Gödel debate. It is hard to overestimate the difficulty of assessing the debate on this topic, knowing that Gödel himself refused to publish his paper "Is Mathematics Syntax of Language",[2] after six versions of this paper. Though believing Carnap's view of his theorems in *The Logical Syntax of Language* mistaken,[3] he was not satisfied by his own argument that his theorems supported mathematical realism rather than the formalism favored by Carnap.[4]

Gödel's incompleteness theorem shows the existence of a statement (called "Gödel sentence", or "G sentence") true but undecidable in Peano arithmetic.

*The first draft of this paper was written for the PILM Conference Nancy University - France (September 30th - October 4th 2002). I am very thankful to Neil Tennant, Jeffrey Ketland and Torkel Franzen for their helpful discussion on this topic and to their friendly help improving this paper. Last, I am very thankful to my colleagues and friends, Scott Walter who corrected the English of this paper, and Manuel Rebuschi who gave to me very useful suggestions and precious corrections. Obviously, remaining errors are mine.
[2] [Gödel, 1995].
[3] [Carnap, 1937].
[4] Gödel was convinced that Mathematics is not purely syntactic, but he felt unable to give a positive reply to the question "what is Mathematics?". [Dubucs 1991], pp. 53–68.

J. van Benthem, G. Heinzmann, M. Rebuschi and H. Visser (eds.), The Age of Alternative Logics: Assessing Philosophy of Logic and Mathematics Today, 51–73.

Thus, at least in formal systems, "somehow truth transcends proof". But transcendence of truth is difficult to pin down philosophically. In Gödel's opinion, the incompleteness theorem gives a picture of independent, unlimited, and always new mathematical facts which are irreducible to conventions based on axioms; at the opposite, Carnap denies that mathematics are independent from language, the incompleteness being only the expression of formal hierarchy of mathematical systems.[5] Obviously the philosophical question of the meaning of the truth in mathematics is behind the question of the reading of the incompleteness theorem, and the broader opposition between realism and anti-realism is standing further. According to Tennant (in *The Taming of the True*), the debate about realism concerns the tenability of a realist view of language, thought and the world. He begins his book by citing Russell:

> On what my be called the realist view of truth, there are 'facts', and there are sentences related to these facts in ways which makes the sentences true or false, quite independently of any way of deciding the alternative. The difficulty is to define the relation which constitutes truth if this view is adopted.[6]

This realist view of the truth expressed by Russell is the first basic assertion of any realism (from Plato's to Quine's) and Gödel's incompleteness theorems seem to give strong reasons to believe that this philosophical standpoint is the right one. In this discussion about the significance of Gödel's proof, the realist holds that the burden of the proof is on the anti-realist who must show that any truth-predicate independent of proof is involved in the incompleteness theorems. semantic anti-realism claims that "any thing worthy of the name true in mathematics or natural science is in principle knowable" and consequently denies that, absolutely, truth transcends proof. That is why Tennant aims to show in his paper that a deflationist reading of Gödel's theorems is licit and that the "non conservativeness argument" is not logically apt to throw it overboard. If philosophical interpretations of Gödel's incompleteness theorems have not changed from the Carnap-Gödel debate, logicians have recently shed more light on the problem: the point is to know if the "non-conservativeness argument" can be removed.

It is well known that Hilbert's program of giving finitist proofs of the consistency of mathematics collapsed when Gödel proved the impossibility of such a project in arithmetic. But my goal in this paper is not to develop this historical point but to clarify the contemporary terms of this debate on Gödel's proof and to try to explain why, from a logical and philosophical point of view, there is no scientific refutation of the deflationary theories of truth *via* Gödel's proof.

In the first part of this paper I show briefly the logical argument involved in Gödel's incompleteness theorems. I will explain in a second step the relations

[5][Carnap, 1937], § 60d., p. 222.

[6][Russell, 1940] quoted by Tennant, [Tennant, 1997], p. 1.

between the deflationist account of the truth and the conservativeness in the one hand, and the "non-conservativeness argument" about Gödel's theorems on the other hand. I will try to show that the question about deflationism and the Gödel phenomena is polemical from a philosophical point of view but not really from a logical point of view. I will compare in conclusion contemporary realist and anti-realist interpretations of Gödel's incompleteness theorems, and wondering about their respective stability I will plead on the latter. More precisely, I point out to contemporary realist that if he is free to reject a deflationary-anti-realist account of Gödel's proof, he is in trouble to avoid the genuine Platonist philosophy of Mathematics and not to betray a sort of discrepancy between his realist ontology and his empiricist epistemology.

5.2 The logic of Gödel's incompleteness theorem

5.2.1 Quine's informal explanation

Maybe the easiest and the more elegant way of exposing Gödel's incompleteness theorem is Quine's. But Quine's way of explaining Gödel's result is not only interesting for its pedagogical virtue, but also for Quine's philosophical conclusions.The reminder of the famous Epimenides paradox of the "I'm lying" in its Quinean version is necessary to understand Quine's account of Gödel's proof:

> "Yields a falsehood when appended to its own quotation" yields a falsehood when appended to its own quotation.

In Quine's opinion, Gödel's proof is akin to Epimenides paradox, at least on first sight:

> Gödel's proof may conveniently be related to the Epimenides paradox or the *pseudomenon* in the 'yields a falsehood' version. For 'falsehood' read 'non-theorem' thus: ' "Yields a non-theorem when appended to its own quotation" yields an non-theorem when appended to its own quotation.'
>
> This statement no longer presents any antinomy, because it no longer says of itself that it is false. What it does say of itself is that it is not a theorem (of some deductive theory that I have not yet specified). If it is true, here is one truth that that deductive theory, whatever it is, fails to include as a theorem. If the statement is false, it is a theorem, in which event that deductive theory has a false theorem and so is discredited.
>
> [...] [Gödel] shows how the sort of the talk that occurs in the above statement - talk of non-theoremhood and of appending things to quotation - can be mirrored systematically in arithmetical talk of integers. In this way, with much ingenuity, he gets a sentence purely in the arithmetical vocabulary of number theory that inherits that crucial property of being true if and only if it is not a theorem of number theory. And Gödel's trick works for any deductive system we may choose as defining 'theorem of number theory'.[7]

[7][Quine, 1966], p. 17.

The analogy between Epimenides paradox and the Gödel's sentence G can be confusing. The self-referentiality of the Liar's sentence cannot give the exact structure of Gödel's theorem which, contrarily to the Liar paradox, avoids carefully every flaw in the logical reasoning. Let us sketch more formally how the proof works in "arithmetic talk of integers".

5.2.2 Formal sketch of Gödel's proof

Gödel's proof shows an uncontroversial logico-mathematical truth. It is based on the coding of the syntax of formal system of arithmetic S and on the representability theorem asserting that it is always possible to represent in the arithmetical object language every metamathematical expressions denoted by an arithmetic formula in S. Thanks to the coding, every arithmetical formula φ can be associated with its Gödelian expression $\overline{\varphi}$. One calls "recursive" such a reasonable base theory where every formula can be expressed by its Gödel's number. Thus every logical relation between metamathematical sentences is, thanks to the coding, perfectly "internalized" by the corresponding arithmetical expressions. In the proof, S is assumed recursively axiomatizable, which implies that the provability predicate of S can be defined so that (1) and (2) below hold. So, if S is a system expressed in first order language and its intended interpretation is over the set of natural number, then we can define the provability predicate for S via the coding:

$$\text{If } n \text{ is a proof in } S \text{ of } m, \text{ then } S \vdash Proof_S(\underline{n}, \underline{m}) \tag{5.1}$$

$$\text{If } n \text{ is not a proof in } S \text{ of } m, \text{ then } S \vdash \neg Proof_S(\underline{n}, \underline{m}) \tag{5.2}$$

where n and m are the respective names of $\underline{n}$ and $\underline{m}$, the Gödel numbers identifying proofs and sentences. Now, thanks to the representability theorem ("every (total) recursive function is representable"), it is possible to define via the coding, in the set of primitive recursive functions of S, the scheme of a crucial formula:

$$S \vdash G \Leftrightarrow \neg\exists y\, Proof_S(y, \overline{G}) \tag{5.3}$$

which represents an arithmetical sentence G which says of itself, in the metamathematical language, that it is not provable in S. To get Gödel's proof one needs to understand that G and $\neg\exists y\, Proof_S(y, \overline{G})$ are in S *inter-deductible via* the coding and that this inter-deductibility *is proved* on the base of the representability theorem. G is a *fixed point* for the negation of the S-provability predicate. (The fixpoint theorem says that for any given formula $\varphi(x)$, with one free variable, there is a sentence α such that the biconditional $\alpha \leftrightarrow \varphi(\overline{\alpha})$ is provable in S, where $\overline{\alpha}$ is α's Gödel number.) So (5.3) is constructed with thinking of G as a sentence x that, referring to itself via its Gödel number, is saying of itself that it has the property φ which is expressed by the formula $\neg\exists y\, Proof_S(y, \overline{x})$.

Now the question is "is G a S-theorem?" the reply is "no" and Gödel's proof of first incompleteness theorem runs as follows:

Consistency and ω-consistency. If we can prove in a system S both $\exists x Fx$ and $\forall x \neg Fx$, then S is inconsistent. S is said to be ω-inconsistent if both $\exists x \neg \psi x$ and $\psi \underline{0}$, $\psi \underline{1}$, ..., $\psi \underline{n}$, ... are provable in S. A system can be ω-inconsistent without being inconsistent. A ω-consistent system is a system which is not ω-inconsistent. Gödel's original proof is " if Peano arithmetic is ω-consistent, then it is incomplete".

If S is consistent, there is no S-proof of G. Suppose that any m codes a proof of G in S, then by (5.3) one gets:

$$S \vdash G \Leftrightarrow \neg Proof_S(\underline{m}, \overline{G}) \tag{5.4}$$

But the surprise lies in the fact that m coding a proof of G codes $\neg Proof_S$ $(\underline{m}, \overline{G})$ but that contradicts (5.1) from which one can infer $Proof_S(\underline{m}, \overline{G})$ and violates the consistency of S.

If S is ω-consistent, there is no S-refutation of G. If G is S-refutable, then $\neg G$ is S-provable, and then, by the defining property of G (saying of itself not being provable), $\exists y\, Proof_S(y, \overline{G})$ is provable, so at least one m coding a refutation of G in S exists: $Proof_S(\underline{m}, \overline{G})$. But we have just proved that G is not S-provable, if S is consistent. That means for every y , $\neg Proof_S(y, \overline{G})$, consequently $\neg Proof_S(\underline{0}, \overline{G})$ is true, $\neg Proof_S(\underline{1}, \overline{G})$ is true, ..., $\neg Proof_S(\underline{k}, \overline{G})$ is true, etc. But there if $\neg G$ is S-provable, then $\exists y\, Proof_S(y, \overline{G})$ is provable, and consequently S is ω-inconsistent.

Conclusion. Gödel's original first theorem of incompleteness shows that if S is a formal system of arithmetic, there is an S-undecidable statement G in S, if S is ω-consistent.[8]

The second theorem of incompleteness proves that if the assertion of the existence of of a S-proof of the S-consistency is substituted for G, one gets, provided that S is consistent:

$$\nvdash_S Con(S) \tag{5.5}$$

It means that no consistent formal system of arithmetic can prove its own consistency, and the proof can be sketched as follows. The following formula says

[8]Rosser [Rosser, 1936] has proved that it is possible to get the incompleteness theorem with the weaker hypothesis of S-consistency. But it requires the construction of a more sophisticated formula R, formalizing "if this sentence has a proof, then there is a smaller proof of its negation". See [Smullyan, 1992] chap. 6, § 4.

that it there is no S-proof of the S inconsistency:

$$\neg\exists y\, Proof_S(y, \neg Cons(S)) \tag{5.6}$$

If one takes now the following formula as an instance of:

$$\neg\exists y Proof_S(y, \neg(0 = 0)) \tag{5.7}$$

It is intuitively clear that (5.7) expresses via the coding the consistency of S: no consistent system of arithmetic can prove that $(0 \neq 0)$ and to say *via* the coding, that there is no proof of "$(0 \neq 0)$" is S is also to assert the consistency of S.

That there is no proof of (5.7) is proved like the first incompleteness theorem. So every formal system of arithmetic cannot derive the assertion of its own consistency, provided that it is consistent:

$$Con(S) \rightarrow \neg\exists y\, Proof_S(y, \neg Con(S)) \tag{5.8}$$

If $\nvdash_S Con(S)$ then there are there are models of S where Con(S) is satisfied, as the set of natural numbers, and there are models of S where $\neg$ Con(S) is also satisfied (non standard models). That explains why the Gödel sentence must be true but unprovable. The formal system S being defined with an intended interpretation over the universe of natural numbers, the Gödel sentence has to be satisfied but not proved in the intended models of arithmetic, provided the consistency of S, and that explains why one asserts that Gödel's incompleteness theorem has demonstrated that there are arithmetical sentences, not being logical consequences of a reasonable base theory of numbers, are true but not analytic.

Why do *G* and the consistency of *S* have to be asserted? The most important point of that demonstration is that G as well as the consistency of S cannot be asserted in the formal system S but from a stronger system of which S is a proper subset, say $S*$. Maybe the more intuitive manner to present this fact is to deal with the condition of ω-consistency used by Gödel's original proof.

At this stage, it is helpful to base the argument on the second incompleteness theorem, and we remind that it can be done thanks to a formula like (5.7). That does not exist a numeral coding the proof of an inconsistency in the arithmetic seems obvious. Following the natural progression of integers, the instantiations in (5.7) prove:

$$\vdash (0 = 0), \vdash (1 = 1), \vdash (2 = 2), \ldots \tag{5.9}$$

and (by soundness):

$$(0 = 0) : \text{true}, (1 = 1) : \text{true}, (2 = 2) : \text{true}, \ldots \tag{5.10}$$

At one level above S, say $S*$, one is justified to tell that $\forall y \neg Proof_S(y, \underline{0} \neq \underline{0})$ is true, but it is impossible to infer it from S and it is precisely what Gödel's theorem proves. The explanation is that each standard integer satisfies in S the sentence $\neg\exists y\, Proof_S(y, \neg(\underline{0} = \underline{0}))$ but it does not entail that such a sentence is a theorem of *S for all* integers. The sentence $\neg\exists y\, Proof_S(y, \neg(\underline{0} = \underline{0}))$ asserts its own unprovability and we know from the first incompleteness theorem that if S is ω-consistent, G is undecidable. Suppose that we could get an ω-proof of G showing recursively that $\neg\exists y\, Proof_S(y, \underline{\omega} \neq \underline{\omega})$, it would also disprove the assertion of its unprovability, so S would be ω-inconsistent and we could prove $\neg G$. Last, if $\neg G$ could be proved, S would be ω-inconsistent because it would involve to accept, at the ω-level, the existence of a Gödel number which codes the proof of a contradiction. But this possibility of ω-inconsistency of S is not to be confused with inconsistency: a formal system can be ω-inconsistent without being inconsistent. That explains why the condition of ω-consistency can only be done above S, from $S*$, where, given the axiom scheme of mathematical induction and a primitive truth predicate, we can rightly infer that, if S is consistent, then *G is true*.

At the end of this section, we are now able to understand what is at stake in the Gödel phenomena: to wonder if it is possible to grasp the meaning of the incompleteness without a notion of truth transcending the base theory S. Shapiro and Ketland have replied "no" to that question and have found in the incompleteness a logical argument against a contemporary theory of truth called "deflationism". Field and Tennant have differently replied to that anti-deflationist argument, on behalf of deflationism. I am going to analyze now how these logical arguments are philosophically motivated.

5.3 Realism against deflationary theories: the argument of non-conservativeness

Quine's account of Gödel's incompleteness theorem is useful for understanding its main logical point and its logical consequences; but the question of the certainty of Quine's philosophical conclusions about Gödel's proof remains open:

> Gödel's discovery is not an antinomy but a veridical paradox. That there can be no sound and complete deductive systematization of elementary number theory, much less of pure mathematics generally, is true. It is decidedly paradoxical, in the sense that it upsets crucial preconceptions. We used to think that mathematical truth consisted in provability.

That mathematical truth *does not* consist in provability is to be understood in Quine's analysis of Gödel's incompleteness theorem. Gödel's negative theorems prove that no mathematical axiomatic system can include every mathematical truth. Consequently, in realist Quine's opinion, because truth is a relation of sentence to facts, undecided mathematical sentences do express our lack of knowledge of mathematical facts existing independently of proof systems. For example, Wiles has proved that Fermat was right in his conjecture because Fermat supposed a property of integers which is independent of knowledge: "if $n \geq 3$ there is no positive integer *x, y, z* such as $x^n + y^n = z^n$."

Gödel's incompleteness theorem proves, from the *non conservativeness argument* that there are truth sentences in every mathematical theory which cannot be known as true without a notion of truth which is transcendent with respect to the base theory. Such an argument is obviously seducing for realist philosophers and it is presented as a logical refutation of deflationary theories of truth. It is now necessary to show clearly the relation between disquotation, deflationism, and conservativeness.

Disquotation is, after Tarski's work, commonly used to define truth. The disquotation scheme is:

The sentence *"p"* is true if and only if *p*.

The canonical example given by Tarski is well known: "snow is white" is true if and only if snow is white. Truth is disquotation.

Deflationism is a philosophical interpretation of truth which is often based on the disquotation scheme: truth predicate is nothing else than a logical device for "disquoting" expressions and for expressing in finite sentences a infinite list of true sentences ("God knows every truth"). Both Ketland and Tennant recognize that it is not easy to give a unambiguous definition of deflationism: Ketland says that "deflationism about truth is a pot-pourri" and Tennant describes it as "a broad church".[9] They themselves illustrate the ambiguity of deflationism: reading their respective papers, it is possible to give at least two versions of deflationism, both being expressed on the base of the disquotational theory of Truth. The first one can be called the "strong deflationism". Ketland describes it perfectly:

> the concept of truth [...] is redundant and "dispensable": [...] we need to "deflate" the correspondence notion that truth expresses a substantial or theoretically significant language-world relation.[10]

In my opinion, Tennant adopts a weaker version of deflationism, or more precisely, a version of deflationism akin to his semantic anti-realism. The following quotation makes that point quite clear:

> Deflationism has its roots in Ramsey's contention that to assert that ϕ is true is to do no more than assert ϕ, unadorned. Truth is not a substantial property whose metaphysical essence could be laid bare. It has no essence; it is as variegated as the grammatical declaratives that would be its bearers. There would therefore appear to be no gap, on the deflationist's view, between claims that are true and

[9][Ketland, 1999], p. 69; and [Tennant, 1997], p. 558.
[10][Ketland, 1999].

> assertions that are warranted; or, generally, between *truth* on the one hand, and, on the other hand, grounds for assertion, or *proof*.[11]

We can notice that in Tennant's version of deflationism truth is not, strictly speaking, dispensable, and that it is not the language-world relation that we need to "deflate", but the idea of substantial existence of truth, as if truth could be more than the label that we put on all sentences that we can check. If truth is always knowable (i.e. checkable) in principle, then to disquote *"p"* must mean that '*"p"* is a justified belief and that the truth of *"p"* is the verified relation between the sentence *"p"* and the fact x denoted by p.'

To wonder if truth is substantial or non substantial seems maybe obscure and needs to be clarify. The semantic anti-realism advocated by Tennant does not claim that truth is neither substantial nor objective, but that truth does not lie as a substantial property in absolute unprovable sentences: the only licit notion of truth is always epistemologically constrained.[12] Relations between truth and proof and the role of Bivalence are on this point crucial to get the difference between the realist and the semantic anti-realist. Suppose truth as independent from proof, then truth can be imagined as a substantial property of some sentences which are true even though nobody is apt to verify them. Bivalence universally assumed, by definition every declarative sentence is determinately true of false, independently of our means of coming to know whether it is true, or false, and then it would be odd to claim that truth is not a substantial property of sentences. For example, thanks to Bivalence, the Megarians philosophers had constructed a system of Logical Fatalism: if every affirmation or negation about a future is true or false it is necessary or it will be impossible for the corresponding state of affairs to have to exist: the sentence "Jacques Chirac will resign in September, 1st, 2005" has already in itself a truth value before the mentioned date. My point is not to develop the philosophical subtleties which have been made to solve such a puzzle,[13] but to make clear that deflationism *à la Tennant* tries to dissolve the truth property in verification process: the universality of Bivalence must be rejected.

It is now possible to understand why Tennant tries to find in his paper a strategy to save a deflationary reading of Gödel's theorem. From the semantic anti-realism point of view every mathematical truth must be in principle provable. But a widespread view about Gödel's incompleteness theorem leads to assume, as Quine does, that mathematical truth is not provability, because Gödel has succeeded to show in the syntax of number theory a sentence G which, under the hypothesis of the consistency of the number theory, must be *true* and *unprovable* in the number theory. Of course such a demonstration does

[11] [Tennant, 2002], p. 552, emphasis in original.
[12] [Tennant, 1997], p. 15.
[13] See [Vuillemin, 1996].

not affect the general equivalence between mathematical truth and provability in principle, because a proof of G can be done in a meta-theory including the number theory. In other words, Gödel's incompleteness does not affect the tenet of the semantic anti-realism. But an argument about the conservativeness of deflationary theories of truth seems to show that they collapse because of Gödel's result. My point is to show that deflationism in Ketland's meaning effectively collapses, but it is not the case if deflationism is intended in Tennant's meaning.

Ketland has given some important technical results about the conservativeness theories of truth.[14] The main point can be understood without logical formula, and, in order to avoid technical developments, I refer only to the Disquotational Theory (*DT*) described by Ketland and to the result he has proved: *DT* added to any (non-semantical) theory S is *conservative* over S. Intuitively it means only that the truth predicate of *DT* adds *nothing new* to true sentences of S, or that any model of S may be expanded to a model of $S \cup DT$, or that $(S \vdash \varphi) \Leftrightarrow (S \cup DT \vdash \varphi)$. Thus *DT* holds a metaphysically "thin" notion of truth, to repeat Shapiro's word.[15]

The goal of Ketland's demonstration is to show that because of the property of conservativeness, the union of any deflationary theory of truth to the Arithmetic of Peano (PA) is unable to show that G is true. On the contrary, it is the union of PA with a Satisfaction theory of truth which is able to prove that G is true: $PA(S) \vdash G$. That theory S is Tarski's inductive definition of truth expressed by this list of four axioms:

1 $(T_{A_t})(t = u)$ is true if and only if the value of $t =$ the value of u.

2 $(T_{\neg})\neg\varphi$ is true if and only if φ is not true.

3 $(T_{\wedge})(\varphi \wedge \psi)$ is true if and only if φ is true and ψ is true.

4 $(T_{\forall})(\forall\varphi)$ is true if and only if, for each n, $\varphi(n)$ is true.

S added to PA, PA(S) is a "truth-theoretic" *non-conservative* extension of PA. From that union of PA and S, the most important results are that PA(S) *proves* that anything provable in PA is true, *proves* also that PA is consistent, and *proves* that the Gödel sentence G constructed in the syntax of PA is both *true* and not provable in PA. Thus, the non conservativeness argument about Gödel's proof could be called "The Master Argument of non-conservativeness" against deflationism. I let Ketland sum up the argument:

> Stewart Shapiro and I have introduced an innovation in relation to understanding the notion of deflationism about truth. We have proposed that we define "substantial" (for a theory of truth) to mean "non-conservative" (over base theories).

[14][Ketland, 1999], p. 74-79.
[15][Shapiro, 1998].

> We proved that disquotation is conservative, and Tarski's inductive definition is sometimes non-conservative.[16]

It means, to speak very generally, that in order to recognize the truth of G sentence constructed *via* the coding in a theory S, a *non conservative* theory of truth over S is required. This fact leads, in Shapiro's and Ketland's opinion, to a "thick" or a "substantial" notion of truth. That is why, invoking a full Tarskian theory of truth (i.e. a non conservative theory of truth) Ketland writes:

> If I am right, our ability to recognize the truth of Gödel sentences involves a theory of truth (Tarski's) which *significantly transcends the deflationary theories*.[17]

Last but not least, Ketland is convinced that this objection from non-conservativeness is a logico-mathematical refutation of deflationism: because every deflationary theory of truth is conservative (by definition) and because the deflationary philosophy of truth pretends that a theory of truth adds "no content" to a non-truth theoretic base theory, Ketland concludes that, because it is proved that a *non-conservative* truth-theoretic extension of PA is necessary to show that all theorems of PA are true, that PA is consistent and that every Gödel sentence in PA is true, then "deflationism is false".[18]

Ketland's conclusion is not only interesting from a logico-philosophical point of view; it is also an interesting attitude from what one could call a "meta-philosophical" point of view. Assuming that the non-conservativeness objection is a refutation of deflationism, Ketland presupposes that philosophy can be refuted by science. But even if deflationism is not a philosophical system but only a philosophical opinion about the nature of truth, it is undoubtful that deflationism can be attractive for every anti-Platonist (or anti-realist) philosopher. That is why Ketland uses the non-conservativeness of Tarski's theory of truth as a polemical argument against deflationary view of Field's nominalism for example. Ketland concludes his paper in making an analogy between the indispensability of mathematics for the knowledge of the physical world, and the indispensability of a full Tarskian theory of truth to prove some important sentences.

Believing in nominalism, Field replied to Shapiro that non-conservativeness lies with the notion of natural number and with the notion of mathematical induction, not with the notion of truth:

> [...] this point about the extension of schemas has nothing to do with truth: we are committed in advance to extending our schemas to all new predicates, not just "true".[19]

---

[16] Private correspondence.

[17] [Ketland, 1999], emphasis in original; quoted also by Tennant, [Tennant, 2002], p. 566.

[18] [Ketland, 1999], p. 92.

[19] [Field, 1999].

We will find again later on Field's remark. But we can notice how much uneasy the strict nominalist position becomes with respect to Gödel's theorems. The proper extension of Peano arithmetic involved in the assertion of its consistency as well as the convenient set-theoretic point of view appear as very strong realist arguments. But it is the deflationist theory of truth which is the topic of that discussion. Let see how Tennant has tried to show that deflationism can give a licit reading of Gödel's theorems.

5.4 Tennant's deflationist solution

5.4.1 The deflationist use of reflection principles

To justify his philosophical position from a logical point of view, Tennant uses Feferman's reflection principles, proposing a uniform reflection principle compatible with Gödel's proof and anti-realist goals. Feferman's reflection principles are "axiom schemata [...] which express, insofar as is possible without use of the formal notion of truth, that whatever is provable in S is true".[20] The soundness of S for *primitive recursive* sentences could be expressed by the following reflection principle:

> (pa)[21] If $\overline{\varphi}$ is a primitive recursive sentence and $\overline{\varphi}$ is provable-in-S, then φ.

Tennant's principle of "uniform primitive recursive reflection" is:

> $(UR_{p.r.})$ Add to S all sentences of the form $\forall n\, (Prov_S(\overline{\psi(n)}) \rightarrow \forall m\, \psi(m)$ where ψ is *primitive recursive*.[22]

Tennant insists on the point that, producing the consistency extension, $(UR_{p.r})$ has exactly the logical strength to formalize faithfully the reasoning in Gödel's semantical argument, and that is why, thanks to this reflection principle, Tennant is able to give meta-proofs, in S^*, without mention of truth predicate (even in S) that there is neither S-proof nor S-refutation of the G sentence asserting "there is no S-proof of the inconsistency of S". Such meta-proofs explicitly appeal in S^* to the consistency of S, to the S-provability of assertable primitive recursive statements, and to the representability of S-proof. It aims at showing that Gödel's proof can be done in "truth-predicate-free" theories.

To Shapiro's demand that the deflationist do justice to the soundness of S, Tennant suggests that he would express the soundness of S by being prepared to assert, in the extending system S^*, every instance of the reflection schema:

$$Prov_S(\overline{\varphi}) \rightarrow \varphi \tag{5.11}$$

[20][Feferman, 1991].

[21]*pa* is used to abbreviate 'provable then assertable'.

[22][Tennant, 2002], p. 573. I add the emphasis.

So, Tennant , as every logician, recognizes the logical necessity of an extended system to express the soundness of S, but he denies that it is necessary to mention a truth predicate as he denies that Gödel's proof leads to a substantial notion of truth:

> One can agree with Shapiro that the 'deflationist cannot say that all of the theorems of [S] are true'. But the deflationist can instead express (in S^*) his willingness, via the soundness principle, to assert any theorem of S. The anti-deflationist desires to go one step further and embroider upon the same willingness by explicitly using a truth-predicate.[23]

The sentence "all theorems of S are true" is called "the adequacy condition". And it is an important point is that the adequacy condition cannot be asserted in the base theory itself:

> Löb's Theorem ensures that this soundness principle ($Prov_S(\overline{\varphi}) \rightarrow \varphi$) could not be derived in S without making S inconsistent. But here we are contemplating adopting the soundness principle in the extension S^* of S; and this adverts that danger of inconsistency.[24]

When Tennant suggests that the deflationist, instead of saying that all S-theorems are true could be prepared to accept, in S^*, every theorem of S, he finds again the Wittgensteinian distinction between "to say" and "to show":

> When the deflationist adopts the soundness principle above, he is allowing that it may have infinitely many instances.[25]

I have put aside technical details of Tennant's demonstration in order to stress on the main logical-philosophical issues of the " Gödel phenomena". One could believe that Tennant's work on Gödel's proof consists only in switching the truth-predicate for the proof-predicate, and that such a trick cannot be a refutation of the realist interpretation of Gödel's incompleteness theorems. But such a feeling, in my opinion, betrays a misunderstanding of the philosophical disagreement between Tennant and Ketland and, especially, a misunderstanding of the philosophical meaning of Tennant's defense of deflationism. I will show at the end of this paper that the difference between Ketland and Tennant is not scientific but a strict difference of philosophy.

5.4.2 Ketland's reply: reflection principles are truth-theoretically justified

Ketland has written a reply to Tennant's paper.[26] The first important step of his reply is the definition of what he calls a "conditional epistemic obligation":

---

[23][Tennant, 2002], p. 574.
[24][Tennant, 2002], p. 574.
[25][Tennant, 2002], p. 575.
[26]Unpublished paper.

Conditional epistemic obligation: *If* one accepts a mathematical base theory S, then one is committed to accepting a number of *further statements* in the language of the base theory (and one of these is the Gödel sentence G.)[27]

The "further statements" mentioned by Ketland are those which are not provable by the theory S, but which are nevertheless proved as true thanks to a non-conservative extension of S, like the extension produced by the Tarskian theory of satisfaction. Ketland insists on the idea that it is using the notion of *truth* that we can explain the conditional epistemic obligation. His claim is that the objection of non-conservativeness made against the deflationary theories of truth is in fact a "Reflection Argument": informally we can say that by the use of the notion of truth we can arrive at statements as "all sentences proved in S are true", "G expressed in S is non provable in S and true", "all theorems of S are true", etc. The crucial point of Ketland's argument is that the conditional epistemic obligation can be explained by a non-conservative (or substantial) use of the notion of truth leading the deflationist to the following dilemma:

1 Either abandon the conservative constraint [i.e. the deflationist gives up his claim that truth is conservative or dispensable], thereby becoming *some sort* of substantialist about truth;

2 Or abandon the adequacy condition [expressed by the sentence "all theorems of S are true"]. And furthermore, offer some *non-truth-theoretic* analysis of the conditional epistemic obligation.[28]

After a quick and superficial reading of Ketland's and Tennant's paper, it could be surprising to see that, in his reply to Tennant, Ketland uses and mentions Feferman's reflection principles as his own anti-deflationist argument. One can sum up his argument in saying that it is true that G is a theorem of the union of PA with the principle of uniform primitive recursive reflection proposed by Tennant. But he argues against Tennant's deflationist position that the use of Feferman's reflection principles can only being *truth-theoretically* justified: that is only the use of the notion of truth, or a theory of truth as a Tarskian theory of truth, which can explain and justify the reflection principles. The superiority of the anti-deflationist or the substantialist theory of truth, in Ketland's opinion, is even that it is able to *prove* the principles of reflection when the deflationist is only able to assume them, without justification, as Tennant seems to do. And then Ketland shows that philosophy is essentially polemical when he writes:

Part of the point of the articles by Feferman, Shapiro and myself was to show how to *prove* reflection principles, which, "ought to be accepted if one has accepted the basic notions and schematic principles of that theory" (Feferman, 1991, p. 44). On Tennant's proposal, instead of *proving* the reflection principles in the

[27] [Ketland, 2005], an unpublished Reply to Tennant.
[28] ibid.

> manner proposed by Fefeferman, Shapiro and myself, the deflationist may simply *assume* the reflection schemes. As far as I can see, in the absence of the sort of truth-theoretic justification given by Feferman, Shapiro and myself, Tennant's idea is that deflationist may assume these principles *without argument*. No reason, argument or explanation for adopting the reflection principles is given by Tennant . If we avoid explaining why acceptance of mathematical theory S rationally obliges further acceptance of reflection principles, then this is apparently "philosophically modest". I find this curious. It is rather like saying that if we avoid explaining a phenomenon, we achieve "philosophical modesty". Presumably, the ideal way to achieve such "modesty" in the scientific arena would be to abandon scientific explanation altogether.[29]

I aim at showing with modest means - I mean non technically sophisticated from a logico-mathematical point of view - that Ketland is wrong to believe in a *scientific* difference between his substantialist explanation of Gödel's theorem and Tennant's deflationist reading. In reality, they are twins from a scientific point of view but these twins do not have the same philosophical beliefs, and even if their philosophical beliefs are about science, I am convinced that there is no scientific proof that one is wrong and the other is right. Ketland thinks that a deflationary theory of truth cannot be a good theory of truth because, being logically conservative, it is inadequate to explain why the Gödel phenomena in a theory S requires a non-conservative theory of truth over S (if S is a reasonable base theory). Tennant considers the conservativeness constraint of deflationary theories of truth as a reasonable base for a philosophical understanding of truth and he has consequently to propose a strategy to accommodate the proper extension involved by the truth of the independent sentence G and the conservativeness requirement for deflationism.

The only way I see to understand the philosophical difference between two logicians discussing about the same uncontroversial theorem is to suggest that they do not interpret in the same way the logical hierarchy of object language and metalanguage involved in Gödel's theorem. We have to keep in mind that the topic of their discussion is to wonder if a deflationist reading of Gödel's theorem is licit or not, provided that deflationism is defined by the conservativeness constraint. My position is on behalf of Field and Tennant: deflationism is not at all disproved by the non-conservativeness argument. My position is based on two types of arguments. The first one is logical. I want to stress on the fact that the non-conservativeness argument is based on some hierarchy of formal systems and that involves what Quine called a "semantic ascent" in which we can find truth-predicates among other predicates (to join Field's remark). In my opinion, every reasonable deflationism must respect that semantic ascent and consequently the objection from non-conservativeness is pointless. The second type of argument concerns the philosophical meaning of the deflationary

[29]ibid.

theories of truth (especially the disquotational theory of truth) which Ketland depreciates for philosophical reasons. I will sketch in conclusion my understanding of relations between science and philosophy, trying to show also the specific difficulties of the contemporary mathematical realism inspiring the argument of non-conservativeness.

5.5 The non-conservativeness of truth or the semantic ascent

I am going to develop in this section a deflationist reply to the non-conservativeness objection. My first point is that it is possible, quoting Field, to "deflate" the argument of non-conservativeness from the Tarskian solution of the Liar paradox: when the Liar say, "I am lying", does his sentence involves a non-conservative theory of true? The anti-deflationist would reply "yes" without hesitation, but the deflationist would be certainly more cautious. We have seen Quine making an analogy between the Liar paradox and Gödel's proof, so, I will just develop that analogy going back to the solution of the former.

It is well known that the solution of the Liar paradox is in the distinction between object language and metalanguage. If the Liar says that he is lying about some other sentence uttered five seconds ago, the paradox vanishes as it does in common language: when one confesses lying one does not mean that the confession itself is false, but the confession is normally understood as a true_1 sentence speaking about false_0 sentences that one asserted as true_0. Quine comments the Tarskian solution and suggests a finely-shaded opinion about the non-conservativeness of the truth theory involved in the Liar paradox:

> For the ith level, for each i, the variables 'x_i', 'y_i', etc. range over that levels and lower ones; thus 'x_0', 'y_0', etc. range only over sets. Predicates 'true_0',' true_1', and so on are then all forthcoming by direct definition. For each i, 'true_i' is dependably disquotational in application to sentences containing no bound variables beyond level i. We get a self-contained language with a hierarchy of better and better truth predicates but no best. Truth_0 is already good enough for most purposes, including classical mathematics.[30]

Quine's opinion seems here in agreement with the anti-deflationist position about truth. As Ketland Quine could assume the idea that the substantiality of truth in a language L is related to the undefinability of truth in L. Ketland has pointed out that it would be misleading to describe Quine as a deflationist.[31] But I would now suggest an explanation of how deflationism works inside to the Tarskian hierarchy in order to see how deflationism can avoid the non-conservativeness objection.

[30][Quine, 1992], p. 89.
[31][Ketland, 1999], p. 70, n.1.

In my opinion, the anti-deflationist reflection argument in unfair with the deflationary (disquotational) theory of truth: the latter has never claimed one and only one level for every truth. On the Foundations of Mathematics electronic list of discussion,[32] Ketland and Franzen have challenged Tennant to give a non-truth-theoretic explanation of the use of the principle of reflection producing the consistency proper extension of PA. But I believe that this challenge is based on a misunderstanding of what means a reasonable version of deflationism. A reasonable deflationary theory of truth, as Tennant seems to advocate, does not claim that we can really throw overboard the notion of truth. Following Quine's lesson we can say that the disquotational theory of truth teaches us only that for every sentences φ with no bound variables of level higher than i we are allowed to disquote them when we say that they are $true_i$.

But what is the relevance of my remark on the semantic ascent with Gödel's proof showing that in every consistent arithmetical base theory without predicate we can construct *true* but unprovable sentences? I mean only that Gödel's proof of S-incompleteness is finally significant from the point of view of a theory S^* of higher order than S. The anti-deflationist replies that here is precisely his argument against deflationism: that we can get thanks to the Tarskian theory of truth "a hierarchy of better and better truth predicates but no best." But the deflationist replies that we have to take care of what sort of axioms we need and not precisely to truth-predicates which can be correctly deleted thanks to recursive definitions.

The anti-deflationist reflection argument seems consequently a metaphysical use of logical principles which does not only deal with *truth*: principles of reflections are logical means to construct different hierarchies of theories and to compare their respective strength. So, when Tennant proposes the principle of uniform primitive recursive reflection to give a meta-proof that G has neither S-proof nor S-refutation, he describes exactly the logical strength, no more, that is needed for the theorem of incompleteness, and, contrarily to Ketland's claim, he has not to *prove* that principle, because, if I am not mistaken, he would need a stronger one for such a job.

Finally, I remain convinced that the anti deflationist argument about our ability to "recognize the truth of Gödel sentences" does not pay attention to levels of "truth". It is nevertheless easy to imagine the following situation: being "inside" PA it is impossible to "see" its consistency and so it is impossible to conclude that G is true. On the contrary, every formula φ which is a theorem or the negation of a theorem of the arithmetic can be checked thanks to an effective algorithm. One concludes usually that G is a truth of PA but not provable in PA, hence the realist and widespread view about the distinction between truth

[32] http://www.cs.nyu.edu/mailman/listinfo/fom/.

and proof (the former transcending the latter.) But G being equivalent, modulo PA, to PA consistency statement, it seems obvious that the assertion of G, or the assertion of PA consistency, needs more strength from axioms than PA's and if one want to adorn that assertion of a truth-predicate that one needs an index higher than PA-theorems. The deflationary theory of truth can be understood as the philosophical expression of the attention that one must pay to the level of our truth predicates. Using of truth predicate without index is to be in danger of collapsing in contradiction.

In a very stimulating private correspondence Ketland has replied to me that considerations on a hierarchy of truth-predicates are irrelevant vis-à-vis his argument: PA is a base theory without truth-predicate, and PA(S) proves theorems which are not provable in PA. But I reply that it means also that there are theorems which are proved in PA and consequently these proofs are "truth-theoretically explained", to speak in Ketland's language, by a truth-predicate with a level which is lower than theorems proved by PA(S) only. I persist in seeing an interesting analogy (deserving clarification) between that necessary semantic ascent involved in Gödel's proof and the hierarchy of truth-predicates solving the Liar paradox. Nevertheless it seems to me clear that disquotational process is irrespective of types or orders and, consequently, that the non-conservativeness argument collapses.

5.6 Philosophical developments

I said in the introduction of this paper that the philosophical debate about the meaning of Gödel's incompleteness theorem had not changed a lot from the controversy between Carnap and Gödel himself. Vuillemin has pointed out that this discussion was purely philosophical and not strictly scientific (i.e. logico-mathematical) and I agree with him. I see the debate about deflationism and the objection of non-conservativeness as a motivated philosophically one and nobody can prove to be right, nobody can disprove the rejected thesis. The deflationist theory of truth in its disquotational version is in fact a philosophical anti-metaphysical position wearing logical clothes ample enough to avoid the bullet from the non-conservativeness objection. It is more surprising, but no irrational, that such a bullet has been shot by logicians who have learned the lessons of Logical Positivism. The argument of non-conservativeness against deflationism is motivated by mathematical realism. The undefinability of truth seems to stand for transcendence in Ketland's philosophy of mathematics. But the undefinability of truth could be also calculability's (via Church's conjecture), or specifiability's (via set theory), and so one could doubt about the metaphysical significance of the former.

One of the merits of Tennant's position in his paper, is that it does not pretend to "disprove" the anti-deflationist interpretation of Gödel's incompleteness theorem, but only to show that a deflationary account of these famous theorems is licit. This point of his paper seems to involve a pluralist theory of meta-philosophy: there could be several licit philosophical reading of a same scientific result. Such a meta-philosophical pluralism could mean giving up the confusion of philosophy with science.

I believe that the use of reflection principles is philosophically meaningful from Tennant's anti-realist point of view, which denies that truth is significant if truth is not checkable at least in principle. In other words, Tennant holds explicitly an intuitionist notion of truth according to which truth is *always* epistemically constrained: it makes no sense to say that p is true if we do not have at hand a method of checking the truth value of p. This intuitionist conception of truth is an old and respectable notion of truth adopted by Epicurus, Descartes and Kant, before Brouwer and other logicians.[33]

Now one can wonder if the claim that truth is always epistemically constrained is consistent with deflationism: if deflationism is expressed by the disquotational theory of truth, an intuitionist theory of truth says more than "truth is disquotation", but insists on the idea that disquotation in itself is not enough without at least a way of proving the truth. On the contrary, a disquotational theory of truth can also fit with realism and with the thesis that truth is the property of true sentences expressing what is the case, independently of any way of checking the so-called correspondence.[34]

The only way to reply to this aporia, is to remind that the disquotational theory is philosophically neutral, and that the discussion is on the conservativeness or the non-conservativeness of the truth theory. Shapiro and Ketland hold that the non-conservativeness of truth involved in Gödel's proof implies that truth is transcendent to proof, which is uncontroversial if truth and proof are related to a formal system S. But the question if truth remains meaningful when it is absolutely transcendent to proof is a philosophical question where Gödel's incompleteness theorem has finally little authority. The philosophical debate opposing truth as warranted truth and truth as determined by the universality of Bivalence will remain open and every logician-philosopher can make a free but rational philosophical choice to decide his position. Tennant's solution needs to give up the universality of the principle of Bivalence. From an anti-realist point of view, Bivalence holds only in decidable theories.[35] From a realist point of view like Quine's, to give up Bivalence would complicate uselessly the logic.[36]

---

[33] [Vuillemin, 1981].

[34] In this respect, Tennant's reference to the prosentential theory of truth could be criticized, if the prosentential theory accepts that truth can be recognition-transcendent.

[35] [Tennant, 1997], p. 173-176.

[36] [Quine, 1995], p. 56-57.

5.7 Conclusion: towards a positive and philosophically stable interpretation of Gödel's proof

A philosophical system is "stable" if it is not only consistent, but if any of its theses does not create insuperable difficulties *vis-à-vis* other theses of the system. Quinean realism is, in my opinion, a nice example of an unstable philosophical system: on the one hand empiricism and physicalism stand as dogmas, on the other hand, mathematical entities or mathematical facts belong to our universe, even when the higher parts of set theory have no obvious links with the empirical world. But Gödel shows a philosophical problem starting at the bottom of the mathematical hierarchy, in elementary number theory, where undecidable statements will be always out of reach. There is no strict contradiction between empiricism and mathematical realism, because the former is an epistemological claim and the latter is an ontological one. Nevertheless, even if this distinction between epistemology and ontology leaves a way out, undecidable mathematical statements cause embarrassment: because of the universal application of Bivalence, their truth value is depending on facts which are definitely beyond empirical evidence. The instability of the comes from its effort to both naturalize mathematical knowledge (hence to deny transcendence), and to assume Bivalence and Gödel's proof that mathematics does not consist in provability but in relation to mathematical facts.[37] The only way for the Quinean of not to feel uneasy in front of this twofold goal is to believe naïvely that Plato had only religious motivations to reject empiricism.

At the end of this analysis, I believe that the Gödel phenomena shows that there are only two stable philosophical interpretations of incompleteness theorems. The first is the orthodox realist position, i.e. the genuine Platonism in philosophy of Mathematics which claims that truth is bivalent and transcendent. In order to get stability, the Platonist has to deny empiricism, even if empiricism is not inconsistent with his ontology. Assuming willingly the transcendent existence of a pure intelligible world incompletely described by mathematical theories and various conjectures, the Platonist gives up knowledge for faith and accepts in his philosophy Faith and Mysticism. The unfathomable mystery lies in the impossibility to give a satisfactory answer to the epistemological question: if mathematical objects are really transcendent to our knowledge, and if they are defined only negatively, being neither spatial, nor temporal, etc., then it is very difficult to explain a causal relation to mathematical objects to our brains. That is also why contemporary realism (Quinean realism for example) is

[37] This difficulty is akin to Gödelian Optimism which denies transcendence but accept bivalence. From Gödelian Optimist's point of view, there is no mathematical proof which is forever beyond proof. But the difficulty is, in such a philosophical system, to interpret Gödel's proof into a bivalent theory and to deny transcendence, because even if it is logically possible, it is very uneasy. On this topic, see [Tennant, 1997], pp. 159-244.

not Platonist in the genuine meaning of this term. But, once mystery accepted, the genuine Platonist claims that the burden of the proof is on the anti-realist as well as on the formalist or even on the unauthentic realist (i.e. the Quinean philosopher) who all have to explain what is mathematical content an have to give a plausible explanation of the history of mathematics. Vuillemin, who has finally held the genuine Platonism at the end of his philosophical development, challenged the logical positivism like that.[38]

The anti-realism *à la Tennant* brings an interpretation of Gödel's incompleteness consistent with the spirit of logical empiricism but without the instability of this latter. Contrarily to Quine's contention, Tennant holds that mathematical truth, and truth in general, consists in provability and he succeeds in showing that it is logically possible to explain Gödel's proof without assuming the substantialist notion of truth but a deflationary-anti-realist account of truth. Then it is possible to explain the history of mathematics as being the history of systems of proofs, without claiming the existence of a transcendent intelligible world that every mathematician brain strives to discover.

To conclude on this point, a sketch of philosophical analysis of Wiles' demonstration of Fermat's conjecture will shed more light on the anti-realist way out. There are differences and analogy between Fermat conjecture and G-sentences. First, Fermat's conjecture is intuitively clear and simple, when the Gödel sentence is weird. Even if Gödel's theorem is an exploit and appears as one of the most important theorems of Mathematical Logic, it is obviously shorter and less difficult from a technical point of view than Wiles' proof which is developed in two hundred pages and can be checked by very few mathematicians in the world. But the main point is that Wiles has proved that there is no solution to the equation $x^n + y^n = z^n$ when $n \geq 3$, that there is no associated curve to this diophantine equation, basing his demonstration on the contemporary Taniyama's conjecture — which is broader than Fermat's — and other mathematical theories taking place later than Fermat. Wiles has turned Fermat's conjecture into a *true* mathematical statement. There is no need of great mathematical knowledge to see that Gödel's theorem proves abstractly the existence of undecidable statements in number theory, when Wiles has proved generally the non-existence of solution for a defined equation. But an analogy with the lesson of Gödel's theorems can be made: the truth of G can be justified only beyond S, from a $S*$ theory to which S belongs, and, maybe in similar way, the proof of Fermat's conjecture is given from another broader conjecture to which the former appears to be a particular case.

The Platonist is convinced that Wiles' demonstration belongs to the history of mathematical *discoveries*, and Gödel's proof shows *in abstracto* the neces-

[38] [Vuillemin, 1997].

sity of this historical development, because it shows that Mathematics cannot be conceived as pure syntax, but has content. The indefinite development of History of Mathematics shows concretely that new mathematical truths can always be reached, and Gödel has logically proved that such a development has no end because there are always, in every mathematical axiomatic system, undecidable statements.

The anti-realist philosopher can concede that Carnap was wrong in believing that Mathematics is only purely syntactic, and can concede that Gödel's proof, showing the distinction between logical and mathematical axioms, leads one to acknowledge also that there is mathematical content. But he denies transcendence to mathematical content, which is 'reduced' to proof systems: every mathematical truth is asserted inside some theory and gets its meaning therefrom. There are no mathematical facts that are really independent of proof. So, from the anti-realist point of view, Wiles' demonstration is based on relations of mathematical proof systems, and Gödel's proof shows formal properties of every mathematical proof system in which elementary number theory can be expressed.

If I agree with Vuillemin's thesis that philosophy has its origin in a free but rational choice, I disagree with his last Platonism and I throw overboard Myth, Mysticism, and Faith, to hold only positive and rational explanations in philosophy of knowledge. That is why, to all philosophers who are reluctant to adopt the genuine Platonism, the anti-realist-deflationary account of Gödel's incompleteness theorems should seem to be apt to get the last word on this so fascinating and so difficult logical-philosophical topic.

References

Carnap, R.: 1937, *The Logical Syntax of Language*. London: Routledge and Kegan.

Dubucs, J.: 1991, 'La philosophie de Kurt Gödel'. *L'âge de la science* 53–68.

Feferman, S.: 1991, 'Reflecting on incompleteness'. *Journal of Symbolic Logic* **56**, 1–49.

Field, H.: 1999, 'Deflating the Conservativeness Argument'. *Journal of Philosophy* 533–540.

Gödel, K.: 1995, *Unpublished Philosophical Essays*. Basel: Birkhäuser Verlag.

Ketland, J.: 1999, 'Deflationism and Tarski's Paradise'. *Mind* **108** (429), 69–94.

Ketland, J.: 2005, 'Deflationism and the Gödel Phenomena - Reply to Tennant'. *Mind* **114** (453), 78–88.

Quine, W.: 1966, *The Ways of Paradox and Other Essays*, Chapt. 1- The Ways of Paradox, 1–18. Harvard University Press.

Quine, W.: 1992, *Pursuit of Truth*. Cambridge, Mass.: Harvard University Press.

Quine, W.: 1995, *From Stimulus to Science*. Harvard University Press.

Rosser, R.: 1936, 'Extensions of some theorems of Gödel and Church'. *Journal of Symbolic Logic* **1**, 87–91.
Russell, B.: 1940, *An Inquiry into Meaning and Truth*. Allen & Unwin.
Shapiro, S.: 1998, 'Proof and Truth: Trough Thick and Thin'. *Journal of Philosophy* (95), 493–521.
Smullyan, R.: 1992, *Gödel's Incompleteness Theorems*. Oxford University Press.
Tennant, N.: 1997, *The Taming of the True*. Oxford, New York: Oxford University Press.
Tennant, N.: 2002, 'Deflationism and the Gödel Phenomena'. *Mind* **111**, 551–582.
Vuillemin, J.: 1981, 'Trois philosophes intuitionnistes: Epicure, Descartes, Kant'. *Dialectica* **35**, 21–41.
Vuillemin, J.: 1996, *Necessity or Contingency - The Master Argument*, Lecture Notes number 56. Stanford: CSLI Publications.
Vuillemin, J.: 1997, 'La question de savoir s'il existe des réalités mathématiques a-t-elle un sens?'. *Philosophia Scientiae* **2** (2), 275–312.

Chapter 6

TRANSPOSITIONS

Henk Visser
University of Maastricht
h.visser@cs.unimaas.nl

6.1 Introduction

Philosophical discussions on 'the nature of mathematical entities' are only relevant if the adopted points of view influence the way in which mathematicians are actually reasoning (Heyting). But what do practicing mathematicians themselves think of the question what mathematics is about? They do not subscribe to the (ironical) view that 'mathematics may be defined as the subject in which we never know what we are talking about, nor whether what we are saying is true' (Russell). Neither do they hold that mathematical entities 'participate' in a 'sphere' called 'logical reality' (Beth). Mathematicians simply do not pursue the question so deeply; their 'objects' are numbers, points, functions, groups, etc. However, there is a kind of relativity involved: a mathematical theory is not *per se* about, say, points or 'geometrical objects', for it may happen that a certain problem, allegedly about such things, can be better solved by imagining that it is about other things such as numbers or 'arithmetical' objects, and conversely.

It will be argued that such 'transpositions' can have an important heuristic value. Switching from one 'domain' to another, more perspicuous 'field of activity', may facilitate the mathematical problem solving process, accordingly as either the mathematician's 'intuitive' skills, or the computer's 'digital' powers can be better exploited in it. This will be demonstrated by solutions to the problem of finding models for finite (affine and projective) geometries.

The argument supports the view that the 'nature of the mathematical objects' may indeed be relevant to the way in which mathematicians are actually reasoning, though in a mundane interpretation which is totally different from its philosophical meaning.

J. van Benthem, G. Heinzmann, M. Rebuschi and H. Visser (eds.), The Age of Alternative Logics: Assessing Philosophy of Logic and Mathematics Today, 15–86.

6.2 Transpositions for combinatorial problems

Suppose we want to determine a competition scheme for an even number of players in such a way that every two players encounter each other in exactly one round, in which all players participate. This problem can easily be solved for 6 players by a standard search procedure (with lexicographical ordering and depth first search). The solution below shows the notation, and if necessary also the structure, of a 'tournament' consisting of 'rounds', and 'rounds' consisting of 'games' (Figure 1).

However, human beings are not proficient in this procedure for larger numbers, and even computer programs may fail for more than 200 players. Yet there is a simple solution, accomplished by a transposition. Kraitchik gave it, and it will henceforth be called 'Kraitchik's solution'. The idea can be demonstrated for 6 players as follows. Replace or represent each number by a point in such a way that 1 is represented by the center of a circle, and the other numbers by points, regularly distributed on the circle. Then the first round is read off from the figure by taking the line 12 and its perpendiculars 36 and 45, the second round by 13 and its perpendiculars 42 and 56, and so on (Figure 2).

It is immediately clear that this method can be applied to every even number of players. Though this is important in itself, we are more interested in the

12 34 56

13 25 46

14 26 35

15 24 36

16 23 45

Fig. 1.

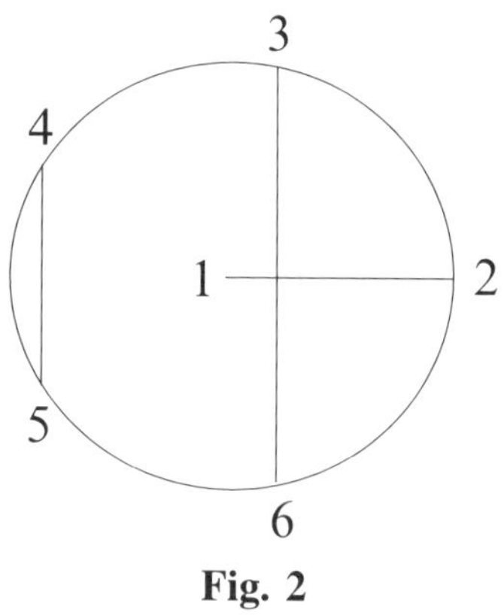

Fig. 2

conditions that make this transposition possible. It is a fact that two 'simple' elements – in this case the players – or the numbers – determine exactly one 'complex' element, in this case the games – or the pairs of numbers. This suggests that a similar transposition will also be successful in the task of finding arithmetical models of elementary geometrical axiomatic systems about points and lines in which the condition holds that two distinct points determine a straight line, or, in other words, that for any two points there is exactly one (straight) line containing them both. Let me explain.

6.3 Transpositions in finite geometry

Suppose we have simple geometrical theories A(n) in which the following axioms hold:

1 For every two distinct points, there is exactly one line containing them both.

2 Through a point not on a given line there is exactly one line that does not meet the given line.

3 Not all points are on the same line.

4 There exists at least one line.

5 Every line contains exactly n points.

As soon as we give the variable n in the fifth axiom a certain value, say 3, we can try to draw points and lines in such a way that these axioms are satisfied, but it is also clear that it is more feasible to regard this as a combinatorial problem. Then the task is facilitated if we apply a transposition by taking numbers for points, and arrays of numbers for lines. This was once done by John Wesley Young, who solved the combinatorial problem for $n = 3$ by a certain reasoning process, with the following result (Figure 3).

This result can also be achieved by a brute force method, using the lexicographical ordering and depth first search. It is easily verified that the condition that every two numbers or 'points' determine exactly one array or 'line', is satisfied. So far there is nothing special about this task. However, things are different for human beings, as soon as larger values for n are taken. Therefore we take advantage of the presence of the above-mentioned condition and try the

123	246	349	478	569
145	258	357		
167	279	368		
189				

Fig. 3

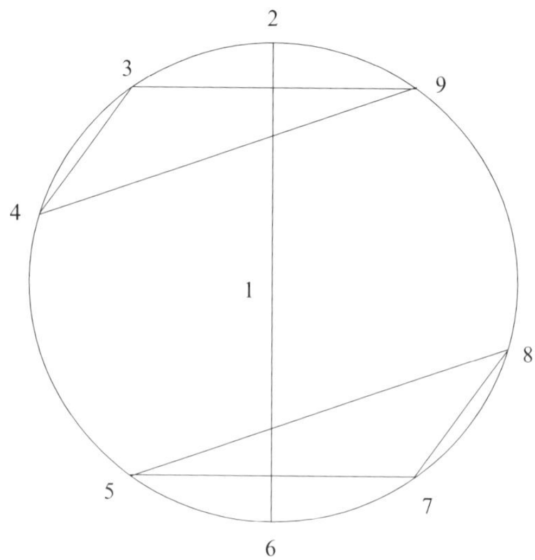

Fig. 4

126	349	578
137	452	689
148	563	792
159	674	823

Fig. 5

hypothesis that solutions can also and more easily be found after a transposition into a domain consisting of a circle with center 1 and containing the other eight points at equal distances from each other. That the answer is positive for $n = 3$ can be seen from the following perspicuous representation, in which *three* 'lines' are pictured, whereas the other nine 'lines' are found by rotations (Figure 4).

Apparently the resulting 'arithmetical' solution is as follows (Figure 5).

That is to say, as soon as the 'geometrical' solution has been found, the rotations are not carried out, but the successive lines are determined by cyclical permutations. In any case, we have found a model of the axiom system A(3), and it is clear that we want to find a similar picture as Figure 4 for axiom system A(4), in which axiom 5 says that every line contains exactly 4 points.

As a matter of fact, Figure 6 shows *two* 'lines', and it is easily seen that the third and fourth 'lines' are found by rotations of the second 'line' over 120 degrees. Then the usual cyclical permutations finish the job. All this results in the following 'arithmetical' model of the axiom system A(4) (Fig. 7 below).

This is not the end of the story. One would also like to find a model for the axiom system A(5), but before tackling this problem, I can pretend that the reader has had enough of the geometrical theories A(n) for the time being. The

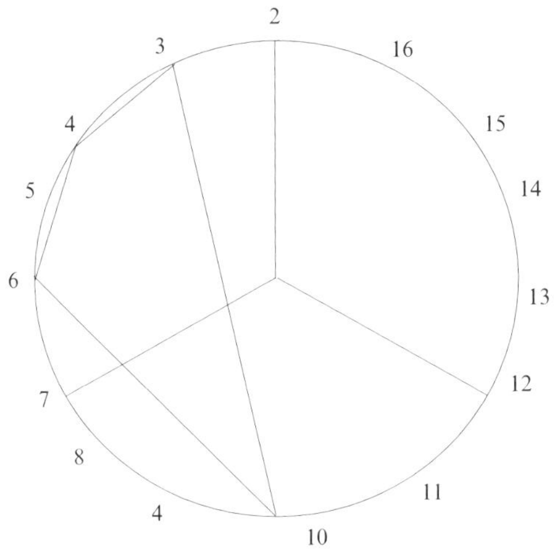

Fig. 6

1	2	7	12	3	4	6	10	8	9	11	15	13	14	16	5
1	3	8	13	4	5	7	11	9	10	12	16	14	15	2	6
1	4	9	14	5	6	8	12	10	11	13	2	15	16	3	7
1	5	10	15	6	7	9	13	11	12	14	3	16	2	4	8
1	6	11	16	7	8	10	14	12	13	15	4	2	3	5	9

Fig. 7

real reason is that the above solution does not give sufficient indications about how to find a solution for the problem of A(5) in a systematic way. Therefore I switch over to the even simpler geometrical theories P(n), in which the following axioms hold:

1 For every two distinct points, there is exactly one line containing them both.

2 For every two distinct lines, there is exactly one point contained by both.

3 Not all points are on the same line.

4 There exists at least one line.

5 Every line contains exactly n points.

Again we can find an arithmetical model for P(3) – the case in which every line contains exactly three points – standard brute force procedure leads to Figure 8.

The similarity with the foregoing cases suggests that here also a transposition is possible, or, more precisely, that a perspicuous representation can be found

123 246 347
145 257 356
167

Fig. 8

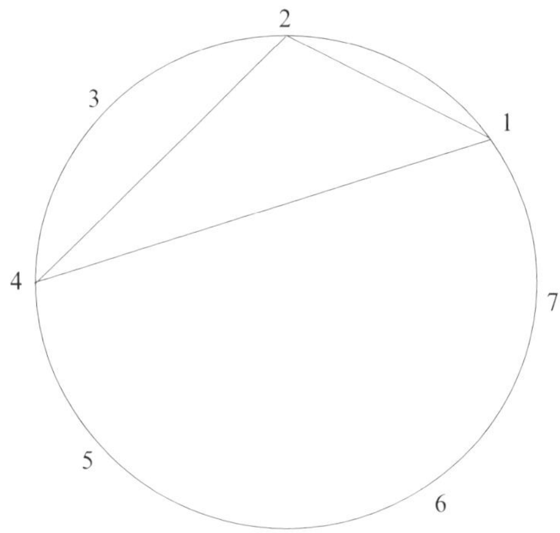

Fig. 9

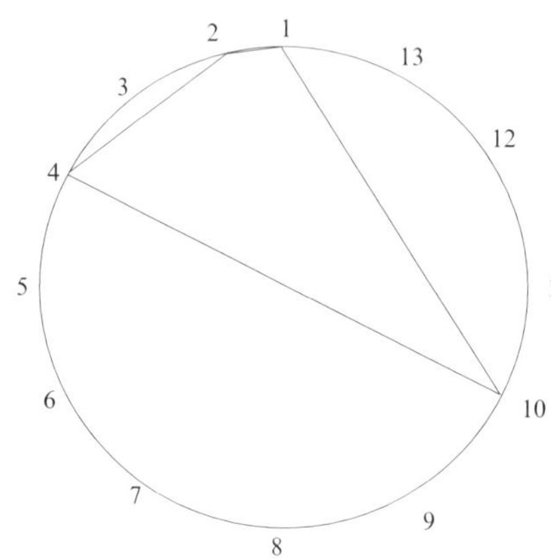

Fig. 10

in the form of a circle, so that rotations can do the work. It appears that we succeed as soon as we place all points on a circle, and choose a triangle in such a way that it has exactly one point in common with each of its rotations around the center of the circle (cf. Fig. 9). Similarly, a model for P(4) was found with a suitable quadrangle (cf. Fig. 10).

Obviously, the corresponding numerical model begins with 1 2 4 10, followed by 2 3 5 11, and so on, until 13 1 3 9 has been reached. But now the question can be asked what is so special about the division of the points on the circle,

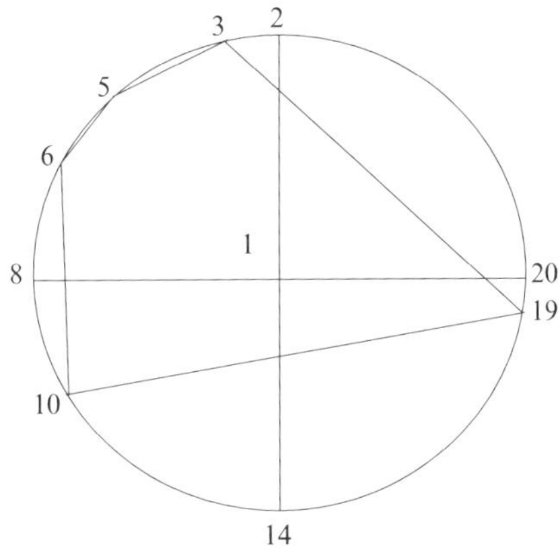

Fig. 11

that every two of these quadrangles have exactly one point in common. The answer is that the successive distances of the vertices of the quadrangle are, respectively, 1, 2, 6 and 4, and this implies that the sums of adjacent distances on the circle are all different. In arithmetical terms: $13 = 1 + 2 + 6 + 4$ is such that each number under 13 appears just once as a partial sum in this equation, in the sense that $1 = 1, 2 = 2, 3 = 1 + 2, 4 = 4, 5 = 4 + 1, 6 = 6, 7 = 4 + 1 + 2, 8 = 2 + 6, 9 = 1 + 2 + 6, 10 = 6 + 4, 11 = 6 + 4 + 1$ and $12 = 2 + 6 + 4$. As soon as this is seen, the construction of models for P(5) and P(6) is easy. Without much effort, the partitions $21 = 1 + 3 + 10 + 2 + 5$ and $31 = 1 + 14 + 4 + 2 + 3 + 7$ can be found. This is a remarkable result, and the hypothesis can be formulated that the problem of finding models for the axiom systems A(n) can also be solved by a transposition to the arithmetical domain in which suitable partitions are found.

As a matter of fact, inspection of the geometrical model of Figure 4 for A(3) yields the equation $8 = 1 + 5 + 2$, in which all numbers under 8 except 4 occur as a partial sum. Similarly, Figure 6 for A(4) gives $15 = 1 + 2 + 4 + 8$. But now we know how to proceed in the as yet unsolved case of A(5), if the search for a suitable pentagon would get stuck: by a transposition! Although I still found a solution by means of the geometrical representation, resulting in Figure 10, the 'modern' way of finding a model-forming partition is leaving the tedious work to the computer.

So after I had found the above solutions, I asked the computer scientist Jeroen Donkers for help. He wrote two computer programs, one for the partition problem that resulted after the three (!) transpositions of the problem of finding a model for axiom systems A(n), and another for the same problem for axiom systems P(n). It appeared that the partition problem for A(5) has two different solutions. The first is given by the equation $24 = 1 + 2 + 8 + 9 + 4$, which can already be read from Figure 10, and the second by the equation $24 = 1 + 3 + 5 + 2 + 13$. However, the partition problem for A(6) appeared to have no solution at all, and this means that this axiom system has no models. On the other hand, the partition problem for A(7) has four different solutions, and the one for A(8) has three.

Similar results were found for axiom systems P(n). Donkers' program showed that the partition problem has two different solutions for P(4), for P(5) only one, for P(6) five, but for P(7) none. It follows that the axiom system for P(7) has no models. There is no need to pursue this subject further, since it may be assumed that such results are corollaries of theorems in affine or projective geometry. My purpose was to show how productive problem solving could arise from a series of transpositions, from geometrical representations to arithmetical representations, from these arithmetical representations to other geometrical representations, and from these geometrical representations to other arithmetical representations. Human beings and computers can help each other in finding solutions to problems, depending on the types of representation each of them can manage best.

We have seen that the choice of a particular geometrical representation for a certain problem can be guided by the success that was achieved with it in the case of a similar problem. But the choice of another medium is one thing, and the selection of certain representations in a chosen medium another. In Kraitchik's solution of the competition problem, a circle plus center was selected instead of only a circle. Then a particular representation, consisting of a diameter with perpendiculars, gave the key to the general solution.

It is easily seen that the choice of a circle without center would have yielded simple solutions for the competition problem with 6, 8, and 10 players. One has only to find suitable combinations of sides and diagonals of (regular) polygons, but their different structures already show that this is not the right road to a general solution.

The question remains how a generalizable representation could be found. To answer this question, I return to the numerical solution of Figure 1. We can transpose each of the rounds to the figure of a circle with center 1 and the other numbers regularly divided over the circle, and then conclude that the third round, 14 26 35, leads to a configuration that leaves nothing to be desired (cf. Fig. 12).

That Figure 12 contains, in a sense, a complete solution of the composition problem for six players, if not for every even number of players, is clear to any well-trained mathematician. Nevertheless it does not require a great effort to find this figure, at least if the above procedure is followed: as soon as the right transpositions are carried out, the solution of the problem is ready to hand. The same holds for the preceding examples. But we have also seen that the choice of promising transpositions was guided by the idea that they were

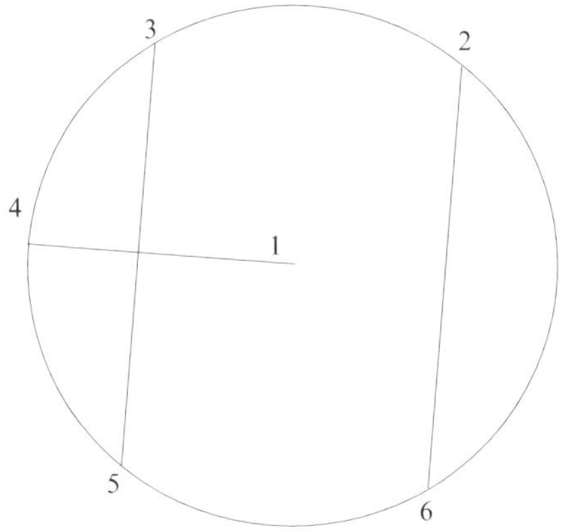

Fig. 12

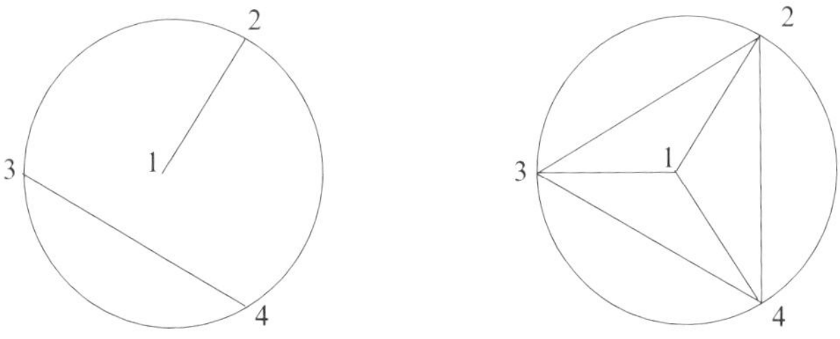

Fig. 13

effective under similar conditions. Thus there are two psychologically relevant aspects of the procedure of transposition: first, the idea that another medium might further the solution, second, the insight that the solution is contained in a particular representation. But there is also a philosophically relevant aspect of transposition, because mathematicians have a certain 'freedom' in choosing the medium of their liking.

6.4 The philosophical relevance of transpositions

The given geometrical figures show, in a sense, how models of finite geometries can arise, but they are themselves not complete models. It seems that performing the rotations would result in pictures that are not perspicuous anymore, as this is the case for A(2) (cf. Fig. 13) and P(2) (cf. Fig. 14).

This is, of course, nothing new. At most we may notice that the complete model of A(2), as pictured in Figure 16, can be seen as a three-dimensional figure that is in principle not different from the well-known tetraeder model below (cf. Fig. 15).

But now there is a remarkable difference between the last picture that shows the model of P(3), and Fano's famous 'projective plane' (Fig. 18).

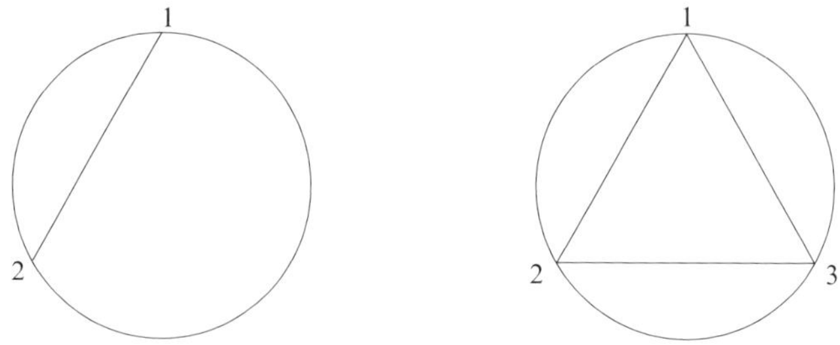

Fig. 14

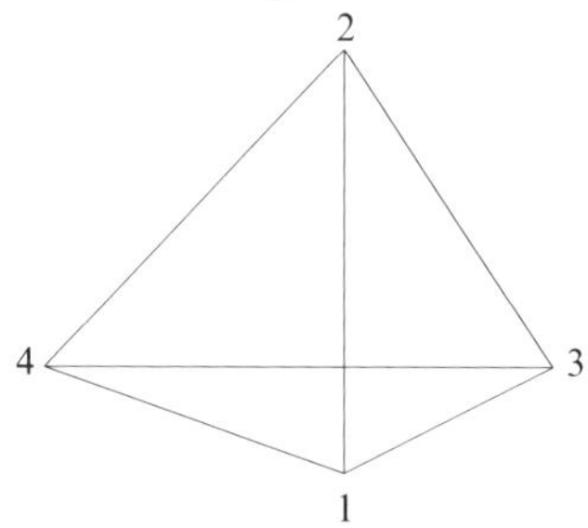

Fig. 15

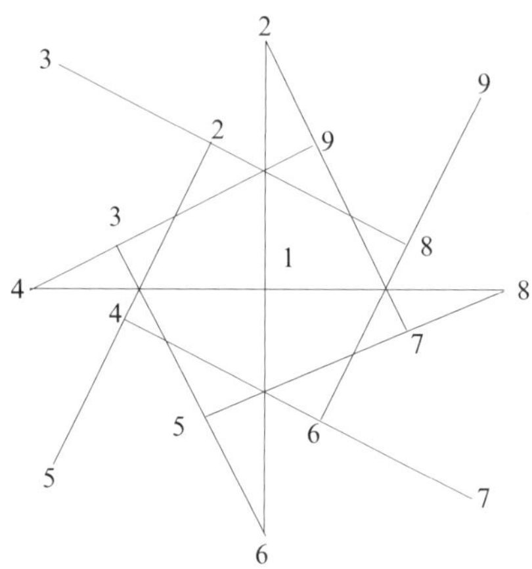

Fig. 16

Yet there is a possibility to depict complete models of A(3) and P(3) too, if one allows that every point gets two 'locations'. The results are sketched in Fig. 16 and Fig. 17.

Figure 17 shows the cyclical structure of the model, whereas the form of Figure 18 suggests something quite different.

It is also possible to construct rotation symmetrical representations of models for A(4) and P(4), provided that every point is given three locations. In order to do this in an easy way for P(4), we start with the line containing the points

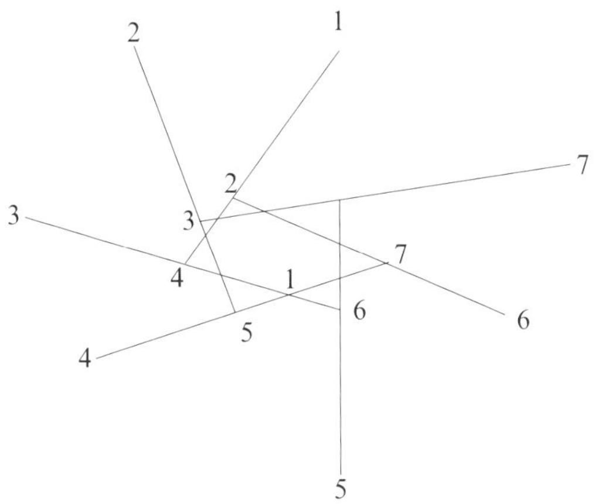

Fig. 17

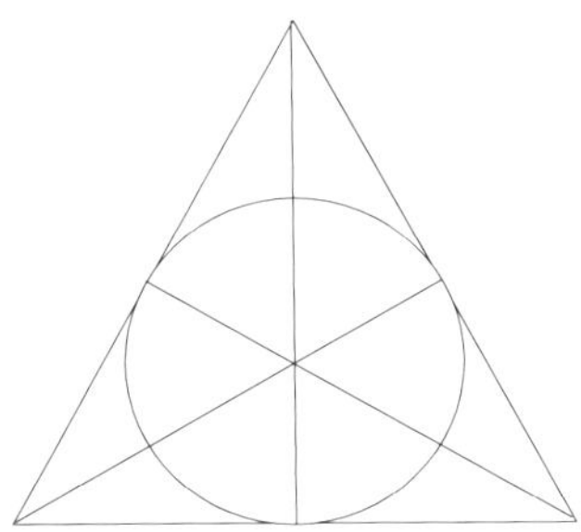

Fig. 18

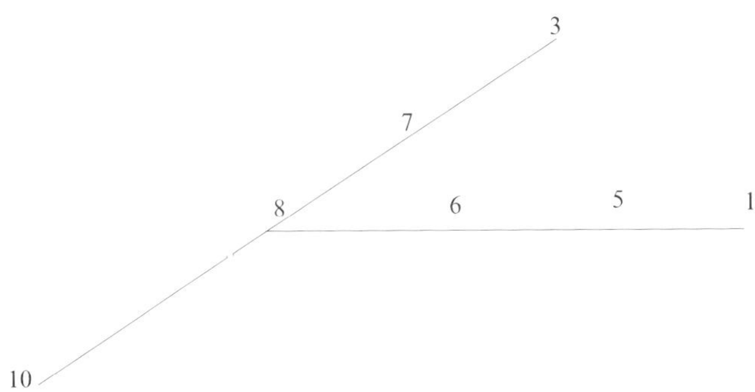

Fig. 19

1, 5, 6 and 8 and continue with the line containing the points 3, 7, 8 and 10 as follows (Fig. 19).

Then we rotate this figure in such a way that the first line 1, 5, 6, 8 coincides with the second line and the second line produces a third line with the points 5, 9, 10 and 12. By repeating this procedure, we successively get the other lines

7, 11, 12, 1; 9, 13, 1, 3; 11, 2, 3, 5; 13, 4, 5, 7; 2, 6, 7, 9; 4, 8, 9, 11; 6, 10, 11, 13; 8, 12, 13, 2; 10, 1, 2, 4 and 12, 3, 4, 6. This results in a figure that is similar to Figure 17 and also has a nice rotational symmetry.

There is no need to pursue the subject further. Only the fact that points can get more 'locations' might disturb some 'naïve' mathematicians. Yet it shows that it makes no sense to speak of 'points' as if the 'nature' of such mathematical objects is 'fixed', let alone 'predetermined': it is the creative mathematician who decides how to represent mathematical objects. He is free to depict them in the medium of his liking, and when he wants to represent one and the same point by more than one 'dot', he may do this and describe it according to his preference. He is not interested in the nature of mathematical entities, but only in the nature of representations of mathematical entities, as long as they are subservient to his aim, that is, solving mathematical problems and requiring insight into the obtained solutions.

II

TRUTH VALUES BEYOND BIVALENCE

Chapter 7

MANY-VALUED AND KRIPKE SEMANTICS

Jean-Yves Béziau*
Institute of Logic
University of Neuchâtel
jean-yves.beziau@unine.ch

7.1 Introduction

Many-valued and Kripke semantics are generalizations of classical semantics in two different "opposite" ways. Many-valued semantics keep the idea of homomorphisms between the structure of the language and an algebra of truth-functions, but the domain of the algebra may have more than two values. Kripke semantics keep only two values but a relation between bivaluations is introduced.

Many-valued semantics were proposed by different people among whom Peirce, Łukasiewicz, Post, Bernays. In fact all these people are also considered as founders of the semantics of classical zero-order logic (propositional logic). And from their work it appears that the creation of many-valued semantics is almost simultaneous to the creation of the bivalent two-valued semantics. From this point of view we cannot say that many-valued semantics are an abstract meaningless generalization developed "après coup", as suggested by Quine in ([Quine 1973], p. 84). However it is true that the meaning of the "many" values is not clear. As Quine and other people have noticed, the division between distinguished and non distinguished values in the domain of the algebra of truth-functions of many-valued semantics is clearly a bivalent feature. So, in some sense many-valued semantics are bivalent, in fact they can be reduced, as shown for example by Suszko, to bivalent (non truth-functional) semantics. Suszko was also against the terminology "logical values" for these many values.

*Work supported by the Swiss Science Foundation.

J. van Benthem, G. Heinzmann, M. Rebuschi and H. Visser (eds.), The Age of Alternative Logics: Assessing Philosophy of Logic and Mathematics Today, 89–101.

He thought that Łukasiewicz was seriously mistaken to consider the third value of his logic as possibility (see [Suszko 1977] and also [da Costa, *et al.* 1996], [Tsuji 1998]). I don't share Suszko's criticism on this point. It seems to me that the many values can be conceived as degrees of truth and degrees of falsity and that we can consider a four-valued semantics in which the two distinguished values can be called "possibly true" and "necessary true", and the two non distinguished values can be called "possibly false" and "necessary false". With this intuition we can develop a four-valued modal logic [Dugundji 1940]. The use of many-valued semantics for the development of modal logic has been completely left out. This can be explained by two reasons: on the one hand the negative results proved by Dugundji showing that S5 and other standard modal logics cannot be characterized by finite matrices [Dugundji 1940], on the other hand the rise of popularity of Kripke semantics.

Today many people identify Kripke semantics with modal logic. Typically a book called "modal logic" nowadays is a book about Kripke semantics (cf. e.g. the recent book by [Blackburn *et al.* 2001]). But modal logic can be developed using other kinds of semantics and Kripke semantics can be used to deal with many different logics and it is totally absurd to call all of these logics "modal logics". Kripke semantics are also often called "possible worlds semantics", however this is quite misleading because the crucial feature of these semantics is not the concept of possible world but the relation of accessibility. Possible worlds can easily be eliminated from the definition of Kripke semantics and then the accessibility relation is defined directly between the bivaluations. For this reason it seems better to use the terminology "relational semantics". Of course, if we want, we can call these bivaluations "possible worlds", this metaphor can be useful, but then why using this metaphor only in the case of relational semantics? In fact in the *Tractatus* Wittgenstein used the expression "truth-possibilities" for the classical bivaluations. Other concepts of the semantics of classical zero-order logic were expressed by him using a modal terminology: he said that a formula is necessary if it holds for all truth-possibilities, impossible if it holds for none, and possible if it holds for some. But Wittgenstein was against the introduction of modal concepts inside the language as modal operators.

Many-valued and Kripke semantics may be philosophically controversial, anyway they are very useful and powerful technical tools which can be fruitfully used to give a mathematical account of basic philosophical notions, such as modalities. It seems to me that instead of focusing on the one hand on some little philosophical problems and on the other hand on some developments limited to one technique, one should promote a better interaction between philosophy and logic developing a wide range of techniques, as for example the combination of Kripke semantics (extended as to include the semantics Jaskowski) and Many-valued semantics (extended as to include non truth-functional many-valued semantics). My aim is this paper is to give a hint of how these techniques can be developed by presenting various examples.

7.2 Many-valuedness and modalities

Many people have nowadays forgotten that the first formal semantics for modal logic was based on many-valuedness. This was first proposed by Łukasiewicz in 1918 and published in [Łukasiewicz 1920]. Moreover many-valued logic was developed by Łukasiewicz in view of modalities, he introduced a third value which was supposed to represent possibility. Although there is no operator of possibility in the standard version of Łukasiewicz's three-valued logic $Ł_3$, at first there was one, eliminated after Tarski showed that it was definable in terms of other non modal connectives.

Łukasiewicz's logic was dismissed as a modal logic by many people, since it has strange features like the validity of the formulas: $\diamond a \wedge \diamond b \rightarrow \diamond(a \wedge b)$. Later on, in 1940, the negative result of Dugundji showing that some of the famous Lewis's modal systems like S4 and S5 cannot be characterized by finite matrices was another drawback for the many-valued approach to modal logic. Nevertheless Łukasiewicz insisted in this direction and in 1953 he presented a four-valued system of modal logic [Łukasiewicz 1953]. This system is also full of strange features and was never taken seriously by modal logicians. At the end of the 1950s the rise of Kripke semantics put a final colon to the love story between many-valuedness and modalities. Nowadays the many-valuedness approach to modal logic is considered as prehistory.

However I think it is still possible to develop in a coherent and intuitive way many-valued systems of modal logic. A possible idea is to consider a set of four-values, two non distinguished values, 0^- and 0^+, and two distinguished values, 1^- and 1^+. These values are ordered by the following linear order: $0^- \prec 0^+ \prec 1^- \prec 1^+$. A possible interpretation is to say that 0^- means necessary false, 0^+ possibly false, 1^- means possibly true and 1^+ means necessary true.

The basic laws for modalities are the following:

$\Box a \vdash a$ $\qquad a \vdash \diamond a$

$a \nvdash \Box a$ $\qquad \diamond a \nvdash a$

$\Box a \vdash \diamond a$ $\qquad \diamond a \nvdash \Box a$

In order for these laws to be valid the tables defining possibility and necessity *must* obey the conditions given by the following table:

a	$\Box a$	$\diamond a$
0^-	0	0
0^+	0	1
1^-	0	1
1^+	1	1

Table 1.

In this table 0 means 0^- or 0^+ and 1 means 1^- or 1^+.

We have many possibility Nevertheless all systems obeying the conditions given by Table 1 obey the involution laws:

$$\Box a \dashv\vdash \Box\Box a$$

$$\Diamond a \dashv\vdash \Diamond \Diamond a$$

the De Morgan laws for modalities:

$$\Box a \wedge \Box b \dashv\vdash \Box(a \wedge b)$$

$$\Diamond a \vee \Diamond b \dashv\vdash \Diamond(a \vee b)$$

as well as Kripke law, considering that implication is defined classically as $\neg a \vee b$ and that disjunction is standardly defined with the operator min:

$$\Box(\neg a \vee b) \vdash \neg\Box a \vee \Box b.$$

One possibility for the minus/plus choice is to reduce the four values to two values 0^- and 1^+. We get then the following table:

a	$\Box a$	$\Diamond a$
0^-	0^-	0^-
0^+	0^-	1^+
1^-	0^-	1^+
1^+	1^+	1^+

Table 2. M4-Red

With this idea we get the collapse of compound modalities:

$$\Diamond a \dashv\vdash \Box \Diamond a$$

$$\Diamond\Box a \dashv\vdash \Box a$$

We are getting therefore very close to S5, although we know, due to Dugundji's theorem that this table cannot define S5. So what are the laws of S5 which are not valid in M4-Red? It depends on the way that we define the non modal connectives. We can reduce the four values to two values 0^- and 1^+ for these connectives or not.

If we do not operate the reduction, we have the standard definitions for conjunction and disjunction with the operators min and max defined on the linear order, and we define the negation in the following logical way:

a	$\neg a$
0^-	1^+
0^+	1^-
1^-	0^+
1^+	0^-

Table 3.

In this case the rule of necessitation

if $\vdash a$ then $\vdash \Box a$

is not valid, as shown by the following table:

p	$\neg p$	$p \vee \neg p$	$\Box(p \vee \neg p)$
0^-	1^+	1^+	1^+
0^+	1^-	1^-	0^-
1^-	0^+	1^-	0^-
1^+	0^-	1^+	1^+

Table 4.

The fact that the rule of necessitation is not valid can be seen as a serious defect. However, Łukasiewicz has argued at length against the validity of such rule (see [Łukasiewicz 1954]).

Another possibility is to operate a reduction of two values for all molecular formulas. In this case, we get a logic in which the law of necessitation is valid but in which self-extensionality

if $a \dashv\vdash b$ then $\Box a \dashv\vdash \Box b$

if $a \dashv\vdash b$ then $\Diamond a \dashv\vdash \Diamond b$

does not hold.

7.3 Possible worlds semantics without possible worlds

It seems that possible worlds are, as stressed by the name, essential in possible worlds semantics.

In possible worlds semantics we have possible worlds and this would be the difference with classical semantics or many-valued semantics. So an expression like "possible worlds semantics without possible worlds" sounds a bit paradoxical like "orange juice without orange", etc. But in fact, as we will see, possible worlds can easily be eliminated from the standard definition leading to a definition which is equivalent in the sense that it defines the same logics.

There are several presentations of possible worlds semantics, let us take a standard one, close to the one given by Johan van Benthem (cf. [van Benthem 1983]).

We consider a Kripke structure $K =< W, R, V >$, as a set W of objects called *possible worlds*, a binary relation R between these worlds called *accessibility relation*, and a function V assigning a set of possible worlds to each atomic formula. Then we give the following definition:

DEFINITION PWS.

(0) $\models_w p$ iff $w \in V(p)$

(1) $\models_w \neg a$ iff $\not\models_w a$

(2) $\models_w a \wedge b$ iff $\models_w a$ and $\models_w b$

(3) $\models_w a \vee b$ iff $\models_w a$ or $\models_w b$

(4) $\models_w a \to b$ iff $\not\models_w a$ or $\models_w b$

(5) $\models_w \Box a$ iff for every $w' \in W$ such that wRw', $\models_{w'} a$

(6) $\models_w \diamond a$ iff for some $w' \in W$ such that wRw', $\models_{w'} a$

What does mean this definition? What does this definition define? It defines a binary relation between the worlds of W of K and formulas, badly expressed by the notation $\models_w a$. This can be read as "the formula a is true in the world w". From this definition, we then define what it means for a formula a to be true in the Kripke structure K: a is true in K iff it is true in every world w of K.

As we see, in these definitions, the nature of the worlds is never used, they can be anything. Why then calling them worlds? What is used is the relation of accessibility: different properties of this relation lead to different logics.

The second important point is that the definition defines a binary relation between the worlds of W of K and formulas by *simultaneous* recursion: in clauses (4) and (5), to define the relation between a world w and a formula, we use the relation defined between another world w' and formulas. In classical semantics and many-valued semantics, we only use *simple* recursion.

Let us now transform this definition into a worldless definition Instead of considering a Kripke structure, we consider a Ipke structure

$I =< D, R >$

as a set D of functions called *distributions of truth-values* assigning to every atomic formula a the values 0 (false) or 1 (true), and a binary relation R between these distributions called *accessibility relation*.

We now extend these distributions into bivaluations, i.e. function assigning to every formula (atomic or molecular) the values 0 (false) or 1 (true).

DEFINITION PWS-W.

(0) $\beta_\delta(p) = 1$ iff $\delta(p) = 1$

(1) $\beta_\delta\neg(a) = 1$ iff $\beta_\delta(a) = 0$

(2) $\beta_\delta(a \wedge b) = 1$ iff $\beta_\delta(a) = 1$ and $\beta_\delta(b) = 1$

(3) $\beta_\delta(a \vee b) = 1$ iff $\beta_\delta(a) = 1$ or $\beta_\delta(b) = 1$

(4) $\beta_\delta(a \to b) = 1$ iff $\beta_\delta(a) = 0$ or $\beta_\delta(b) = 1$

(5) $\beta_\delta(\Box a) = 1$ iff for every $\beta'_{\delta'}$ such that $\delta R\delta'$, $\beta'_{\delta'}(a) = 1$

(6) $\beta_\delta(\diamond a) = 1$ iff for some $\beta'_{\delta'}$ such that $\delta R\delta'$, $\beta'_{\delta'}(a) = 1$

Using the above definition, we can then define, what it means to be true in the Ipke structure I: a formula a is true iff it is true for every bivaluation.

EQUIVALENCE OF THE TWO DEFINITIONS. *It is the same to be true in a Kripke structure or to be true in an Ipke structure.*

This claim means more precisely that given a Kripke structure, we can construct an Ipke structure which leads to the same notion of truth and vice-versa. The construction is very simple. Given a Kripke structure, we transform a possible world w into a distribution δ_w by putting $\delta_w(p) = 1$ iff $w \in V(p)$. Given an Ipke structure, we transform a distribution δ into a possible world w_δ obeying the condition: $w \in V(p)$ iff $\delta(p) = 1$. This condition in fact defines the function V.

In both cases the accessibility relation is transposed from worlds to distributions and vice-versa.

Someone may claim that possible worlds are nice tools, they help imagination, they are *heuristical*. But we may call bivaluations in DEFINITION PWS-W, possible worlds. We still get the heuristics, but keep a low ontological cost. In fact some people even call possible worlds, the bivaluations of the standard semantics of classical propositional logic, following the first idea of Wittgenstein.

In some recents advances in possible worlds semantics (Dutch trend), possible worlds may be useful, but they are totally useless for the standard semantics of $S5$, etc. On the other hand to work without possible worlds can simplify further constructions as the ones presented in the next sections.

7.4 Combining many-valued and Kripke semantics

If we consider possible worlds semantics without possible worlds, i.e., given by DEFINITION PWS-W, it is easy to combine them with many-valued semantics: instead of considering bivaluations, we consider functions into a finite set of values divided into two sets, the sets of distinguished values and the set of non-distinguished values. We will call such combined semantics *Many-valued Kripke semantics*.

Sometimes people talk about impossible worlds or incomplete worlds (see e.g. the volume 38 (1997) of *Notre Dame Journal of Formal Logic*). An impossible world is a world in which a formula and its negation can both be true, an incomplete world is a world in which a formula and its negation can both be false. These impossible worlds (or incomplete worlds) semantics can be described more efficiently by Many-valued Kripke semantics.

Let us give an example of many-valued relational semantics, we consider a many-valued Ipke structure $MI =< D, R >$, where D is a set of distributions assigning to every atomic formula a, the values 0, $\frac{1}{2}$ or 1 and where R is a binary relation of accessibility between these distributions. We now extend these distributions into three valuations, i.e. function assigning to every formula (atomic or molecular) the values 0, $\frac{1}{2}$ or 1.

DEFINITION MI.

(0) $\theta_\delta(p) = \delta(p)$

(1) $\theta_\delta\neg(a) = 1$ iff $\theta_\delta(a) = 0$

(2) $\theta_\delta(a \wedge b) = min(\theta_\delta(a), \theta_\delta(b))$

(3) $\theta_\delta(a \vee b) = max(\theta_\delta(a), \theta_\delta(b))$

(4) $\theta_\delta(a \rightarrow b)$ is distinguished iff $\theta_\delta(a)$ is non distinguished or $\theta_\delta(b)$ is distinguished.

(5) $\theta_\delta(\Box a) = 1$ is distinguished iff for every $\theta'_{\delta'} \in W$ such that $\delta R \delta'$, $\theta'_{\delta'}(a)$ is distinguished.

(6) $\theta_\delta(\diamond a) = 1$ is distinguished iff for some $\theta'_{\delta'} \in W$ such that $\delta R \delta'$, $\theta'_{\delta'}(a)$ is distinguished.

At first this definition seems quite the same as DEFINITION PWS-W of the preceding section, but since we have a third value, things change. From clause (1), we deduce that

(2') $\theta_\delta\neg(a) = \frac{1}{2}$ iff $\theta_\delta(a) = \frac{1}{2}$.

If we consider that 1 is distinguished and the values 0 and $\frac{1}{2}$ are non-distinguished, then the principle of contradiction expressed by the formula $\neg(p \wedge \neg p)$ is not true in MI, provided we standardly define "true in MI" by "distinguished for every three-valuations": we have some three-valuations in which both values of p and $\neg p$ are $\frac{1}{2}$, and therefore in which the value of $\neg(p \wedge \neg p)$ is $\frac{1}{2}$, i.e. non-distinguished. This is nothing very knew and this is what happens in Łukasiewicz three-valued logic Ł$_3$, where we have:

$$\nvdash \neg(a \wedge \neg a)$$

We are just *combining* different semantics. What happens here is that, at the level of modalities, we don't either have the principle of non contradiction:

$$\nvdash \neg(\Box a \wedge \neg\Box a) \qquad \nvdash \neg(\diamond a \wedge \neg \diamond a)$$

If we take $\frac{1}{2}$ and 1 as distinguished and only 0 as non-distinguished and provided we define the consequence relation in the usual way, then the formulas above are valid but the formulas below expressing the *ex-falso sequitur quod libet* which are valid with only 1 as distinguished are not valid anymore:

$$\nvdash (p \wedge \neg p) \rightarrow q \qquad p, \neg p \nvdash q$$
$$\nvdash (\Box p \wedge \neg \Box p) \rightarrow q \qquad \Box p, \neg \Box p \nvdash q \;\; ll$$
$$\nvdash (\diamond p \wedge \neg \diamond p) \rightarrow q \qquad \diamond p, \neg \diamond p \nvdash q$$

These two possible Many-valued Kripke semantics show that the principle of contradiction is independent of the *ex-falso sequitur quod libet* in its two forms, consequential or implicational.

7.5 JKL semantics

Following some ideas of Jaskowski, we can change the definition of truth in a Kripke structure K, by saying that a formula a is true in K iff it is true at *some* world, i.e. there is *some* valuation in which it is true. In case we are working with Many-valued Kripke semantics, this means: there is *some* valuation for which the value of this formula is distinguished.

We will call many-valued with this definition of truth, "JKL-semantics". Such Semantics were introduced in [Béziau 2001].

If we consider the JKL Semantics corresponding to the Semantics MI of the preceding section, with only 1 as distinguished, we have:

$$a, \neg a \nvdash b \qquad \Box a, \neg \Box a \nvdash b \qquad \diamond a, \neg \diamond a \nvdash b$$

but

$$\vdash (a \wedge \neg a) \rightarrow b \qquad \vdash (\Box a \wedge \neg \Box a) \rightarrow b \qquad \vdash (\diamond a \wedge \neg \diamond a) \rightarrow b$$

and

$$\nvdash \neg(a \wedge \neg a) \qquad \nvdash \neg(\Box a \wedge \neg \Box a) \qquad \nvdash \neg(\diamond a \wedge \neg \diamond a).$$

Something that would be interesting is a logic in which the principle of contradiction and the *ex-falso sequitur quod libet* in its two forms are not valid only for modalities. This fits well for example for a logic of beliefs, where someone may have contradictory beliefs without "exploding", but where contradictions explode at the factual level. For this, we need a more sophisticated construction.

7.6 Non truth-functional Kripke semantics

Many-valued semantics are generally truth-functional, that means that they are matrices (see [Béziau 1997] for a detailed account on this question). But it is also possible to introduce non truth-functional many-valued semantics. I have introduced these kind of Semantics in [Béziau 1990] and developed furthermore the subject in [Béziau 2002].

To understand what it means, let us first explain the difference between truth-functional semantics and non truth-functional semantics at the level of bivalent semantics. The set of bivaluations of the semantics of propositional classical logic is the set of homomorphisms from the algebra of formulas and the matrix of truth-functions defined on $\{0, 1\}$. Since the algebra of formulas is an absolutely free algebra, this set can be generated by the set of distributions, i.e. functions assigning 0 or 1 to atomic formulas. A *non truth-functional bivalent semantics* is a semantics where the bivaluations cannot be reduced to homomorphisms between the algebra of formula and an algebra of truth-functions defined on $\{0, 1\}$.

The semantics of classical logic can be presented in two different ways which are equivalent: the usual way with the distributions and the matrix, or by defining directly a set of bivaluations (functions from the whole set of formulas into $\{0, 1\}$ obeying the following conditions:

(1) $\beta\neg(a) = 1$ iff $\beta(a) = 0$
(2) $\beta(a \wedge b) = 1$ iff $\beta(a) = 1$ and $\beta(b) = 1$
(3) $\beta(a \rightarrow b) = 0$ iff $\beta(a) = 1$ and $\beta(b) = 0$

We have a fairly simple example of non truth-functional bivalent semantics, if we replace the condition (1) by the conditions (1'):

(1') if $\beta\neg(a) = 1$ then $\beta(a) = 0$

In this logic, we may have $\beta\neg(a) = \beta(a) = 0$. The logic generated by this condition has been studied in [Béziau 1999a]. Another example of non truth-functional bivalent semantics can be found in [Béziau 1990b]. A general study of logics from the viewpoint of bivalent semantics (truth-functional or non truth-functional) has been developed in [da Costa, *et al.* 1994].

The definition of non truth-functional many-valued semantics is a straightforward generalization: A *non truth-functional many-valued semantics* is a semantics where the valuations cannot be reduced to homomorphisms between the algebra of formula and an algebra of truth-functions defined on a given set of values.

A very simple is the following: we replace Łukasiewicz's condition for negation by the following:

(1') if $\beta\neg(a) = \frac{1}{2}$ then $\beta(a) = \frac{1}{2}$

Now we will construct a non truth-functional many-valued semantics. As in the case of the bivalent semantics for classical propositional logic, truth-

In general this is presented in a rather informal way, where the matrix does not really appear but is described indirectly by means of truth-tables, see [Béziau 2000].

functional bivalent (or many-valued) semantics can be presented in two different way. For example, instead of DEFINITION PWS-W, we can consider an Ipke structure $I = < B, R >$ as a set B of bivaluations assigning to every formula (atomic or molecular) 0 or 1 and a relation R of accessibility between bivaluations.

Then we stipulate that these bivaluations should obey the following conditions:

DEFINITION PWS-W GLOBALIZED

(1) $\beta\neg(a) = 1$ iff $\beta(a) = 0$
(2) $\beta(a \wedge b) = 1$ iff $\beta(a) = 1$ and $\beta(b) = 1$
(3) $\beta(a \vee b) = 1$ iff $\beta(a) = 1$ or $\beta(b) = 1$
(4) $\beta(a \to b) = 1$ iff $\beta(a) = 0$ or $\beta(b) = 1$
(5) $\beta(\Box a) = 1$ iff for every $\beta' \in B$ such that $\beta R \beta'$, $\beta'(a) = 1$
(6) $\beta(\Box a) = 1$ iff for every $\beta' \in B$ such that $\beta R \beta'$, $\beta'(a) = 1$

Now we replace condition (1) by the following set of conditions:
(1.1.1.) if $\beta(a) = 0$ then $\beta\neg(a) = 1$
(1.2.2.) if $\beta\neg\neg(a) = 1$ then $\beta\neg(a) = 0$
(1.2.3.) if $\beta\neg(a \wedge b) = 1$ then $\beta(a \wedge b) = 0$
(1.2.4.) if $\beta\neg(a \vee b) = 1$ then $\beta(a \vee b) = 0$
(1.2.5.) if $\beta\neg(a \to b) = 1$ then $\beta(a \to b) = 0$

This semantics is non truth-functional. In the logic defined by this semantics, we have:

$\nvdash \neg(\Box a \wedge \neg\Box a)$ $\quad$ $\nvdash \neg(\diamond a \wedge \neg \diamond a)$
$\Box a, \neg\Box a \nvdash b$ $\quad$ $\Box a, \neg\Box a \nvdash b$
$\nvdash (\Box a \wedge \neg\Box a) \to b$ $\quad$ $\nvdash (\diamond a \wedge \neg \diamond a) \to b.$

but

$\vdash \neg(a \wedge \neg a)$ $\quad$ $a, \neg a \vdash b$ $\quad$ $\vdash (a \wedge \neg a) \to b$

provided there are no modalities in a.

7.7 Conclusion: Many possibilities

We have presented different way to generalize and to combine many-valued and Kripke semantics, and in fact there are still some other possibilities like the semantics developed by Buchsbaum and Pequenos (see e.g. [Buchsbaum *et al.* 2004]) or like the semantics of possible translations developed by Carnielli and Marcos (see e.g. [Carnielli, *et al.* 2002]).

All these tools may be very useful both from an abstract viewpoint of a general theory of logics (see e.g. [Béziau 1994]) and from applications to philosophical problems. For example they can be used, as we have shown, to construct models showing the independency of some properties of negation relatively to some other ones. This is very useful in the field of paraconsistent logic.

References

Béziau, J.-Y., (1990), "Logiques construites suivant les méthodes de da Costa", *Logique et Analyse*, **131-132**, 259–272.

Béziau, J.-Y., (1994), "Universal Logic", in *Logica'94 - Proceedings 8th International Symposium*, Czech Academy of Science, Prague, 73–93.

Béziau, J.-Y., (1997), "What is many-valued logic?", in *Proceedings of the 27th International Symposium on Multiple-Valued Logic*, IEEE Computer Society, Los Alamitos, 117–121.

Béziau, J.-Y., (1999), "What is paraconsistent logic ?", in *Frontiers of paraconsistent logic*, Research Studies Press, Baldock, 2000, 95–112.

Béziau, J.-Y., (1999), "Classical negation can be expressed by one of its halves", *Logical Journal of the IGPL*, **7**, 145–151.

Béziau, J.-Y., (1999b), "A sequent calculus for Łukasiewicz's three valued logic based on Suszko's bivalent semantics", *Bulletin of the Section of Logic*, **28**, 89–97.

Béziau, J.-Y., (2000), "The philosophical import of Polish logic", in *Methodology and Philosophy of Science at Warsaw University*, M. Talsiewicz (ed), Semper, Warsaw, 109–124.

Béziau, J.-Y., (2001), "The logic of confusion", in *Proceedings of the International Conference of Artificial Intelligence IC-AI'2002*, H.R. Arbnia (ed), CSREA Press, Las Vegas, 2001, 821–826.

Béziau, J.-Y., (2002), "Non truth-functional many-valued semantics", paper presented at the WoLLiC, Rio de Janeiro, August 1–3, 2002, submitted.

Béziau, J.-Y., (2004), "A new four-valued approach to modal logic", *Logika*, **22**.

Béziau, J.-Y. and D. Sarenac, (2001), "Possible worlds: a fashionable nonsense?", submitted.

Blackburn, P., M. de Rijke and Y. Venema, (2001) *Modal Logic*, Cambridge University Press.

Buchsbaum, A., T. Pequeno and M. Pequeno, (2004). *Game semantics for four players paraconsistent modal logic*, to appear.

Carnielli, W. A. and J. Marcos, (2002). "A taxonomy of C-systems", in W. A. Carnielli, M. E. Coniglio, and I. M. L. D'Ottaviano (eds), *Paraconsistency: The logical way to the inconsistent. Proceedings of the II World Congress on Paraconsistency (WCP'2000)*, Marcel Dekker, 1–94.

Da Costa, N.C.A. and J.-Y. Béziau, (1994), "Théorie de la valuation", *Logique et Analyse*, **146**, 95–117.

Da Costa, N.C.A., J.-Y. Béziau, O.A.S. Bueno, (1996), "Malinowski and Suszko on many-valuedness : on the reduction of many-valuedness to two-valuedness", *Modern Logic*, **6**, 272–299.

Dugundji, J., (1940), "Note on a property of matrices for Lewis and Langford's calculi of propositions", *Journal of Symbolic Logic*, **5**, 150–151.

Font, J.M. and P. Hajek, (2002), "On Łukasiewicz's four-valued logic", *Studia Logica*, **70**, 157–182.

Goldblatt, R., "Mathematical modal logic: a View of its Evolution". to appear in *A History of Mathematical Logic*, edited by D. van Dalen, J. Dawson and A. Kanamori.

Łukasiewicz, J., (1920), "O logice trójwarto±ciowej", *Ruch Filozoficny*, **5**, 170–171.

Łukasiewicz, J., (1953), "A system of modal logic", *Journal of Computing Systems*, **1**, 111–149.

Łukasiewicz, J., (1954), "On controversial problem of Aristotle's modal syllogistic", *Dominican Studies*, **7**, 114–128.

Proc. Acta. Phil. (1963), *Proceedings of a Colloquium on Modal and Many-valued Logics* , Acta Philosophica Fennica, **16**, Helsinki, 1963.

Quine, W.V.O., (1973), *Philosophy of logic*, Prentice-Hall, New Jersey, 1973.

Suszko, R., (1977), "The Fregean axiom and Polish mathematical logic in the 1920s", *Studia Logica*, **36**, 377–380.

Tsuji, M. , (1998), "Many-valued logics and Suszko's Thesis revisited", *Studia Logica*, **60**, 299–309.

van Benthem, J., (1983), *Modal logic and classical logic*, Bibliopolis, Naples.

Chapter 8

THE LOGIC OF COMPLEMENTARITY

Newton C. A. da Costa,
Décio Krause
Federal University of Santa Catarina
Department of Philosophy (both authors)
ncacosta@terra.com.br
d.krause@pop.com.br

> "We must, in general, be prepared to accept the fact that a complete elucidation of one and the same object may require diverse points of view which defy a unique description."
>
> —Niels Bohr, 1929

8.1 Introduction

J. Kalckar, the editor of volume 6 of Bohr's Collected Works [Bohr, 1985] pp. 26-27, suggested that the first reference to the notion of complementarity is to be found in a manuscript written by Bohr on July 10, 1927, where we read that "... the theory exhibited a duality when one considered on one hand the superposition principle and on the other hand the conservation of energy and momentum (...) Complementary aspects of experience that cannot be unified into a space-time picture on the classical theories" (cf. *ibid.*, pp. 26-27). Roughly speaking, the idea involves something like this: complementarity means the possibility of unifying aspects which cannot be put together from a 'classical' perspective. Kalckar also keeps off the view claimed by some writers who have sustained that Bohr was motivated by sources outside physics, like the

J. van Benthem, G. Heinzmann, M. Rebuschi and H. Visser (eds.), The Age of Alternative Logics: Assessing Philosophy of Logic and Mathematics Today, 103–120.

readings of Kierkegaard or of the Danish philosopher H. Høffding. According to Kalckar, the very origins of such an idea came from physics itself and, to reinforce his claim, he recalls L. Rosenfeld's words, which interest us also here: "Bohr's conception of complementarity in quantum mechanics is not the expression of a 'specific philosophical position', but an inherent *part of the theory* which has the same validity as its formal aspect and is inseparable from it." (*apud ibid.*, p. 28, italics ours).

It should be remarked that it seems to exist a discrepancy between Rosenfeld and Bohr in what concerns the way of understanding complementarity. This difference may justify Bohr's refuse to accept that von Weizsäcker had described 'the logic of complementarity', as we shall see below. Apparently, Bohr envisaged his ideas on complementarity as forming part of a general epistemological principle, which could guide us not only in physics (from which the ideas really came), but in any other field of science as well; as he said, "... the lessons taught us by recent developments in physics regarding the necessity of a constant extension of the frame of concepts appropriate for the classification of new experiences leads us to a general epistemological attitude which might help us to avoid apparent conceptual difficulties in other fields of science as well" ([Bohr, 1937]). In other words, we might say that, according to Bohr, complementarity may be viewed as a kind of a general regulative methodological principle. On the other hand, there are positions sustained by people like Rosenfeld (and von Weizsäcker), who see such ideas as making part of the (physical) theory itself. What is the difference?

The difference lies in what we consider as a meta-theoretical principle of science and what is to be considered as a strict principle of a particular (say, axiomatized) scientific theory. The former may be viewed as a meta-principle, while the latter is something to be 'internalized' within the object language of the theory itself. In what follows, we shall try to push this distinction a little bit in relation to the concept of complementarity. This is of particular importance for, as we shall see below, the very idea of complementarity resembles that of the existence of contradictions; keeping it as a meaning principle, it seems easier to understand how it may help us in accepting that "[t]he apparently incompatible sorts of information about the behavior of the object under examination which we get by different experimental arrangements can clearly not be brought into connection with each other in the usual way, but may, as equally essential for an exhaustive account of all experience, be regarded as 'complementary' to each other" [Bohr, 1937].

In this paper, we shall consider how 'complementary ideas' can be seen from both perspectives, that is, as standing both for a general regulative meaning principle and also as a law that can be internalized in the language of the theory proper. As we shall see, although resembling contradictions (but see a way of better specifying them below), the concept of complementary propositions can be put within a certain object language (so keeping it as an 'inherent part of the theory' as Rosenfeld has claimed) without risk of trivializing the whole theory. This will enable us to discuss also the role played by logic in the context of the physical sciences.

We begin firstly by describing the main features connected with the idea of complementarity. We note that there is no general agreement among historians and philosophers (and even among physicists) about the precise meaning of Bohr's Principle of Complementarity (henceforth, PC), what makes the historical analysis quite problematic [Beller, 1992], [Jammer, 1966], [Jammer, 1974]. Even so, after revising some of the main references made by Bohr himself and by various of his commentators on complementarity, we arrive at a characterization of 'complementary propositions' from a strict logical point of view (that is, as defined in a suitable formal language). Then, we shall sketch the main norms of the logic of such propositions, by evidencing that it is a kind of paraconsistent logic, termed *paraclassical logic* (see [da Costa and Vernengo, 1999], [Souza, 2000]). Nonetheless, complementarity, for us, also encompasses meta-theoretical meaning principles imposing some limitations on theories; in addition, it sanctions the use of incompatible approaches in physics. Complementarity, as a meaning principle, plays the role of a kind of normative rule.

Secondly, we insist that the relevance of this kind of study is neither merely historical nor an exercise of logic. In addition to the necessity of a philosophical distinction between meaning principles and strict physical laws, we believe that this discussion has a profound philosophical significance also in showing some of the relationships that there exist between certain non-classical logics and the empirical sciences, in particular to physics. Of course, although in this paper we neither have pursued the historical details on complementarity in deep, though we have mentioned some of the main references one finds in the literature, nor have investigated the logical system we propose in all its formal aspects (a task we hope to accomplish in the near future), we hope to make clear the general underlying idea of the paper. It was and continue partially motivated by Bohr's own way of accepting both the particle and the wave pictures of reality. We believe that the understanding of a wide field of knowledge, like quantum physics, may gain in much if we accept a pluralistic view according to which there are several and eventually non equivalent ways of looking at it (perhaps some of them based on non-classical logics), each one being adequate from its particular perspective, and showing details which cannot be seen from the other points of view, analogously to the different drawings of an engineer in descriptive geometry, *à la* Monge, of a given object.

8.2 Complementarity

The concept of 'complementarity' was introduced in quantum mechanics by Niels Bohr in his famous 'Como Lecture', in 1927, although the basic ideas go back to 1925 [Bohr, 1927], [Bohr, 1985]. The consequences of his view were fundamental, particularly for the development of the Copenhagen interpretation of quantum mechanics and constitutes, as it is largely recognized in the literature, one of the most fundamental contributions to the development of quantum theory (see also [Beller, 1992], [Jammer, 1966; Jammer, 1974]).

In this section we make clear in what sense we understand the word 'complementarity'. The quotations taken from Bohr and from other important commentators aim at to reinforce our view, although we are of course aware that a few isolated quotations cannot provide evidence for the full understanding of concepts, especially regarding the present (and difficult) case. Even so, we hope we can convince the reader that complementarity can be interpreted as a more general principle related to 'incompatibility' in some sense (the 'sense' being explained in the next sections) than to some kind of impossibility of 'simultaneously measuring'.

In what concerns this point, we remark that we find Bohr speaking about complementary concepts which cannot be used *at the same time* (as we see in several of his papers listed in our references [Bohr, 1985, p. 369]). Though this way of talking should be viewed as a way of speaking, for it stands for situations which, according to Bohr himself, demand specific analysis; as he says, "[o]ne must be very careful, therefore, in analyzing which concepts actually underly limitations" (ibid., p. 370). Really, there are several ways of looking at complementarity. Pauli, for instance, claimed that "[if] the use of a classical concept excludes of *another*, we call both concepts (. . .) *complementary* (to each other), following Bohr" ([Pauli, 1980, p. 7], quoted in [Cushing, 1994, p. 33]). By the way, J. Cushing also stressed his own view, in saying that "[w]hatever historical route, Bohr did arrive at a doctrine of mutually exclusive, incompatible, but necessary classical pictures in which any given application emphasizing one class of concepts *must* exclude the other" (op. cit., pp. 34-5).

This idea of complementary propositions as 'excluding' each other (what appears to mean something like 'incompatibility') is reinforced by Bohr himself in several passages, as the following ones:

> "The existence of different aspects of the description of a physical system, seemingly incompatible but both needed for a complete description of the system. In particular, the wave-particle duality." (*apud* [French and Kennedy, 1985, p. 370])

> "The phenomenon by which, in the atomic domain, objects exhibit the properties of both particle and waves, which in classical, macroscopic physics are mutually exclusive categories." (ibid., pp. 371–372.)

> "The very nature of the quantum theory thus forces us to regard the space-time co-ordination and the claim of causality, the union of which characterizes the

classical theories, as complementary but exclusive features of the description, symbolizing the idealization of observation and definition respectively." [Bohr, 1927, p. 566.]

"The apparently incompatible sorts of information about the behavior of the object under examination which we get by different experimental arrangements can clearly not be brought into connection with each other in the usual way, but may, as equally essential for an exhaustive account of all experience, be regarded as 'complementary' to each other." ([Bohr, 1937, p. 291]; [Scheibe, 1973, p. 31.])

"Information regarding the behaviour of an atomic object obtained under definite experimental conditions may, however, according to a terminology often used in atomic physics, be adequately characterized as *complementary* to any information about the same object obtained by some other experimental arrangement excluding the fulfillment of the first conditions. Although such kinds of information *cannot be combined into a single picture* by means of ordinary concepts, they represent indeed equally essential aspects of any knowledge of the object in question which can be obtained in this domain." ([Bohr, 1938, p. 26], *apud* [Scheibe, 1973, p. 31], second italic ours.)

E. Scheibe also says that

"... which is here said to be 'complementary', is also said to be 'apparently incompatible', the reference can scarcely be to those classical concepts, quantities or aspects whose *combination* was previously asserted to be characteristic of the classical theories. For 'apparently incompatible' surely means incompatible on classical considerations alone." [Scheibe, 1973, p. 31.]

In other words, it is perfectly reasonable to regard complementary aspects as *incompatible*, in the sense that their *combination* into a single description may lead to difficulties. But in a theory grounded on standard logic, the conjunction of two theses is also a thesis; in other words, if α and β are both theses or theorems of a theory (founded on classical logic), then $\alpha \wedge \beta$ is also a thesis (or a theorem) of that theory. This is what we intuitively mean when we say that, on the grounds of classical logic, a 'true' proposition cannot 'exclude' another 'true' proposition. In this sense, the quantum world is rather distinct from the 'classical', for although complementary propositions are to be regarded as acceptable, their conjunction seems to be not.

This corresponds to the fact that, in classical logic, if α is a consequence of a set Δ of statements and β is also a consequence of Δ, then $\alpha \wedge \beta$ (α *and* β) is also a consequence of Δ. If β is the negation of α (or vice-versa), then this rule implies that from the set of formulas Δ we deduce a contradiction $\alpha \wedge \neg\alpha$ (or $\neg\beta \wedge \beta$). In addition, when α and β are in some sense incompatible, $\alpha \wedge \beta$ constitutes an impossibility.

Therefore, as we shall show below, part of a natural procedure to surmount the problem is to restrict the rule in question. But before that, let us make some few additional remarks on complementarity.

8.3 Recent results

As it is well known, Bohr and others like P. Jordan and F. Gonseth have suggested that complementarity could be useful not only in physics but in other areas as well, in particular in biology and in the study of primitive cultures (see [Jammer, 1974, pp. 87ff], where still other fields of application, like psychology, are mentioned). Although these applications may be interesting, they are outside the scope of this paper. Keeping within physics, it should be recalled that in 1994 Englert et al. argued that complementarity is not simply a consequence of the uncertainty relations, as advocated by those who believe that "two complementary variables, such as position and momentum, cannot simultaneously be measured to less than a fundamental limit of accuracy", but that

> "(...) uncertainty is not the only enforce of complementarity. We devised and analyzed both real and thought experiments that bypass the uncertainty relation, in effect to 'trick' the quantum objects under study. Nevertheless, the results always reveal that nature safeguards itself against such intrusions –complementarity remains intact even when the uncertainty relation plays no role. We conclude that complementarity is deeper than has been appreciated: it is more general and more fundamental to quantum mechanics than is the uncertainty rule." [Englert *et al.*]

If Englert *et al.* are right, then it seems that the paraclassical logic we shall describe below may in fact be useful.

Recently (1998), some experiments developed in the Weizmann Institute in Israel indicated that the Principle of Complementarity has been verified also for fermions (electrons) [Buch *et al.*, 1998]. Through nano-technology devices created in low-temperature scales, the scientists developed measuring techniques which have enabled them to show that in a certain two-slit experiment, the wave-like behaviour occurs when the possible paths a particle can take remain indiscernible, and that a particle-like behaviour occurs when a 'which-path' detector is introduced, determining the actual path taken by the electron. These recent experiments show that the ancient intuitions and some *Gedankenexperimente* performed by Bohr and others were in the right direction, so sustaining Bohr's position that complementarity is in fact a characteristic trait of matter. So, to accommodate this idea within a formal description of physics is in fact an important task.

A still more recent (2001) 'experimental proof' of Bohr's principle came from Austria, where O. Nairz and others have reported that Heisenberg uncertainty principle, which is closely related to complementarity, was demonstrated for a massive object, namely, the fullerene molecule C_{70} at a temperature of 900 K. In justifying their work, they said that "[t]here are good reasons to believe that complementarity and the uncertainty relation will hold for a sufficiently well isolated object of the physical world and that these quantum properties are generally only hidden by technical noise for large objects. It is therefore interesting to see how far this quantum mechanical phenomenon can be experimentally extended to the macroscopic domain" [Nairz, 2001].

This apparently opens the road for the acceptance of the validity of complementarity also in the macroscopic world. The analysis of these applications should interest not only physicists and other scientists, but philosophers as well. We believe that Bohr's intuitions that complementarity is a general phenomenon in the world deserves careful examination in the near future. But let us go back to logic.

8.4 Logics of complementarity

The expression 'logic of complementarity' has been used elsewhere to designate different logical systems, or even informal conceptions, which intended to provide a description of Bohr's ideas of complementarity from a 'logical' point of view.

As a historical remark, we recall that some authors like C. von Weizsäcker, M. Strauss and P. Février have already tried to elucidate Bohr's principle from such a logical point of view (cf. [Février, 1951], [Jammer, 1974, pp. 377ff], [Strauss, 1973]). Jammer mentions Bohr's negative answer to von Weizsäcker's attempt of interpreting his principle, and observes that this should be taken as a warning for analyzing the subject (ibid. p. 90). As shown by Jammer, Bohr explained that his rejection was due to his conception that "[t]he complementary mode of description (...) is ultimately based on the communication of experience, [quoting Bohr] 'which has to use the language adapted to our usual orientation in daily life' "; Jammer continues by recalling that, to Bohr, "objective description of experience must always be formulated in [quoting Bohr again] 'plain language which serves the needs for practical life and social intercourse' " ([Jammer, 1974, p. 379]). These points reinforce our emphasis that Bohr ascribed to complementarity the role of a meaning principle. So, maybe Bohr's rejection of accepting a 'logic of complementarity' could be due to the discrepancies (or 'divergent conceptions') related to the way of understanding complementarity. In his 1966 book, Jammer also suggested something analogous [Jammer, 1966, p. 356].

Another tentative of building a 'logic of complementarity' was P. Février's. She began by considering Heisenberg uncertainty relations not simply as something which can be derived in the formalism of quantum theory, but attributed to them a distinctive fundamental role as being the very basic principle on which quantum theory should be built on. She distinguished (yet not explicitly) between propositions which can and which cannot be composed. The last are to stand for complementary propositions; in her logic, a third value is used for the conjunction of complementary propositions to mean that their conjunction is 'absolutely false'. Other connectives are presented by the matrix method, so that a 'logic of complementarity' is proposed, yet not detailed in full

([Février, 1951]; for further details on her system, see [da Costa and Krause, to appear]).

Strauss' logic is based on his conception that the complementarity principle "excludes simultaneously decidability of incompatible propositions" [Jammer, 1974, p. 356]; then, he proposed a different theory in which conjunctions and disjunctions of complementary propositions were to be excluded. So, in a certain sense, although described in probabilistic terms, we may say that his intention was to develop a logic in which two propositions, say α and β (which stand for complementary propositions) may be both accepted, but not their conjunction $\alpha \wedge \beta$ (op. cit., p. 335). It is interesting to remark that Carnap declared that Strauss' logic was 'inadvisable' [Carnap, 1995, p. 289]. Today, by using another kind of paraconsistent logic, termed Jaskowski logic, we think that perhaps Strauss' position can be sustained.

Leaving these historical details aside, we shall proceed as follows. After introducing the concept of a theory which admits a *Complementarity Interpretation* (to use Jammer's words – see below), we shall argue that under a plausible definition of complementarity, the underlying logic of such a theory is *paraclassical* logic. In the sequel we shall sketch the main features of this logic.

8.5 Complementarity theories

Bohr's view provides the grounds for defining a very general class of theories, which we have elsewhere termed 'complementarity' ($\mathcal{C}$-theories; see [da Costa and Krause, to appear]). Here, we generalize the concept of a $\mathcal{C}$-theory, by defining 'complementarity theories with meaning principles' (termed $\mathcal{C}_{mp}$-theories), more or less paraphrasing Carnap (but without any compromise with his stance), in which some meta-rules are considered in order we know about the possibility of accepting (or not accepting) certain propositions. Before characterizing these theories, let us see how some authors read complementarity; this will guide our definition of $\mathcal{C}_{mp}$-theories.

To begin with, let us quote Max Jammer:

> "Although it is not easy, as we see, to define Bohr's notion of *complementarity*, the notion of *complementarity interpretation* seems to raise fewer definitory difficulties. The following definition of this notion suggests itself. A given theory T admits a complementarity interpretation if the following conditions are satisfied: (1) T contains (at least) two descriptions D_1 and D_2 of its substance-matter; (2) D_1 and D_2 refer to the same universe of discourse U (in Bohr's case, microphysics); (3) neither D_1 nor D_2, if taken alone, accounts exhaustively for all phenomena of U; (4) D_1 and D_2 are mutually exclusive in the sense that their combination into a single description would lead to logical contradictions."
>
> "That these conditions characterize a complementarity interpretation as understood by the Copenhagen school can easily be documented. According to Léon Rosenfeld, (...) one of the principal spokesmen of this school,

> complementarity is the answer to the following question: What are we to do when we are confronted with such situation, in which we have to use two concepts that are mutually exclusive, and yet both of them necessary for a complete description of the phenomena? "Complementarity denotes the logical relation, of quite a new type, between concepts which are mutually exclusive, and which therefore cannot be considered at the same time – that would lead to logical mistakes – but which nevertheless must both be used in order to give a complete description of the situation." Or to quote Bohr himself concerning condition (4): "In quantum physics evidence about atomic objects by different experimental arrangements (...) appears contradictory when combination into a single picture is attempted." (...) In fact, Bohr's Como lecture with its emphasis on the mutual exclusive but simultaneous necessity of the causal (D_1) and the space-time description (D_2), that is, Bohr's first pronouncement of his complementarity interpretation, forms an example which fully conforms with the preceding definition. Bohr's discovery of complementarity, it is often said, constitutes his greatest contribution to the philosophy of modern science." [Jammer, 1974, pp. 104-5.]

Jammer's quotation is interpreted as follows. We take for granted that both D_1 and D_2 are sentences formulated in the language of a complementary theory T, so that items (1) and (2) are considered only implicitly. Item (3) is understood as entailing that *both* D_1 and D_2 are, from the point of view of T, *necessary* for the full comprehension of the relevant aspects of the objects of the domain; so, item (3) is asserted on the grounds of a certain meaning principle; so, we take both D_1 and D_2 as 'true' sentences, that is, $T \vdash D_1$ and $T \vdash D_2$. Important to remark that here the concept of truth is taken in a syntactical way: a sentence is true in T if it is a theorem of T, and false if its negation is a theorem of T. If neither the sentence nor its negation are theorems of T, then the sentence (so as its negation) is said to be independent.

Item (4) deserves further attention. Jammer (*loc. cit.*) says that 'mutually exclusive' means that the "combination of D_1 and D_2 into a single description would lead to logical contradictions", and this is reinforced by Rosenfeld's words that the involved concepts "cannot be considered at the same time", since this would entail a "logical mistake". Then, we informally say that *mutually exclusive*, *conjugate propositions*, or *complementary propositions*, are sentences which lead (by classical deduction) to a contradiction; in particular, their conjunction yields a contradiction.

So, following Jammer and Rosenfeld, we shall say that a theory T admits complementarity interpretation, or that T is a $\mathcal{C}$-theory, if T encompasses 'true' formulas α and β (which may stand for Jammer's D_1 and D_2 respectively) which are 'mutually exclusive' in the above sense, for instance, that their conjunction yields to a strict contradiction if classical logic is applied. In other words, if $\vdash$ is the symbol of deduction of classical logic, then, α and β being complementary, we have $\alpha, \beta \vdash \gamma \wedge \neg\gamma$ for some γ of the language of T (see [Mendelson, 1997, pp. 34-5]).

The problem with this characterization of complementarity is that if the underlying logic of T is classical logic, then T, involving complementary propositions in the above sense, is contradictory or inconsistent. Apparently, this is precisely what Rosenfeld claimed in the above quotation. Obviously, if we intend to maintain the idea of complementary propositions as forming part of the theory and being expressed in the object language without trivialization, one solution (perhaps the only one) is to employ as the underlying logic of T a logic such that the admission of both α and β would not entail a strict contradiction (i.e., a formula of the form $\gamma \wedge \neg\gamma$). One way to do so is to modify the classical concept of deduction, obtaining a new kind of logic, called *paraclassical* logic, as we shall do in what follows.

That kind of logic is the underlying logic of what we have termed complementary theories; here we call complementary theory or $\mathcal{C}_{mp}$-theory, a $\mathcal{C}$-theory with meaning principles. For instance, as we have seen, Heinsenberg uncertainty relations were taken by Février as the starting point for quantum physics. According to her, these relations should not be a simple result obtained within the formalism of quantum theory, but should be the *base* of quantum mechanics. Meaning principles, as we have said before, are here understood as assumptions which sanction either some restrictions of 'classical' procedures or the utilization of certain 'classical' incompatible schemes in the domain of scientific theories. The word 'classical' refers to classical physics (Bohr strongly believed that all the discourse involving quantum phenomena should be done in the language of classical physics).

8.6 The underlying logic of $\mathcal{C}_{mp}$-theories

We shall restrict our explanation to the propositional level, although it is not difficult to extend our system to encompass quantifiers (and set theory) as it would be necessary if we intend to construct a possible logical basis for physical theories. Let us call C an axiomatized system for the classical propositional calculus. The concept of deduction in C is taken to be the standard one; we use the symbol $\vdash$ to represent deductions in C (see [Mendelson, 1997]). Furthermore, the formulas of C are denoted by Greek lowercase letters, while Greek uppercase letters stand for sets of formulas. The symbols $\neg$, $\rightarrow$, $\wedge$, $\vee$ and $\leftrightarrow$ have their usual meanings, and standard conventions in writing formulas will be also assumed without further comments.

Definition 1. *Let Γ be a set of formulas of C and let α be a formula (of the language of C). Then we say that α is a (syntactical)* P*-consequence of Γ, and write $\Gamma \vdash_{\mathsf{P}} \alpha$, if and only if*

For a derivation of Heisenberg relations within the Hilbert space formalism, see [Redhead, 1992, pp. 59ff].

(P1) $\alpha \in \Gamma$, *or*

(P2) α *is a classical tautology, or*

(P3) *There exists a consistent (according to classical logic) subset* $\Delta \subseteq \Gamma$ *such that* $\Delta \vdash \alpha$ *(in classical logic).*

We call $\vdash_{\mathsf{P}}$ the relation of P-*consequence.*

Definition 2. P *is the logic whose language is that of* C *and whose relation of consequence is that of* P-*consequence. Such a logic will be called paraclassical.*

It is immediate that, among others, the following results can be proved:

Theorem 1.

1. *If* α *is a theorem of the classical propositional calculus* C *and if* Γ *is a set of formulas, then* $\Gamma \vdash_{\mathsf{P}} \alpha$*; in particular,* $\vdash_{\mathsf{P}} \alpha$.
2. *If* Γ *is consistent (according to* C*), then* $\Gamma \vdash \alpha$ *(in* C*) iff* $\Gamma \vdash_{\mathsf{P}} \alpha$ *(in* P*).*
3. *If* $\Gamma \vdash_{\mathsf{P}} \alpha$ *and if* $\Gamma \subseteq \Delta$*, then* $\Delta \vdash_{\mathsf{P}} \alpha$ *(The defined notion of* P-*consequence is monotonic.)*
4. *The notion of* P-*consequence is recursive.*
5. *Since the theses of* P *(valid formulas of* P*) are those of* C*,* P *is decidable.*

Definition 3. *A set of formulas* Γ *is* P-*trivial iff* $\Gamma \vdash_{\mathsf{P}} \alpha$ *for every formula* α*. Otherwise,* Γ *is* P-*non-trivial. (Similarly we define the concept of a set of formulas being trivial in* C*).*

Definition 4. *A set of formulas* Γ *is* P-*inconsistent if there exists a formula* α *such that* $\Gamma \vdash_{\mathsf{P}} \alpha$ *and* $\Gamma \vdash_{\mathsf{P}} \neg\alpha$*. Otherwise,* Γ *is* P-*consistent.*

Theorem 2.

1. *If* α *is an atomic formula, then* $\Gamma = \{\alpha, \neg\alpha\}$ *is* P-*inconsistent, but* P-*non-trivial.*
2. *If the set of formulas* Γ *is* P-*trivial, then it is trivial (according to classical logic). If* Γ *is non-trivial, then it is* P-*nontrivial.*
3. *If* Γ *is* P-*inconsistent, then it is inconsistent according to classical logic. If* Γ *is consistent according to classical logic, then* Γ *is* P-*consistent.*

A semantical analysis of P, for instance a completeness theorem, can be obtained without difficulty [da Costa and Vernengo, 1999]. We remark that the set $\{\alpha \wedge \neg\alpha\}$, where α is a propositional variable, is trivial according to classical logic, but it is not P-trivial. Notwithstanding, we are not suggesting that complementary propositions should be understood necessarily as pairs of contradictory sentences. This is made clear by the following definition:

Definition 5 (Complementarity Theories or $\mathcal{C}_{mp}$-theories). *A $\mathcal{C}$-theory is a set of formulas T of the language of* C *(the classical propositional calculus) closed by the relation of* P*-consequence, that is, $\alpha \in T$ for any α such that $T \vdash_{\mathsf{P}} \alpha$. In other words, T is a theory whose underlying logic is* P*. A $\mathcal{C}_{mp}$-theory is a $\mathcal{C}$-theory subjected to meaning principles.*

Of course the definition of a $\mathcal{C}_{mp}$-theory is a little bit vague. However, for instance in the case of a meaning principle that introduces restrictions in the acceptable statements of the theory, the hypothesis and axioms used in deductions have to satisfy such restrictive conditions. For instance, if a meaning principle of a theory T is formulated as Heisenberg Uncertainty Principle, this circumstance will impose obvious restrictions to certain statements of T.

Theorem 3. *There exist $\mathcal{C}$-theories and $\mathcal{C}_{mp}$-theories that are inconsistent, although are* P*-non-trivial.*

Proof: Immediate consequence of Theorem 2. □

Finally, we state a result (Theorem 4), whose proof is an immediate consequence of the definition of P-consequence. However, before stating the theorem, let us introduce a definition:

Definition 6 (Complementary Propositions). *Let T be a $\mathcal{C}_{mp}$-theory (in particular, a $\mathcal{C}$-theory) and let α and β be formulas of the language of T. We say that α and β are T-complementary (or simply complementary) if there exists a formula γ of the language of T such that:*

1. $T \vdash_{\mathsf{P}} \alpha$ *and* $T \vdash_{\mathsf{P}} \beta$
2. $T, \alpha \vdash_{\mathsf{P}} \gamma$ *and* $T, \beta \vdash_{\mathsf{P}} \neg\gamma$ *(in particular,* $\alpha \vdash_{\mathsf{P}} \gamma$ *and* $\beta \vdash_{\mathsf{P}} \neg\gamma$*).*

Theorem 4. *If α and β are complementary theorems of a $\mathcal{C}_{mp}$-theory T and $\alpha \vdash_{\mathsf{P}} \gamma$ and $\beta \vdash_{\mathsf{P}} \neg\gamma$, then in general $\gamma \wedge \neg\gamma$ is not a theorem of T.*

Proof: Immediate, as a consequence of Theorem 2. □

In other words, T is inconsistent from the point of view of classical logic, but it is P-non-trivial.

It should be emphasized that our way of characterizing complementarity does not mean that complementary propositions are always contradictory, for α and β above are not necessarily one the negation of the other. However, as complementary propositions, we may derive from them (in classical logic) a contradiction; to exemplify, we remark that 'x is a particle' is not the direct negation of 'x is a wave', but 'x is a particle' *entails* that x is not a wave. This reading of complementarity as not indicating strict contradiction, as we have already made clear, is in accordance with Bohr himself; let us quote him once more to reinforce this idea. Bohr says:

> "In considering the well-known paradoxes which are encountered in the application of the quantum theory to atomic structure, it is essential to remember, in this connection, that the properties of atoms are always obtained by observing their reactions under collisions or under the influence of radiation, and that the (...) limitation on the possibilities of measurement is directly related to the apparent contradictions which have been revealed in the discussion of the nature of light and of the material particles. In order to emphasize that we *are not concerned here with real contradictions*, the author [Bohr himself] suggested in an earlier article the term 'complementarity'." [Bohr, 1929b, p. 95] (italics ours).

Let us give a simple example of a situation involving a $\mathcal{C}_{mp}$-theory. Suppose that our theory T is a fragment of quantum mechanics admitting Heisenberg relations as a meaning principle and having as its underlying logic paraclassical logic. If α and β are two incompatible propositions according to Heisenberg's principle, we can interpret this principle as implying that α entails $\neg\beta$ (or that β entails $\neg\alpha$). So, even if we add α and β to T, we will be unable to derive, in T, $\alpha \wedge \beta$. Analogously, Pauli's Exclusion Principle has also an interpretation as that of Heisenberg's.

As we said before, the basic characteristic of $\mathcal{C}_{mp}$-theories is that, in making P-inferences, we suppose that some sets of statements we handle are consistent. In other words, $\mathcal{C}_{mp}$-theories are closer to those theories scientists *actually* use in their everyday activity than those theories with the classical concept of deduction. In other words, paraclassical logic (and paraconsistent logics in general) seems to fit more accurately the way scientists reason when stating their theories.

8.7 The paralogic associated to a logic

As we noted in [da Costa and Krause, to appear], the technique used above to define the paraclassical logic associated to classical logic can be generalized to other logics $\mathcal{L}$ (including logics having no negation symbol, but we will not deal with this case here), as well as the concept of a $\mathcal{C}_{mp}$-theory. More precisely, starting with a logic $\mathcal{L}$, which can be seen as a pair $\mathcal{L} = \langle \mathcal{F}, \vdash \rangle$, where $\mathcal{F}$ is an abstract set called the set of formulas of $\mathcal{L}$ and $\vdash \subseteq \mathcal{P}(\mathcal{F}) \times \mathcal{F}$ is the deduction relation of $\mathcal{L}$ (which is subjected to certain postulates depending on the particular logic $\mathcal{L}$) [Béziau, 1994], we can define the $\mathsf{P}_{\mathcal{L}}$-logic associated to $\mathcal{L}$ (the 'paralogic' associated to $\mathcal{L}$) as follows.

Let $\mathcal{L}$ be a logic, which may be classical logic, intuitionistic logic, some paraconsistent logic or, in principle, any other logical system. By simplicity, we suppose that the language of $\mathcal{L}$ has a symbol for negation, $\neg$. Then,

Definition 7. *A theory based on $\mathcal{L}$ (an $\mathcal{L}$-theory) is a set of formulas Γ of the language of $\mathcal{L}$ closed under $\vdash_{\mathcal{L}}$ (the symbol of deduction in $\mathcal{L}$). In other words, $\alpha \in \Gamma$ for every formula α such that $\Gamma \vdash_{\mathcal{L}} \alpha$.*

Definition 8. *An $\mathcal{L}$-theory Γ is $\mathcal{L}$-inconsistent if there exists a formula α of the language of $\mathcal{L}$ such that $\Gamma \vdash_{\mathcal{L}} \alpha$ and $\Gamma \vdash_{\mathcal{L}} \neg\alpha$, where $\neg\alpha$ is the negation of α. Otherwise, Γ is $\mathcal{L}$-consistent.*

Definition 9. *An $\mathcal{L}$-theory Γ is $\mathcal{L}$-trivial if $\Gamma \vdash_{\mathcal{L}} \alpha$ for any formula α of the language of $\mathcal{L}$. Otherwise, Γ is $\mathcal{L}$-non-trivial.*

Then, we define the $\mathsf{P}_{\mathcal{L}}$-logic associated with $\mathcal{L}$ whose language and syntactical concepts are those of $\mathcal{L}$, except the concept of deduction, which is introduced as follows: we say that α is a $\mathsf{P}_{\mathcal{L}}$-*syntactical consequence* of a set Γ of formulas, and write $\Gamma \vdash_{\mathsf{P}_{\mathcal{L}}} \alpha$ if and only if:

(1) $\alpha \in \Gamma$, or

(2) α is a provable formula of $\mathcal{L}$ (that is, $\vdash_{\mathcal{L}} \alpha$), or

(3) There exists $\Delta \subseteq \Gamma$ such that Δ is $\mathcal{L}$-non-trivial, and $\Delta \vdash_{\mathcal{L}} \alpha$.

For instance, we may consider the paraconsistent calculus C_1 [da Costa, 1974] as our logic $\mathcal{L}$. Then the paralogic associated with C_1 is a kind of 'para-paraconsistent' logic.

It seems worthwhile to note the following in connection with the paraclassical treatment of theories. Sometimes, when one has a paraclassical theory T such that $T \vdash_{\mathsf{P}} \alpha$ and $T \vdash_{\mathsf{P}} \neg\alpha$, there exist *appropriate* propositions β and γ such that T can be replaced by a classical consistent theory T' in which $\beta \to \alpha$ and $\gamma \to \neg\alpha$ are theorems. If this happens, the logical difficulty is in principle eliminable and classical logic maintained.

8.8 Logic and physics

When we hear something about the relationships between logic and quantum physics, we usually tend to relate the subject with the so called 'quantum logics', a field that has its 'official' birth in Birkhoff and von Neumann's well known paper from 1936 (see [Dalla Chiara and Giuntini, 2001]). This is completely justified, for their fundamental work caused the development of a wide field of research in logic. Today there are various 'quantum logical systems', although they have been studied specially as pure mathematical systems, far from applications to the axiomatization of the microphysical world and also far from the insights of the forerunners of quantum mechanics (for a general and updated account on this field, see [Dalla Chiara and Giuntini, 2001]).

Of course an axiomatization of a given empirical theory is not always totally determinate, and the need for a logic distinct from the classical as the underlying logic of quantum theory is still open to discussion. In fact, the axiomatic basis of a scientific theory depends on the several aspects of the theory, explicitly or implicitly, appropriate to take into account its structure. So, for example, (Ludwig, 1990) studies an axiomatization of quantum mechanics based on classical logic. All the stances, that of employing a logic like paraclassical logic (or other kind of system as one of those mentioned above), and that of Ludwig, are in principle acceptable, since they treat different perspectives of the same domain of discourse, and different 'perspectives' of a domain of science may demand for distinct logical apparatuses; this is a philosophical point of view radically different from the classical. As we said in another work (see [da Costa and Krause, to appear]), the possibility of using non-standard systems in the foundations of physics (and in general of science) does not necessarily entail that classical logic is wrong, or that (in particular) quantum theory *needs* another logic. Physicists probably will continue to use classical (informal) logic in the near future. But we should realize that other forms of logic may help us in the better understanding of certain features of the quantum world as well, not easily treated by classical devices, such as complementarity. Only the future of physics will perhaps decide what is the better solution, a decision that involves even pragmatic factors.

To summarize, we believe that there is no just one 'true logic', and distinct logical (so as mathematical and perhaps even physical) systems, like paraclassical logic, may be useful to approach different aspects of a wide field of knowledge like quantum theory. The important point is to be open to the justifiable revision of concepts, a point very lucidly emphasized by Niels Bohr, who wrote:

> "For describing our mental activity, we require, on one hand, an objectively given content to be placed in opposition to a perceiving subject, while, on the other hand, as is already implied in such an assertion, no sharp separation between object and subject can be maintained, since the perceiving subject also belongs to our mental content. From these circumstances follows not only the relative meaning of every concept, or rather of every word, the meaning depending upon our arbitrary choice of view point, but also we must, in general, be prepared to accept the fact that a complete elucidation of one and the same object may require diverse points of view which defy a unique description. Indeed, strictly speaking, the

conscious analysis of any concept stands in a relation of exclusion to its immediate application. The necessity of taking recourse to a complementarity, or reciprocal, mode of description is perhaps most familiar to us from psychological problems. In opposition to this, the feature which characterizes the so-called exact sciences is, in general, the attempt to attain to uniqueness by avoiding all reference to the perceiving subject. This endeavour is found most consciously, perhaps, in the mathematical symbolism which sets up for our contemplation an ideal of objectivity to the attainment of which scarcely any limits are set, so long as we remain within a self-contained field of applied logic. In the natural sciences proper, however, there can be no question of a strictly self-contained field of application of the logical principles, since we must continually count on the appearance of new facts, the inclusion of which within the compass of our earlier experience may require a revision of our fundamental concepts." [Bohr, 1929a, pp. 212-213].

References

Beller, M., (1992). "The birth of Bohr's complementarity: the context and the dialogues", *Stud. Hist. Phil. Sci.* **23** (1), 1992, 147–180.

Béziau, J.-Y., (1994). *Recherches sur la Logique Universelle (excessivité, négation, séquents)*, Thesis, Un. Denis Diderot, U. F. R. de Mathématiques, 1994.

Bohr, N., (1927). "The quantum postulate and the recent development of atomic theory" [3], Atti del Congresso Internazionale dei Fisici, 11–20 Sept. 1927, Como-Pavia-Roma, Vol.II, Zanichelli, Bologna, 1928, reprinted in [Bohr, 1985, 109–136].

Bohr, N. (1928). "The quantum postulate and the recent developments of atomic theory", *Nature* (Suppl.) **121**, 1928, 580–590, reprinted in [Bohr, 1985, 147–158].

Bohr, N., (1929a). "The quantum of action and the description of nature" (1929), in [Bohr, 1985, 208–217].

Bohr, N., (1929b). "Introductory survey" to Bohr 1934, in [Bohr, 1985, 279–302].

Bohr, N., (1934). *Atomic theory and the description of nature*, Cambridge, Cambridge Un. Press, 1934. Reprinted in [Bohr, 1985, 279–302].

Bohr, N. (1937). "Causality and complementarity", *Phil. of Sci.* **4** (3), 1937, 289–298.

Bohr, N., (1938). "Natural philosophy of human cultures" (1938), in *Atomic physics and human knowledge*, New York, John-Wiley, 1958, 23–31 (also in *Nature* **143**, 1939, 268–272).

Bohr, N., (1958). "Quantum physics and philosophy: causality and complementarity" (in) Klibanski, R. (ed.) *Philosophy in the Mid-Century*, I. Firenze, La Nuova Italia, 1958, 308–314.

Bohr, N., (1985). *Collected works*, Rüdinger, E. (general ed.), Vol. 6: Foundations of Quantum Physics I. Kolckar, J. (ed.). Amsterdam, North-Holland, 1985.

Buch, E., Schuster, R. (1998). Heiblum, M., Mahalu, D. and Umanski, V., (1998). "Dephasing in electron interference by a 'which-path' detector", *Nature* **391**, 1998.

Carnap, R., (1995). *An introduction to the philosophy of science*, New York, Dover Pu., 1995.

da Costa, N. C. A. (1974). "On the theory of inconsistent formal systems", *Notre Dame J. Formal Logic* **11**, 1974, 497–510.

da Costa, N. C. A. y Vernengo, R. J., (1999). "Sobre algunaslógicas paraclássicas y el análisis del razonamientojurídico", *Doxa* **19**, 1999, 183–200.

da Costa, N. C. A. and Bueno, O., (2001). "Paraconsistency: towards a tentative interpretation", *Theoria-Segunda Época* 16/1, 2001, 119–145. *Reports on Mathematical Logic* **4**, 7–16.

da Costa, N. C. A. and Krause, D., (to appear). "Complementarity and paraconsistency", in S. Rahman, J. Symons, D. M. Gabbay and J. P. van Bendegem (eds.), *Logic, Epistemology, and the Unity of Science*, Dordrecht, Kluwer Ac. Press, forthcoming: (`www.cfh.ufsc.br/~dkrause/Artigos/CosKraBOHR.pdf`).

da Costa, N. C. A. and Krause, D., (to appear). "Remarks on the applications of paraconsistent logic to physics", forthcoming: (`www.cfh.ufsc.br/~dkrause/Artigos/CosKra03PL.pdf`).

Cushing, J. T., (1994). *Quantum mechanics: historical contingency and the Copenhagen hegemony*, Chicago & London, the univ. of Chicago Press, 1994.

Dalla Chiara, M. L. and Giuntini, R., (2001). "Quantum logic", in: `http://xxx.lanl.gov/list/quant-ph/0101`.

Englert, B.-G., Scully, M. O. and Walther, H., (1994). "The duality in matter and light", *Scientific American* **271** (6), 1994, 56–61.

French, A. P. and Kennedy, P. J. (eds.), (1985). *Niels Bohr, a centenary volume*, Cambridge MA & London, Harward Un. Pres, 1985.

Février, D.-P., (1951). *La structure des théories physiques*, Paris, Presses Un. de France, 1951.

Jammer, M., (1966). *The conceptual development of quantum mechanics*, McGraw-Hill, 1966.

Jammer, M., (1974). *Philosophy of Quantum Mechanics*, New York, John Wiley, 1974.

Ludwig, G., (1990). *Les Structures de base d'une théorie physique*, Berlin, Springer-Verlag, 1990.

Mendelson, E., (1997). *Introduction to mathematical logic*, London, Chapman & Hall, 4th. ed., 1997.

Nairz, O., Arndt, M. and Zeilinger, A., (2001). "Experimental verification of the Heisenberg uncertainty principle for hot fullerene molecules", (`arXiv:quant-ph/0105061`).

Pauli, W., (1980). *General principles of quantum mechanics*, Berlin, Springer-Verlag, 1980.

Redhead, M., (1992). *Incompleteness, non locality, and realism: a prolemenon to the philosophy of quantum mechanics*, Oxford, Clarendon Press, 1992.

Scheibe, E., (1973). *The logical analysis of quantum mechanics*, Oxford, Pergamon Press, 1973.

de Souza , E., (2000). "Multideductive logic and formal unification of physical theories", *Syntheses* **125**, 2000, 253–262.

Strauss, M., (1973). "Mathematics as logical syntax - A method to formalize the language of a physical theory", in Hooker, C. A., (ed.), *Contemporary research in the foundations and philosophy of quantum theory*, Dordrecht, D. Reidel, 1973, 27–44.

Strauss, M., (1975). "Foundations of quantum mechanics", in Hooker, C. A. (ed.) *The Logico-Algebraic Approach to Quantum Mechanics*, Vol. I., Dordrecht, D. Reidel, 1975, 351–364.

Suppes, P., (2002). *Representation and Invariance of Scientific Theories*, Stanford, Center for the Study of Language and information, 2002.

Chapter 9

SEMANTICS FOR NAIVE SET THEORY IN MANY-VALUED LOGICS

Technique and Historical Account

Thierry Libert
Department of Mathematics
Free University of Brussels
tlibert@ulb.ac.be

Introduction

It must be admitted that mathematical investigations in providing *alternative* semantics have carried innovative ideas, and if all have not led to further developments and applications, they have often led to a better understanding of the topic considered. Even within a well-established framework, the use of alternative semantics has proved its fruitfulness. As an example, for independence results in ZF, one may quote the Boolean-valued version of forcing due to Scott and Solovay, in which a set is conceived of as a function which takes its values into a given complete Boolean algebra, no more into the 2-valued one. This concerns classical logic and perhaps this would remind the reader of the primal use of many-valued semantics for proving the independence of axioms in propositional logic. Note that there is no need to be interested in a possible meaning of the additional "truth values" to do that, we would rather say that the understanding is in the application.

That said, one may legitimately ask whether the use of many-valued semantics could not also benefit our understanding of the set-theoretical paradoxes themselves. In any case, to know which logic(s) can support the full comprehension scheme, or some maximal fragments of it, is an interesting question in itself, at least not devoid of mathematical curiosity. So we shall review

J. van Benthem, G. Heinzmann, M. Rebuschi and H. Visser (eds.), The Age of Alternative Logics: Assessing Philosophy of Logic and Mathematics Today, 121–136.

some attempts in that direction, giving an historical account on the subject in connection with Skolem's pioneering work in Łukasiewicz's logics. On the way we will also point out and distinguish two ways of formalizing set theory, namely *comprehension* and *abstraction*, which have been considered by several authors.

Our purpose is to stress the use of *fixed-point arguments* in semantical consistency proofs and thus the role of *continuity* in avoiding the paradoxes. Then it will become apparent how these investigations were actually close to other contemporary ones, namely Kripke's work on the liar paradox and Scott's on models for the untyped λ-calculus.

To make the links between our different sources clearer, we have included in the appendix on page 134 a diagram of some selected references. Throughout the paper, such references are conveniently marked with the symbol '$*$'.

9.1 Moh Shaw-Kwei's paradox

The existence of the Russell set is prohibited in classical logic. By tampering with the negation some alternative logics have proved more tolerant. Nevertheless, there exist other sets that can affect such non-classical systems. This was illustrated in 1954 by Moh Shaw-Kwei [20]* who presented an extended version of Curry's paradox. Here it is.

Let $\rightarrow$ be the *official* implication connective of the logic considered. Precisely, in order that $\rightarrow$ be referred to as an implication connective, it is assumed that *modus ponens* holds, namely

$$MP: \quad p\,, p \rightarrow q \vdash q$$

where $\vdash$ is the consequence relation associated with the logic. To express Moh's paradox, we define the *n-derivative* $\rightarrow^n$ of the implication inductively as follows:

$$p \rightarrow^0 q :\equiv q \qquad \text{and} \qquad p \rightarrow^{n+1} q :\equiv p \rightarrow (p \rightarrow^n q) \qquad (n \in \mathbb{N})$$

Then the implication connective is said to be *n-absorptive* if it satisfies the *absorption rule of order n*, that is,

$$A_n: \quad p \rightarrow^{n+1} q \vdash p \rightarrow^n q$$

Now, assuming that the implication is n-absorptive for some $n > 0$, it is shown that any formula can be derived from the existence of the set $C_n := \{x \mid x \in x \rightarrow^n \bot\}$, where $\bot$ is a "falsum" constant defined in such a way that $\bot \vdash \varphi$, for any φ.

Proof.

$$
\begin{array}{llll}
Comp & \vdash \ \forall x\,(x \in C_n \leftrightarrow (x \in x \rightarrow^n \bot)) & \text{[Comprehension]} & (1)\\
 & \vdash \ C_n \in C_n \leftrightarrow (C_n \in C_n \rightarrow^n \bot) & \text{[Univ. Quant. Elimin.]} & (2)\\
 & \vdash \ C_n \in C_n \rightarrow^n \bot & [(2)^{\rightarrow}, A_n] & (3)\\
 & \vdash \ C_n \in C_n & [(2)^{\leftarrow}, (3), MP] & (4)\\
 & \vdash \ \bot & [(3), (4), MP(n \text{ times})] & (5)
\end{array}
$$

Thus any formula can be derived and the theory is meaningless.

As a particular case we have $C_1 = \{x|\ x \in x \rightarrow \bot\}$, which might be called the Curry set, and then we recover the Russell set $R = \{x|\ x \notin x\}$ if $\neg\varphi$ is defined by $\varphi \rightarrow \bot$, as it is the case in classical logic. Note incidentally that $C_0 = \emptyset$, but this later is not problematic.

9.2 The Łukasiewicz logics

A nice illustration is supplied by the most popular many-valued logics, the Łukasiewicz ones. We shall content ourselves here with recalling the truth-functional characterization of the connectives and quantifiers of these logics.

The set of truth values for the *infinite*-valued Łukasiewicz logic Ł_∞ is taken to be the real unit interval $I := [0,1] \subseteq \mathbb{R}$ with its natural ordering, which will be referred to as the *truth ordering* and denoted by $\leqslant_T$. The only *designated* value is the maximum 1. Here are the truth functions of the logical operators (note we are using the same notation for the operators and their truth functions):

- negation is defined by $\neg x := 1 - x$, for any $x \in I$;
- conjunction and disjunction are respectively minimum and maximum with respect to the truth ordering, i.e., $x \wedge y := \min_{\leqslant_T}\{x, y\}$ and $x \vee y := \max_{\leqslant_T}\{x, y\}$, for any $x, y \in I$;
- quantifiers are thought of as generalized conjunction and disjunction, i.e., for any $A \subseteq I$, $\forall A := \inf_{\leqslant_T} A$ and $\exists A := \sup_{\leqslant_T} A$;
- last but not least, the truth function of the implication is specifically defined by $x \rightarrow y := \min_{\leqslant_T}\{1, 1 - x + y\}$. Observe that this is not $\neg x \vee y$, we only have $\neg x \vee y \leqslant_T x \rightarrow y$. It is also interesting to note that $x \rightarrow y = 1$ iff $x \leqslant_T y$, so that $\rightarrow$ is in some sense a characteristic function of the truth ordering. Accordingly, a very particular property of the biconditional is that $x \leftrightarrow y = 1$ iff $x = y$.

For our purposes, it is to be noticed that if we equip I with the usual topology of the real line, then each of the truth functions of the connectives is continuous. The truth functions of the quantifiers are continuous as well with respect to a reasonable topology on the set of subsets of I. That *non-logical* property of these operators will be of interest to us.

The n-valued Łukasiewicz logic $Ł_n$ ($n \geqslant 2$) is obtained by restricting the set of truth values to $I_n := \{0, \frac{1}{n-1}, \frac{2}{n-1}, \ldots, 1\}$. Particular cases are then the 2-valued logic $Ł_2$, which is nothing but classical logic, and the 3-valued one $Ł_3$, which historically was the first many-valued logic introduced by Łukasiewicz.

It was observed by Moh Shaw-Kwei that the implication is $(n-1)$-absorptive in $Ł_n$, whereas it is not n-absorptive in $Ł_\infty$, for any value of n, whereupon the author asked whether one could develop the naive theory of sets from $Ł_\infty$.

9.3 Skolem's conjecture

This observation was the starting point in the late fifties of a course of papers initiated by Skolem, who conjectured and tried to prove in [21]* the consistency of the full comprehension scheme in $Ł_\infty$.[1] On his road Skolem had also considered and investigated the consistency problem of some fragments of that scheme in $Ł_3$ and $Ł_2$ (see [23; 22; 24]*), on which we are going to dwell later.

Skolem's conjecture was partially confirmed by Skolem himself in [21]* and by Chang and Fenstad in different papers [5; 9]*. For instance, Skolem only showed the consistency of the comprehension scheme restricted to formulae containing no quantifiers, while Chang had quantifiers but no parameters, or parameters but then some restrictions on quantifiers. Anyhow, from the technical point of view, what should be said is that all these first attempts are *semantical* and that their proofs are based on the original method of Skolem, using at some stage a fixed-point theorem, namely *Brouwer's fixed-point theorem* for the space I^m, $m \in \mathbb{N}$ (or even for $I^{\mathbb{N}}$), stating that any continuous function on I^m (or $I^{\mathbb{N}}$) has a fixed point. We will see that another famous fixed-point theorem was also used, but rather implicitly, in Skolem's papers [23; 22]*.

To understand how a fixed-point argument can be involved in a semantical consistency proof of fragments of the comprehension scheme in many-valued logics, let us see what a model of such fragments could be.

9.4 Comprehension

A *structure* $\mathcal{M} :\equiv \langle M ; \in_{\mathcal{M}} \rangle$ for set theory in a given many-valued logic, of which T is the set of truth values (and throughout the paper T shall only be reserved for I or any of the I_n's), is formally defined by a non-empty set M together with a function $\in_{\mathcal{M}} : M \times M \longrightarrow T : (x, y) \longmapsto \in_{\mathcal{M}}`(x, y) := |x \in y|_{\mathcal{M}}$, where this later is to be understood as the truth value of $x \in y$ in M. More generally, and relatively to such a structure $\mathcal{M}$, $|\varphi|_{\mathcal{M}}$ will be denoting the truth

[1] As just mentioned, this suggestion was made by Moh Shaw-Kwei in [20]*. It should be remarked however that Skolem does not cite that work, and so he might have arrived at his conclusions independently.

value of any closed formula φ interpreted in $\mathcal{M}$, which is defined inductively in the usual manner.

In this many-valued setting, we shall say that a structure is a *model* for some set theory if it fulfils some fragment of the comprehension scheme:

$$Comp[\Sigma] :\equiv \left| \begin{array}{l} \text{for any } \varphi(x, \bar{y}) \text{ in } \Sigma, \\ \forall \bar{p}\, \exists y\, \forall x (x \in y \leftrightarrow \varphi(x, \bar{p})) \end{array} \right.$$

where Σ is any given fragment of the language of set theory. We usually denote such a y given by the φ-instance of that scheme by '$\{x|\, \varphi(x, \bar{p})\}$'.

Now it is easily seen how a fixed-point argument can come into play. Indeed, let $\varphi(x) := f(x \in x)$, where $f(\cdot)$ is any propositional function in one variable, and then let $\tau := $ '$\{x|\, \varphi(x)\}$'. If the existence of τ was guaranteed in a structure $\mathcal{M}$, we would have $|\tau \in \tau|_{\mathcal{M}} = |f(\tau \in \tau)|_{\mathcal{M}} = |f|(|\tau \in \tau|_{\mathcal{M}})$, where $|f|$ is denoting the truth function of f, showing that this latter should have a fixed point. It is then not surprising that Brouwer's fixed-point theorem was involved in the first attempts in $Ł_\infty$, as it precisely ensures that any truth function on I defined by means of the Łukasiewicz connectives and quantifiers is continuous and so has a fixed point. More generally, this would suggest that *continuity*, in a very broad sense as we shall see, might be regarded as a kind of *safety* property against the set-theoretical paradoxes.

Thus far nothing guarantees the uniqueness of '$\{x|\, \varphi(x, \bar{p})\}$', so such *set abstracts* may not be properly handled in the theory. As usual, this would require some form of *extensionality* and/or the explicit use of an *abstraction operator* '$\{\cdot|-\}$' in the language, on which we are going to elaborate hereafter.

9.5 Extensionality

The *extensionality principle* asserts that two sets having the same members are in some way indistinguishable. Depending on whether it is considered as a rule or as an axiom, different versions are conceivable. To formulate them, let us adopt the following abbreviations:

$$x \mathrel{\underset{\cdot}{=}} y :\equiv \forall z (z \in x \leftrightarrow z \in y) \qquad x \doteq y :\equiv \forall z (x \in z \leftrightarrow y \in z)$$

Thus, candidates for expressing the extensionality principle are

$$^{\flat}Ext: \; x \mathrel{\underset{\cdot}{=}} y \vdash x \doteq y \qquad ^{\flat}Ext^{\sharp}: \vdash \forall x \forall y (x \mathrel{\underset{\cdot}{=}} y \rightarrow x \doteq y).$$

And if the equality symbol $=$ was available in the language, these should rather appear as

$$Ext: \; x \mathrel{\underset{\cdot}{=}} y \vdash x = y \qquad Ext^{\sharp}: \vdash \forall x \forall y (x \mathrel{\underset{\cdot}{=}} y \rightarrow x = y)$$

Of course, in order that $=$ deserve the status of *equality*, it is to be required at least that it be so interpreted in any structure as to guarantee the principle of *substitutivity* (as a rule or as an axiom scheme). This can be met by simply defining the truth function of $=$ on a structure $\mathcal{M}$ by $=_{\mathcal{M}}`(x,y) := 1$ if $x = y$ in $M$, and $=_{\mathcal{M}}`(x,y) := 0$ if $x \neq y$ in M. It was however observed in [5]* that this *strict* interpretation of $=$ never yields a model of $Comp + Ext^{\sharp}$. Note that a more reasonable definition of the equality relation on a structure might be anyone such that $=_{\mathcal{M}}`(x,y) = 1$ if and only if $x = y$ in M. In any case this would suffice to ensure that $=$ has the substitutivity property (but as a rule, not as an axiom scheme actually). Such a structure will be said to be *normal*.

In any normal structure $\mathcal{M}$, Ext amounts to identifying each set y with its *characteristic function* $[y]_{\mathcal{M}} : M \longrightarrow T : x \longmapsto |x \in y|_{\mathcal{M}}$, so that the "bracket" function thus defined, $[\cdot]_{\mathcal{M}} : M \longrightarrow T^M$ where T^M is the set of all functions from M into T, is injective. Accordingly, the universe M of an extensional normal structure may be identified with a subset of T^M, namely the range of $[\cdot]_{\mathcal{M}}$, which is denoted here by $[M \to T]$:

$$M \simeq [M \to T] \subsetneq T^M \qquad (\diamondsuit)$$

Of course, because of Cantor's theorem, not every function from M into T can represent a set in the model. In other words, not every truth function can be involved in formulae defining sets. Which ones could? As was observed, sets defined by means of truth functions having fixed points are most welcome. Now, by providing T and M with some suitable *topological* structure, and then by taking $[M \to T]$ to be a certain class of *continuous* functions, so as to guarantee the existence of fixed points, there could be some M not only in bijection but *homeomorphic* with $[M \to T]$. The whole universe of an extensional normal structure might then itself be regarded as a fixed point of, say, some functor $\mathcal{F}(\cdot) := [\cdot \to T]$ acting on a suitable category of topological spaces. That is indeed a promising way of looking at set-theoretic models.

9.6 Abstraction

One may consider the use of set abstracts in the language itself with the help of an *abstraction operator* '$\{\cdot \mid -\}$'. Then we shall rather speak of *abstraction* instead of comprehension to stress that set abstracts may already appear as *terms* in the formula φ involved in the corresponding instance of that scheme:

$$Abst[\widehat{\Sigma}] :\equiv \left| \begin{array}{l} \text{for any } \varphi(x,\bar{y}) \text{ in } \widehat{\Sigma}, \\ \forall \bar{p}\, \forall x (x \in \{x \mid \varphi(x,\bar{p})\} \leftrightarrow \varphi(x,\bar{p})) \end{array} \right.$$

where $\widehat{\Sigma}$ stands for any given fragment of the language of set theory extended by an abstraction operator.

A characteristic feature of such a language is that it allows abstraction and quantification over variables occurring free in set abstracts, as certified for instance by the formation of the term '$\{x \mid \{u \mid x \in x\} \in x\}$'.

It was precisely by using an abstraction operator, and by a *proof-theoretic* method in fact, that Skolem's conjecture was finally established much later by White in 1979 (see [25]*). So what he actually proved is the consistency of the full *abstraction* scheme in $Ł_\infty$, without equality in the language. It was noticed by the author himself that his system is actually too weak in order to be able to develop classical mathematics inside (even just classical first-order number theory). He also showed that ${}^\flat Ext^\sharp$ cannot be consistently added, but we ignore whether ${}^\flat Ext$ could be. We do not know either whether extensional normal models of the full *comprehension* scheme could be built.[2] Note that such models would give rise to a full universe of *fuzzy* sets.

We shall now leave the consistency problem of the full comprehension scheme in $Ł_\infty$ to consider the one of some logical fragments of it in $Ł_3$ and $Ł_2$, as initiated by Skolem (see [23; 22; 24]*). We will see that, though the use of set abstracts in the language has been salutary too in order to prove their consistency, it can also be fatal to extensionality in the presence of equality in formulae defining sets.

9.7 Extensionality and abstraction with equality

To illustrate this, we begin by showing that, assuming Ext, the Russell set can be defined in $Ł_2$ without negations or implications, simply by using set abstracts and equality in the language:

Proof. Assume $Abst$ & Ext and define $R := \{x \mid \{y \mid x \in x\} = \{y \mid \bot\}\}$. Thus we have :

$$\begin{array}{llll} Abst \vdash R \in R & \leftrightarrow & \{y \mid R \in R\} = \{y \mid \bot\} & \\ & \leftrightarrow & \forall y(R \in R \leftrightarrow \bot) & \text{(using } Ext\text{)} \\ & \leftrightarrow & R \notin R & \end{array}$$

This sort of paradoxes appeared in Gilmore's work on partial set theory [13]*.[3] Note that Russell's set is no longer contradictory in Gilmore's set

[2] So far we have only obtained partial results, comparable to those of [5; 9]*, by using techniques described in [AME/RUT:89].

[3] It should be said that Gilmore's work was first publicized in 1967, but the inconsistency of extensionality only appeared in 1974.

theory. What Gilmore showed is that, within an extensional universe, a substitute for it can be defined by using set abstracts and equality in the language. Although his motivations were elsewhere, that work by Gilmore could have equally been expressed within the 3-valued Łukasiewicz logic. Besides, Brady in [3]* directly adapted Gilmore's technique to $Ł_3$ to much strengthen Skolem's initial result, showing the consistency of an abstraction scheme in that logic, but without equality in the language (as in Skolem). By this mere fact, a significant departure in Brady's paper is that the author succeeded in proving that his model is extensional and so, though he was not aware of that, he actually proved a complementary result of Gilmore's, namely that one can recover extensionality by dropping equality out of formulae defining sets.

To clearly express those results of Gilmore and Brady, let Π^3 denote the set of formulae containing no occurrences of $\rightarrow$ in the language of set theory.[4] Adding $=$ as subscript means that we may officially use it in formulae defining sets.

Theorem ([13]*). *$Abst[\widehat{\Pi}^3_=]$ is consistent in $Ł_3$, but inconsistent together with Ext.*

Theorem ([3]*). *$Abst[\widehat{\Pi}^3] + {}^\flat Ext$ is consistent in $Ł_3$.*

It should be mentioned that similar results apply as well to the *paraconsistent* counterpart of $Ł_3$, the quasi-relevant logic RM_3 (see [4; 18]*).

Previously, Skolem had only built extensional models of a comprehension scheme in $Ł_3$ restricted to formulae not containing any occurrence of $\rightarrow$, but not any quantifier either (and without equality in the language). Actually his technique of proof in [23; 22]* cannot be extended in order to handle quantifiers. Skolem however shows in [22]* that it can be adapted to $Ł_2$, initiating by the way, as far as we know, the consistency problem for *positive* comprehension principles. Even better, in [22]* and [24]*, he presents different techniques of which one finally leads him to show the consistency of $Comp(\Pi^2) + Ext$ (see [24]*, Theorem 1), where Π^2 is like Π^3 but without $\neg$ (classical logic!). It is the reason why such formulae are commonly said to be *positive*.

Surprisingly, and without any references to Skolem, the consistency problem for positive comprehension principles has been reinvestigated and invigorated much later in the eighties, where it would rather seem to have his source in Gilmore's work on partial set theory (see [11; 14]*). It is then not so surprising that similar results to those stated for $Ł_3$ have been proved for $Ł_2$:

Theorem ([14]*). *$Abst[\widehat{\Pi}^2_=]$ is consistent in $Ł_2$, but inconsistent together with Ext.*

[4] Notice that this restriction is meaningful in view of Moh Shaw-Kwei's paradox.

Theorem ([16]*). *$Abst[\widehat{\Pi}^2] + {}^\flat Ext$ is consistent in $Ł_2$.*

Here much more interesting *extensional* models were actually discovered in order to recover equality in formulae defining sets, and so by giving up the use of set abstracts.

Theorem ([11]*). *$Comp[\Pi^2_=] + Ext$ is consistent in $Ł_2$.*

Subsequently, the technique used to construct them was adapted by Hinnion in [15]* to the partial and the paraconsistent cases with different success (see also [18]* for the paraconsistent version).

The rest of this talk is in a sense devoted to showing how such models can be obtained. To do that, we first have to point out the guiding idea which was already subjacent in the original work by Skolem.

9.8 Monotonicity: a particular case of continuity

The key step in Skolem's proofs of the consistency of a comprehension scheme in $Ł_3$ and $Ł_2$ (see [23]* and [22]*) is again the observation that the truth functions of formulae defining sets have the fixed-point property. Of course, as the set of truth degrees is discrete, it is no longer possible to invoke Brouwer's theorem to see that. As a matter of fact, Skolem contents himself in [23]* with observing that there are exactly eleven propositional truth functions in one variable constructible in $Ł_3$ without using $\rightarrow$, and that each of them really has a fixed point. In [22]*, a similar remark for positive formulae is applied to $Ł_2$. Although this was not noticed by Skolem, it is another famous fixed-point theorem that is hidden behind these observations, namely the *Knaster-Tarski theorem* for ordered sets.

To proceed we would remind the reader that a particular case of continuity is *monotonicity*. Indeed, it is well-known that if any ordered set is equipped with the topology for which the closed subsets are nothing but the lower sets, i.e., A is *closed* if and only if $x \leqslant a \in A \Rightarrow x \in A$, then the continuous maps are exactly the monotonic ones. Now it has been shown that the ordered sets on which any monotonic/continuous function has a fixed point are characterizable, these are the *dcpo*'s: we say that a partially ordered set D is *directed complete*, or is a *dcpo*, if any directed subset in D has a least upper bound; where $A \subseteq D$ is said to be *directed* if $A \neq \emptyset$ and for all $a, b \in A$, there exists $c \in A$ with $a, b \leqslant_D c$.

Theorem (Knaster-Tarski). *Let D be a dcpo and $f : D \longrightarrow D$ be monotonic. Then f has a fixed point, i.e., there exists (a least) x in D such $x = f(x)$.*

And we mention that this theorem has a converse, which is much harder to prove by far: *a partially ordered set (with a least element) on which any monotonic function has a fixed point is necessarily a dcpo.*

Note that any finite chain or any *complete* infinite one is a dcpo. Thus, any of the I_n's or I with the *truth* ordering $\leqslant_T$ is a dcpo. It is then easily seen that all the connectives and quantifiers *except* negation *and* implication are monotonic, whereas we would remind the reader that *all* were continuous with respect to the *usual* topology on I. Of course, if both negation and implication are rejected, there is absolutely no need to add some imaginary truth-values, so this would only make sense for I_2, which incidentally is the strongest logic in the family.

In the case of Ł$_3$, however, the set of truth degrees is naturally endowed with another ordering, the so-called *knowledge/information* ordering $\leqslant_K$, which comes directly from the various attempts at explanation for the middle value $\frac{1}{2}$ as “unknown”, “undefined”, “undetermined”, “possible”, or whatever expressing in some sense a *lack* of truth value.

$$\begin{array}{ccccc} 0 = \{\texttt{false}\} & & & & \{\texttt{true}\} = 1 \\ & \diagdown & & \diagup & \\ & & \frac{1}{2} = \{\,\} & & \end{array}$$

The partially ordered set ‘$\diagdown\!\diagup$’ thus defined is actually the smallest example of a dcpo that is not a chain. With respect to this ordering it is readily seen that all connectives and quantifiers except implication *only* are monotonic. By the fixed-point theorem above, any set that is not defined by means of $\rightarrow$ is acceptable. So we trace back the embryonic use of this theorem in semantical consistency proofs to Skolem's papers.

9.9 Kripke-style models

Interestingly, it is the same theorem that is implicitly invoked again, but at another level, in Gilmore and Brady's work in providing their *terms models*. Roughly, the universe of these models is fixed in advance and made of set abstracts, regarded as *syntactical* expressions of the form $\{x|\,\varphi(x)\}$, for suitable formulae φ (e.g., $\widehat{\Pi}^2_{(=)}$, $\widehat{\Pi}^3_{(=)}$), and then, by a fixed-point argument, the membership relation is determined inductively in such a way that $\{x|\,\varphi(x)\}$ be a solution to the φ-instance of the *abstraction scheme* under consideration. This so called *inductive method* was later popularized by Kripke [17]* in his seminal work on the liar paradox. For a comprehensive description and analysis of the connection between these works, the reader is referred to Feferman's paper [8]*.

Since Section 7 showed us the limit of abstraction, we are not going to dwell on that way of building models. We would rather prefer to sing the praises of another useful technique which has its source in Scott's work on λ-calculus.

9.10 Scott-style models

In providing models for the *untyped* λ-calculus, Scott discovered that the Knaster-Tarski theorem is reflected within suitable subcategories of dcpo's. Roughly, it was proved that for a wide variety of functors $\mathcal{F}(\cdot)$ acting on those categories the reflective equation $X \simeq \mathcal{F}(X)$ has a solution (obtained by *inverse limit*). Such fixed points have naturally proposed themselves as *semantic domains* of programming languages, so the mathematical branch in theoretic computer science that investigates this is called *domain theory*.

Now, in view of ($\diamondsuit$) on page 126 (Section 9.5), when $T \equiv \langle I_2\,; \leqslant_T \rangle$ or $T \equiv \langle I_3\,; \leqslant_K \rangle$, which are both dcpo's, it might be tempting by using techniques of domain theory to try to solve a reflexive equation of the form $X \simeq [X \rightarrow T] \subseteq \langle X \rightarrow T \rangle$, where $\langle X \rightarrow T \rangle$ is the set of all monotonic functions from X into T. Thus, a fixed-point solution M to such an equation, which should be equipped with its own information ordering $\leqslant_M$, would give rise to a universe of sets that are thought of as particular monotonic functions from M into T. Therefore, by virtue of the monotonicity of the connectives and quantifiers, it might be conceivable that M would yield an extensional normal model of $Comp[\Pi^2]$ or $Comp[\Pi^3]$.

Such an attempt is largely explored in [2] for the 3-valued case. However attractive this idea is, it does not enable one to recover equality in formulae defining sets. This is because the equality is *not* monotonic on a normal structure. Let us illustrate this with $T \equiv \langle I_2\,; \leqslant_T \rangle$, for in that case the equality on a normal $\mathcal{M}$ must be strict, i.e., $=_{\mathcal{M}}`(x, y) = 1$ if $x = y$ in $M$, $= 0$ otherwise. But then, as soon as there exist $x, y$ in $M$ with $x <_M y$, we have $=_{\mathcal{M}}`(x, x) = 1 \not\leqslant_T =_{\mathcal{M}}`(x, y) = 0$, showing that $=_{\mathcal{M}}$ is not monotonic. Concerning the 2-valued case, it is proved in [19] that a solution to $X \simeq [X \rightarrow I_2]$ in the category of dcpo's yields a model of positive *abstraction*, a model that is thus not a pure term model.

Still, we are on the right track and the issue actually lies in a previous note. Again we shall only focus on I_2, though the same move can be made for I_3 (see [18]* for the paraconsistent version).

9.11 The issue

As mentioned earlier, monotonicity is a particular case of continuity. Applying that to $I_2 = \{0, 1\}$ with $\geqslant_T$, the *dual* of the truth ordering, for which the closed subsets are $\emptyset, \{1\}, \{0, 1\}$, it is obvious to see that $f : M \longrightarrow I_2$ is continuous/monotone if and only if $f^{-1}\{1\}$ is closed. We would now remind the reader that in any extensional normal structure $\mathcal{M}$ such a function f is the *characteristic function* $[x]_{\mathcal{M}}$ of the set x it represents, which may thus be identified with a *closed* subset of M.

The next step is then to look at sets as *closed* parts of the universe, but closed in a broader topological sense, not only in connection with any given ordering. Notice that demanding that $=_{\mathcal{M}} : M \times M \longrightarrow I_2$ be continuous amounts to requiring that M be Hausdorff, seeing that ${=_{\mathcal{M}}}^{-1}\{1\} = \Delta_M$, where $\Delta_M := \{(x, y) \in M \times M \mid x = y\}$, and it is well known that a topological space M is Hausdorff if and only if Δ_M is closed. This is a singular departure from the preceding attempts, for the order-topology is never Hausdorff (unless $<_M = \emptyset$). Thus, let us consider the following "functor" acting on Hausdorff topological spaces:

$$\mathcal{F}(X) := \{A \subseteq X \mid A \text{ closed}\}$$

A solution to the reflexive equation $X \simeq \mathcal{F}(X)$ can be found by using techniques described in [AME/RUT:89], where it is shown that another famous fixed-point theorem involving *continuous* functions is reflected within a suitable category. This theorem is *Banach's* for *contracting* functions on *complete metric spaces*, which too have been successfully used in modelling programming processes.[5]

Actually what we get is a *compact* complete metric space M such that $M \simeq \mathcal{F}(M)$, compactness being an essential ingredient in order to show that M yields an extensional normal model of $Comp[\Pi^2_=]$, as expected. Hereby the equality relation makes its entrance in formulae defining sets within an extensional universe. Note that in Ł$_2$ the introduction of $=$ coincides with the one of $\dot{=}$, and this is certified by the fact that these models actually fulfil an extended comprehension scheme involving *restricted quantifications* of the form $\forall x\ (x \in y \rightarrow \cdots)$. But a very characteristic property that comes directly from their definition is the following:

> for any predicate $P(x)$, there exists a smallest set a in M such that $\mathcal{M} \models \forall x (P(x) \rightarrow x \in a)$.

Indeed, it is nothing but the unique a in M such that $[a]^{-1}_{\mathcal{M}}\{1\} = \overline{A}$, where $A = \{x \in M \mid \mathcal{M} \models P(x)\}$. Accordingly, the Russell class is not closed so that Russell's paradox is blocked.

The properties of these models give rise to an interesting first-order *topological set theory* which was axiomatized and investigated by Esser in [6; 7], where he showed in particular that it can interpret ZF.[6] On this point, it should

[5] Recently, the framework of so-called "*continuity spaces*", a common refinement of partially ordered sets and metric spaces, has been proposed to develop a general theory of semantics domains; the interested reader is referred to [10].

[6] For historical reasons, Esser called his theory "GPK^+_∞". Needless to say that anything else more fetching would have been luckier.

be remarked that a pertinent formulation of the axiom of infinity is actually required in order to recover it in ZF. Besides, originally in [11]*, it was shown that with a large cardinal assumption, namely the existence of an uncountable *weakly compact* cardinal, the construction of such structures can be so carried out as to fulfil such a relevant axiom of infinity. These structures, subsequently called *Hyperuniverses*, were further studied by Forti, Honsell, Lenisa in several papers, e.g. in [12] where they even proposed hyperuniverses as an universal framework for investigating the semantics of programming languages...

We have thus shown that an alternative and yet expressive way of avoiding the set-theoretic paradoxes in *classical* logic originates in the use of so-called "*deviant logics*".

APPENDIX
Diagram of selected references

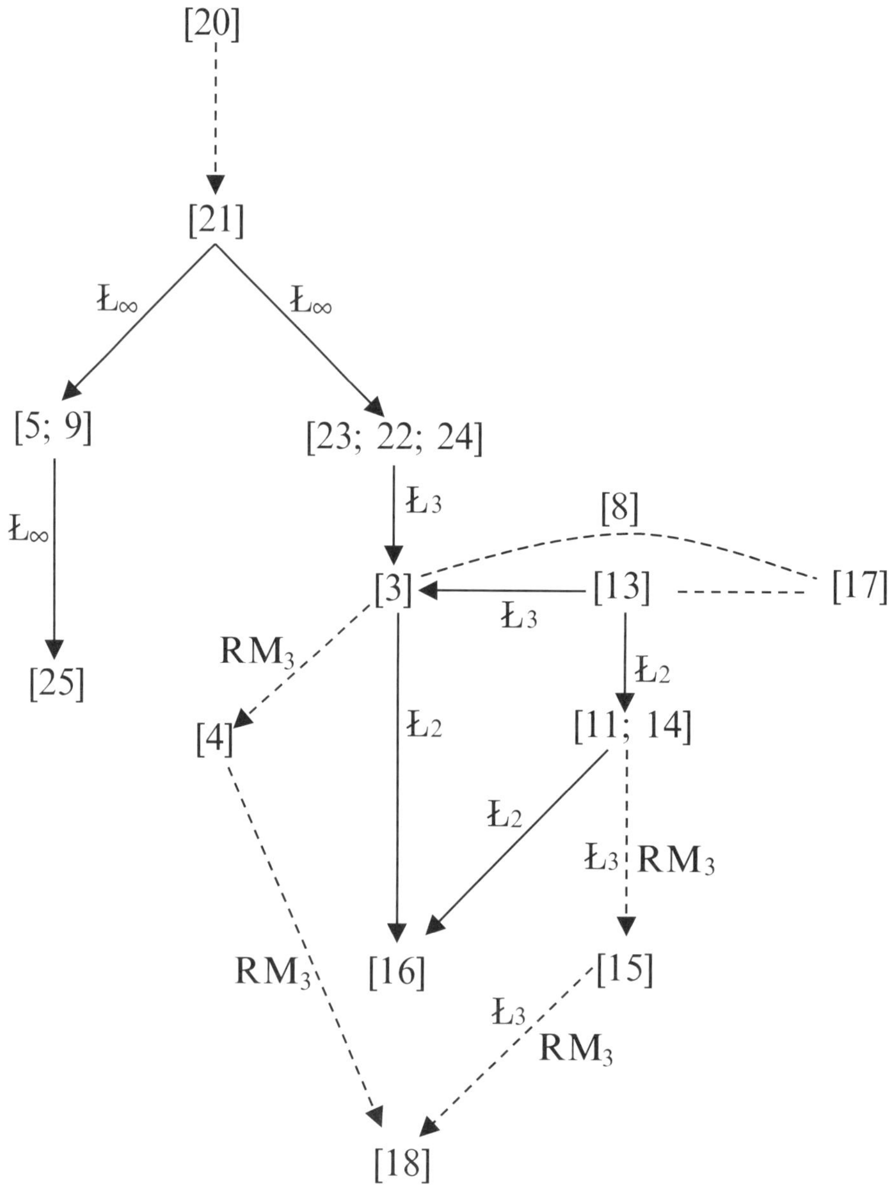

References

[1] America, P. and J. Rutten. "Solving reflexive domain equations in a category of complete metric spaces", *J. Comput. System Sci.* **39** (1989), 343–375.

[2] Apostoli, P. and A. Kanda. "Parts of the continuum: towards a modern ontology of science", forthcoming in *The Poznan Studies in the Philosophy of Science and the Humanities*, L. Nowak (ed.).

[3] Brady, R. T. "The consistency of the axioms of abstraction and extensionality in a three-valued logic", *Notre Dame J. Formal Logic* **12** (1971), 447–453.

[4] Brady, R. T. and R. Routley. "The non-triviality of extensional dialectical set theory", *in* G. Priest, R. Routley, and J. Norman (eds.), *Paraconsistent Logic*, Philosophia Verlag, Munich, 1989, 415–436.

[5] Chang, C. C. "The axiom of comprehension in infinite valued logic", *Math. Scand.* **13** (1963), 9–30.

[6] Esser, O. *Interprétations mutuelles entre une théorie positive des ensembles et une extension de la théorie de Kelley-Morse*, Ph.D. thesis, Université Libre de Bruxelles, 1997, to be published.

[7] Esser, O. "On the consistency of a positive theory", *Math. Logic Quart.* **45** (1999), 105–116.

[8] Feferman, S. "Towards useful type-free theories. I", *J. Symbolic Logic* **49** (1984), 75–111.

[9] Fenstad, J. E. "On the consistency of the axiom of comprehension in the Łukasiewicz infinite valued logic", *Math. Scand.* **14** (1964), 65–74.

[10] Flagg, B. and R. Kopperman, "Continuity spaces: Reconciling domains and metric spaces", *Theoret. Comput. Sci.* **177** (1997), 111–138.

[11] Forti, M. and R. Hinnion. "The consistency problem for positive comprehension principles", *J. Symbolic Logic* **54** (1989), 1401–1418.

[12] Forti, M., F. Honsell, and M. Lenisa, "Processes and hyperuniverses", *in* I. Prívara *et al.* (eds.) *Mathematical Foundations of Computer Sciences, Lectures Notes in Computer Science*, vol. 841, Springer, 1994, 352–361.

[13] Gilmore, P. "The consistency of partial set theory without extensionality", *Axiomatic Set Theory, Proceedings of Symposia in Pure Mathematics*, vol. 13, Part II, *Amer. Math. Soc.*, Providence, R.I., 1974, 147–153.

[14] Hinnion, R. "Le paradoxe de Russell dans des versions positives de la théorie naïve des ensembles", *C. R. Acad. Sci. Paris, Sér. I Math.* **304** (1987), no. 12, 307–310.

[15] Hinnion, R. "Naive set theory with extensionality in partial logic and in paradoxical logic", *Notre Dame J. Formal Logic* **35** (1994), 15–40.

[16] Hinnion, R. and Th. Libert, "Positive abstraction and extensionality", *J. Symbolic Logic* **68** (2003), 828–836.

[17] Kripke, S. "Outline of a theory of truth", *J. Philos. Logic* **72** (1975), 690–716.

[18] Libert, Th. "Models for a paraconsistent set theory", *J. Appl. Log.*, **3** (2005), 15–41.

[19] Libert, Th. *More studies on the axiom of comprehension*, Ph.D. thesis, Université Libre de Bruxelles, 2004, to be published.

[20] Shaw-Kwei Moh. "Logical paradoxes for many-valued systems", *J. Symbolic Logic* **19** (1954), 37–40.

[21] Skolem, Th. "Bemerkungen zum komprehensionsaxiom", *Z. Math. Logik Grundlagen Math.* **3** (1957), 1–17.

[22] Skolem, Th. "A set theory based on a certain 3-valued logic", *Math. Scand.* **8** (1960), 127–136.

[23] Skolem, Th. "Investigations on a comprehension axiom without negation in the defining propositional functions", *Notre Dame J. Formal Logic* **1** (1960), 13–22.

[24] Skolem, Th. "Studies on the axiom of comprehension", *Notre Dame J. Formal Logic* **4** (1963), 162–170.

[25] White, R. W. "The consistency of the axiom of comprehension in the infinite-valued predicate logic of Łukasiewicz", *J. Philos. Logic* **8** (1979), 509–534.

III

CATEGORY-THEORETIC STRUCTURES

Chapter 10

CONTINUITY AND LOGICAL COMPLETENESS: AN APPLICATION OF SHEAF THEORY AND TOPOI

Steve Awodey
Carnegie Mellon University
awodey@cmu.edu

The *main argument* of this paper is as follows:

1 The distinction between the Particular and the Abstract General is present in that between the Constant and the Continuously Variable. More specially, *continuous variation is a form of abstraction.*

2 Higher-order logic (HOL) can be presented algebraically. As a consequence of this fact, it has *continuously variable models.*

3 Variable models are classical mathematical objects; namely, *sheaves.*

4 HOL is *complete* with respect to such continuously variable models. Standard semantics appears thereby as the constant case of "no variation." In this sense, *HOL is the logic of continuous variation.*

The argument will be developed in four sections: (i) the algebraic formulation of HOL is given; (ii) rings of real-valued functions are considered as an example of variable structure; (iii) the idea of continuously variable sets is then discussed; and finally, (iv) it is explained how HOL is the logic of continuous variation.

10.1 Algebraic logic

Categorical logic can be seen as the successful completion of the program of "algebraicizing" logic begun in the 19-century. Everyone is familiar with

J. van Benthem, G. Heinzmann, M. Rebuschi and H. Visser (eds.), The Age of Alternative Logics: Assessing Philosophy of Logic and Mathematics Today, 139–149.

$$\bot \vdash \varphi \qquad \varphi \vdash \top$$

$$\frac{\varphi \vdash \vartheta \text{ and } \psi \vdash \vartheta}{\varphi \vee \psi \vdash \vartheta} \qquad \frac{\vartheta \vdash \varphi \text{ and } \vartheta \vdash \psi}{\vartheta \vdash \varphi \wedge \psi}$$

$$\frac{\vartheta \wedge \varphi \vdash \psi}{\vartheta \vdash \varphi \Rightarrow \psi}$$

$$\frac{\varphi(x) \vdash \vartheta}{\exists x.\varphi(x) \vdash \vartheta} \qquad \frac{\vartheta \vdash \varphi(x)}{\vartheta \vdash \forall x.\varphi(x)}$$

Fig. 1. Adjoint rules for FOL

the boolean algebra approach to propositional logic, but the treatment of quantification in particular has posed a serious obstacle to extending the algebraic treatment. The categorical treatment of quantifiers as adjoint functors — due to F.W. Lawvere in the 1960s — solved this problem, although it has been little appreciated until very recently. Category theory is of course a branch of abstract algebra, but the sense in which the categorical treatment of logic is "algebraic" is deeper than just that. Rather, it is the recognition of the quantifiers — and indeed all of the logical operations — as *adjoint functors* that makes logic algebraic. For it is a general fact about adjoints that they always admit an algebraic description, in a definite, technical sense. This is the same fact that makes possible the equational description of e.g. cartesian products and pairing. Figure 1 (page 140) shows the (two-way) rules of inference for the first-order logical operations expressed as adjoints. [1]

HOL also includes quantification over "higher types" of relations, functions, properties of functions, and so on. Figure 2 indicates the basic ingredients of algebraic HOL, as it results from the adjoint analysis of these operations. The *axioms* consist of a handful of equations of the sort indicated, and the *rules of inference* are essentially substitution of equals for equals, as in elementary algebra.

It may be noted that these are *all* of the logical operations required; the first-order operations are definable from these, as suggested in Figure 3 (which also indicates how even fewer would still suffice). The adjoint rules of Figure 1 on page 140 can then be proven.

In categorical logic we extend the treatment of propositional logic as a boolean algebra to HOL, by introducing the new notion of a *topos*. A topos is a certain kind of algebraic object (a category equipped with a certain adjoint structure) that bears the same relation to HOL as does a boolean algebra to

[1] The quantifier rules require the variable x not occur freely in ϑ. For a full statement see [J. Lambek and P. J. Scott 1986; S. Awodey and C. Butz 2000].

Types: $X \times Y$, Y^X, $\mathcal{P}(X)$, $\mathcal{P}$
Terms: $\langle s, t \rangle$, $\pi_1 t$, $\pi_2 t$, $\lambda x.t$, $t(s)$, $\{x$
$STHV\varphi\}$
Formulas: $s = t$, $s \in t$
Axioms: equations such as:

$$\pi_1\langle s, t \rangle = s$$
$$\lambda x.t(x) = t$$
$$x \in \{x$$
$$STHV\varphi\} = \varphi$$

Rules: substitution of equals for equals.

Fig. 2. Algebraic formulation of HOL

$$\begin{aligned}
\top &=_{df} \{x \mid x = x\} = \{x \mid x = x\} \\
\varphi \wedge \psi &=_{df} \langle \varphi, \psi \rangle = \langle \top, \top \rangle \\
\forall x.\, \varphi(x) &=_{df} \{x \mid \varphi(x)\} = \{x \mid \top\} \\
Y^X &=_{df} \{f \in \mathcal{P}(X \times Y) \mid \forall x \exists y. \langle x, y \rangle \in f\} \\
\mathcal{P}(X) &=_{df} \mathcal{P}^X
\end{aligned}$$

Fig. 3. Logical operations defined

propositional logic:

$$\frac{\text{propositional logic}}{\text{boolean algebra}} = \frac{\text{higher-order logic}}{\textit{topos}}$$

It should be emphasized that this reformulation is still equivalent to standard deductive HOL with respect to the logical formulas and consequences. We do not change the "logical theorems" but only the presentation of the logical system, replacing the machinery of formal deductive systems with elementary algebraic manipulations. It also should be noted that we are making no use of either what the logician calls *standard* or *Henkin* semantics. Instead, from a logical point of view, we are going to specify a *new kind of semantics*. Indeed, the algebraic formulation just given admits continuously variable models, resulting in so-called *topological semantics*. This possibility results from general facts about algebraic objects and continuous variation; so it may be useful to briefly recall how it works in the familiar case of rings, before considering the new one of algebraic logic.

10.2 Rings of $\mathbb{R}$-valued functions

The real numbers $\mathbb{R}$ form a topological space, an abelian group, a commutative ring, a complete ordered field, and much more. We shall consider just the properties expressed in the *language of rings*

$$0, 1, a + b, a \cdot b, -a$$

and first-order logic. For example, $\mathbb{R}$ is a field:

$$\mathbb{R} \models \forall x(x = 0 \vee \exists y.\ x \cdot y = 1)$$

Now consider the *product ring* $\mathbb{R} \times \mathbb{R}$, with elements of the form:

$$r = (r_1, r_2)$$

and the *product operations*:

$$\begin{aligned}
0 &= (0, 0) \\
1 &= (1, 1) \\
(x_1, x_2) + (y_1, y_2) &= (x_1 + y_1, x_2 + y_2) \\
(x_1, x_2) \cdot (y_1, y_2) &= (x_1 \cdot y_1, x_2 \cdot y_2) \\
-(x_1, x_2) &= (-x_1, -x_2)
\end{aligned}$$

Since these operations are still associative, commutative, and distributive, $\mathbb{R} \times \mathbb{R}$ is still a ring. But the element $(1, 0) \neq 0$ cannot have an inverse, since $(1, 0)^{-1}$ would have to be $(1^{-1}, 0^{-1})$. Therefore $\mathbb{R} \times \mathbb{R}$ *is not a field*. In a similar way, one can form the more general product rings $\mathbb{R} \times \ldots \times \mathbb{R} = \mathbb{R}^n$, or $\mathbb{R}^I$ for any index-set I. Elements have the form:

$$r = (r_i)_{i \in I}$$

the *pointwise operations* are defined by:

$$\begin{aligned}
0 &= (0)_i \\
1 &= (1)_i \\
(x_i) + (y_i) &= (x_i + y_i) \\
(x_i) \cdot (y_i) &= (x_i \cdot y_i) \\
-(x_i) &= (-x_i)
\end{aligned}$$

$\mathbb{R}^I$ is again a ring, but with still fewer properties of $\mathbb{R}$. Product rings $\mathbb{R}^I$ are, however, always (von Neumann) *regular:*

$$\mathbb{R}^I \models \forall x \exists y.\ x \cdot y \cdot x = x.$$

For, given x, we can take $y = (y_i)$ with:

$$y_i = \begin{cases} x_i^{-1}, & \text{if } x_i \neq 0 \\ 0, & \text{if } x_i = 0 \end{cases}$$

Then:

$$(x \cdot y \cdot x)_i = x_i \cdot y_i \cdot x_i = \begin{cases} x_i \cdot x_i^{-1} \cdot x_i = x_i, & \text{if } x_i \neq 0 \\ 0 \cdot 0 \cdot 0 = x_i, & \text{if } x_i = 0 \end{cases}$$

The *main point* of these examples is that one can produce rings that violate even more properties of $\mathbb{R}$ by passing to "continuously varying reals". But what is a "continuously varying real number"? Let X be a topological space. A "real number r_x varying continuously over X" is just a continuous function:

$$r : X \to \mathbb{R}$$

We equip these functions with the pointwise operations, as before:

$$(f+g)(x) = f(x) + g(x), \quad \text{etc.}$$

The set $\mathcal{C}(X)$ of all such functions then forms a *subring* of the product ring over the index set of points $|X|$:

$$\mathcal{C}(X) \subseteq \mathbb{R}^{|X|}$$

But unlike the product ring, $\mathcal{C}(X)$ is in general not regular:

$$\mathcal{C}(X) \not\models \forall f \exists g.\ f \cdot g \cdot f = f$$

For take e.g. $X = \mathbb{R}$ and $f(x) = x^2$, then we must have:

$$g(x) = \frac{1}{x^2}, \qquad \text{if } x \neq 0$$

but of course:

$$g(0) = \lim_{x \to 0} g(x) = \lim_{x \to 0} \frac{1}{x^2} = \infty$$

so there can be no *continuous* g satisfying $f \cdot g \cdot f = f$. Summarizing the lesson of these examples, we've seen that the "continuously varying reals" $\mathcal{C}(X)$ have even fewer properties of the field of "constant" reals $\mathbb{R}$ than do the product rings $\mathbb{R}^I$. In that sense, they are closer to a general notion of "quantity". Passing from constants to continuous variation therefore "abstracts away" some properties of the constants. We note by the way that it does so without introducing any new "abstract entities".

10.3 Continuously variable sets

Now let us take stock: we have an algebraic presentation of HOL, and we have seen how continuously varying algebraic structures like rings can violate some properties of the constant ones. We now proceed according to the analogy indicated in Figure 4, comparing real numbers to sets. The similarity rests on regarding reals as *linear* magnitudes, while sets are *extensive* magnitudes. The condition $\forall f \in A^B \exists g \in B^A.\ f \circ g \circ f = f$ on (non-empty) sets corresponding to regularity is actually a form of the axiom of choice (see [F.W. Lawvere 1964]). The notion of a "continuously variable set" that we seek will turn out to be that of a *sheaf*. First, observe that the (algebraically specified) logical operations can be interpreted in other "universes" of sets, e.g. in the universe of "pairs of sets":

$$\mathbf{Sets} \times \mathbf{Sets}$$

The elements have the form:

$$A = (A_1, A_2)$$

and the operations are defined componentwise:

$$(A_1, A_2) \times (B_1, B_2) = (A_1 \times B_1, A_2 \times B_2)$$
$$\mathcal{P}(A_1, A_2) = (\mathcal{P}(A_1), \mathcal{P}(A_2))$$
$$(a_1, a_2) \in (A_1, A_2) \Leftrightarrow a_1 \in A_1 \text{ and } a_2 \in A_2$$
$$\text{etc.}$$

Real numbers	Sets
Algebraic operations $x + y, x \cdot y, x^{-1}, 0, 1$	Algebraic operations $X \times Y, Y^X, \mathcal{P}(X), \emptyset$
Algebraic condition (formula in ring operations) $\forall x \exists y.\ x \cdot y \cdot x = x$	Algebraic condition (formula in HOL operations) $\forall f \in A^B \exists g \in B^A.\ f \circ g \circ f = f$
Variable real number (continuous $\mathbb{R}$-valued function)	Variable set (sheaf)

Fig. 4. Analogy

This interpretation models HOL, but it doesn't satisfy *all* the properties of **Sets**. For example:

$$\mathbf{Sets} \models A \cong 0 \vee \exists x.\, x \in A$$

But in $\mathbf{Sets} \times \mathbf{Sets}$ we can take $(1, 0) \not\cong 0$ as A, and then $a \in (1, 0)$ means $a = (a_1, a_2)$ with $a_1 \in 1$ and $a_2 \in 0$, which is impossible. Next, just as in the case of rings, we can generalize to $\mathbf{Sets} \times \ldots \times \mathbf{Sets} = \mathbf{Sets}^n$, and indeed to $\mathbf{Sets}^I$ for any index set I, to get the "universe" of I-indexed families of sets:

$$A = (A_i)_{i \in I}$$
$$(A_i) \times (B_i) = (A_i \times B_i)$$
$$\text{etc.}$$

These families again model HOL, but they have still fewer properties of **Sets**. However, all such "product universes" do satisfy e.g. the axiom of choice. To find even more general "universes" that violate it, we can consider even more general families of sets:

$$(F_x)_{x \in X}$$

varying continuously over an arbitrary space X. (This generalizes the case where the set I in the previous example is regarded as a discrete space.) But what should a "continuously varying set" be? The problem is that we cannot simply take a "continuous set-valued function":

$$F : X \to \mathbf{Sets}$$

as we did for rings of real-valued functions, since **Sets** is not a topological space! Of course, there are people who already know how to do this sort of thing, so let us look at what the topologists and algebraists do when they need *continuously varying structures*. A "continuously varying space" $(Y_x)_{x \in X}$ over a space X is called a *fiber bundle*. It consists of a space $Y = \sum_{x \in X} Y_x$ and a continuous "indexing" projection $\pi : Y \to X$, with $\pi^{-1}\{x\} = Y_x$, as indicated below.

$$\begin{array}{l} Y = \sum_{x \in X} Y_x \\ \Big\downarrow \pi \\ X \end{array}$$

A "continuously varying group" $(A_x)_{x \in X}$ is a *sheaf of groups*. It consists essentially of a fiber bundle $\pi : A \to X$ as indicated below,

$$\begin{array}{l} A = \sum_{x \in X} A_x \\ \downarrow \pi \\ X \end{array}$$

satisfying the additional requirements:

1 π is a local homeomorphism (see below),

2 each A_x is a group,

3 the operations in the fibers A_x "fit together continuously".

Now, what should a "continuously varying set" be? Clearly, it should be a *sheaf of sets:* an indexed family $(F_x)_{x \in X}$ as indicated below,

$$\begin{array}{l} F = \sum_{x \in X} F_x \\ \downarrow \pi \\ X \end{array}$$

such that each fiber $F_x = \pi^{-1}(x)$ is discrete, and moreover π is a *local homeomorphism*: each point $y \in F$ has some neighborhood U on which π is a homeomorphism $U \xrightarrow{\sim} \pi(U)$. This ensures that the variation over the space is continuous. Some of the logical operations on sheaves can be defined pointwise:

$$(F \times G)_x \cong (F_x \times G_x)$$

Others, however, cannot. The exponential G^F of sheaves F, G is the "sheaf-valued hom" $\hom(F, G)$, defined in terms of *germs of continuous maps* $F \to G$:

$$\begin{aligned} (G^F)_x &\cong \hom(F, G)_x \quad \text{(germs of maps } F \to G) \\ &\ncong G_x^{F_x} \end{aligned}$$

The "universe" $\mathrm{Sh}(X)$ of all sheaves on a space X models HOL, but in general it *violates the axiom of choice*:

$$\mathrm{Sh}(X) \nvDash \mathrm{AC}$$

Indeed, one can find sheaf models of HOL that violate many other properties of sets. Example: germs of continuous $\mathbb{R}$-valued functions Sheaf $\mathcal{R}$ on space X with stalks:

$$\mathcal{R}_x = \{[f, U] \mid x \in U, f : U \to \mathbb{R}\}$$

where

$$[f, U] = [g, V] \Leftrightarrow \exists W_x \subseteq U \cap V.\ f|_W = g|_W$$

i.e. $f(w) = g(w)$ on some sufficiently small neighborhood $W \ni x$. $\mathcal{R}$ is a sheaf of rings. Define a ring structure on $\mathcal{R}$:

$$\begin{aligned}
[f, U] + [g, V] &= [f + g, U \cap V] \\
[f, U] \cdot [g, V] &= [f \cdot g, U \cap V] \\
-[f, U] &= [-f, U] \\
0_{\mathcal{R}} &= [0, X] \\
1_{\mathcal{R}} &= [1, X]
\end{aligned}$$

The ring of continuous functions $\mathcal{C}(X)$ is embedded in $\mathcal{R}$:

$$\begin{aligned}
&\mathcal{C}(X) \to \mathcal{R} \\
&\qquad f \mapsto [f, X]_x \in \mathcal{R}_x, \quad \text{for each } x \in X
\end{aligned}$$

$$\begin{aligned}
\mathcal{R} &= \text{“continuously-varying ring” of “continuously-varying reals”} \\
&= \text{“sheaf of reals”}
\end{aligned}$$

Thus, we've seen that HOL can be modeled in various "universes" other than **Sets**. In particular, the "universe" of all sets varying continuously over a space models HOL, where the notion of a continuously varying set is reasonably taken as that of a sheaf. Moreover, sheaves violate some properties of sets.

10.4 The logic of continuous variation

It's time to be more precise about the notion of a "universe". We've seen that only a few constructions are required to model HOL:

$$0,\ A \times B,\ \mathcal{P}(A),\ a \in A, \ldots$$

A *topos* is defined as a category equipped with adjoint structure corresponding to these operations (see [S. Mac Lane and I. Moerdijk 1992]). In this sense, a topos is a "universe of abstract sets". It's worth noting the following theorem, which just says that we have the definition right.

Theorem. *(topos completeness of HOL)*
A sentence of HOL is provable iff it is true in every topos model.

Given the foregoing discussion, it should come as no surprise to learn that the categories $\mathbf{Sets}$, $\mathbf{Sets} \times \mathbf{Sets}$, $\mathbf{Sets}^I$ are toposes. Moreover, the category $\mathrm{Sh}(X)$ of all sheaves of sets on a space X is also a topos. The topos $\mathrm{Sh}(X)$ of sheaves consists of sets F_x varying continuously in a parameter $x \in X$. The logic of the constant sets is quite strong; the logic of variable sets is much weaker. Fewer things are true of variable sets in general than are true of constant ones (think of the difference between the field of real numbers and the ring of real-valued functions). What is the logic of continuously varying sets? That is, which formulas of HOL are true in all sheaf models? The answer is given by the following theorem from [S. Awodey and C. Butz 2000]:

Theorem. *Logic of sheaves = classical deductive HOL.*

The proof of this fact uses recent, non-trivial results in topos theory.[2] theory The sheaf-theory on which it rests [C. Butz and I. Moerdijk 1999] is rooted in geometry, not logic. It is worth emphasizing that, unlike the preceding theorem, there is no obvious reason why this one needs to be true. Sheaves are classical mathematical objects, and their logical properties depend on continuous variation, not deduction. HOL is a classical deductive system going back to Frege and Russell and having nothing to do with continuity. That these things should coincide is remarkable. Note that the Gödel incompleteness of deductive higher-order logic can be easily understood in these terms:

> Gödel's "true but unprovable"
> involves only "true of all *constant* sets"
> but not "true of all *variable* sets"

A "true but unprovable" Gödel sentence is therefore true only of constant sets, not of all variable ones. Thus, summing up, we see that fewer things are true of all continuously varying sets than of all constant ones. HOL captures just those statements that are "variably true". Precisely: *HOL is deductively complete with respect to topological semantics*, which is the real statement of the second theorem mentioned above.[3]

---

[2]The treatment of classical logic (with the law of excluded middle) is also somewhat delicate, requiring a different interpretation than the usual one in topos.

[3]The author thanks the organizers of PILM 2002 for a stimulating and pleasant meeting.

References

Awodey, S. 1999. “Sheaf representation for topoi”, *Journal of Pure and Applied Algebra* 145, 107–21.

Awodey, S. 2000. “Topological representation of the λ-calculus”, *Mathematical Structures in Computer Science* 10, 81–96.

Awodey, S. and C. Butz 2000. “Topological completeness for higher-order logic”, *Journal of Symbolic Logic* 65(3), 1168–82.

Butz, C. and I. Moerdijk 1999. “Topological representation of sheaf cohomology of sites”, *Compositio Mathematica*, 118, 217–233.

Lambek, J. and P. J. Scott 1986. *Introduction to higher-order categorical logic*, Cambridge University Press.

Lawvere, F.W. 1964. “An elementary theory of the category of sets”, *Proc. Nat. Acad. Sci.* 52, 1506–11.

Mac Lane, S. and I. Moerdijk 1992. *Sheaves in geometry and logic: A first introduction to topos theory*, Springer-Verlag.

Chapter 11

WHAT IS CATEGORICAL STRUCTURALISM?

Geoffrey Hellman*
University of Minnesota
hellm001@umn.edu

In a recent paper Hellman [2003], we examined to what extent category theory ("CT") provides an autonomous framework for mathematical structuralism. The upshot of that investigation was that, as it stands, while CT provides many valuable insights into mathematical structure — specific structures and structure in general —, it does not sufficiently address certain key questions of logic and ontology that, in our view, any structuralist framework needs to address. On the positive side, however, a *theory of large domains* was sketched as a way of supplying answers to those key questions, answers intended to be friendly to CT both in demonstrating its autonomy vis-à-vis set theory and in preserving its "arrows only" methods of describing and interrelating structures and the insights that those methods provide. The "large domains", hypothesized as logico-mathematical possibilities, are intended as suitably rich background universes of discourse relative to which both category-and-topos theory and set theory can be developed side by side, without either emerging as "prior to" the other. Although those domains, as described, resemble natural models of set theory (on an iterative conception) or toposes suitably enriched with an equivalent of the Replacement Axiom, they are defined without set-membership as a primitive, and also without 'function' or 'category' or 'functor' as primitives; all that is required is a combination of 'part/whole' and plural quantification (in effect, the resources of monadic second-order logic). This background logic (including suitable comprehension axioms for wholes and "pluralities") suffices; and ontological commitments are limited to claims of the possibility of indefinitely large domains, any one extendable to a more encompassing one, without end.

*Support of this work by the National Science Foundation (USA, Award SES-0349804) is gratefully acknowledged.

J. van Benthem, G. Heinzmann, M. Rebuschi and H. Visser (eds.), The Age of Alternative Logics: Assessing Philosophy of Logic and Mathematics Today, 151–161.

Two interesting responses to this have already emerged on behalf of CT proponents, one by Colin McLarty [2004] and the other by Steve Awodey [2004]. Here we take the opportunity to come to terms with these and to assess their bearing on our original assessment and proposal. We will begin with a brief review of the main critical points of Hellman [2003]; then we will take up the responses of McLarty and Awodey in turn; and finally, we'll try to draw appropriate conclusions.

11.1 "Category Theory" and structuralist frameworks

The first point to stress is that the very term "category theory" is ambiguous, and the ambiguity follows closely on the heels of another, more basic one, that of "axiom" itself. On the one hand, axioms traditionally are conceived as basic truths simpliciter, as in the traditional conception of Euclidean geometry, or the axioms of arithmetic, or the axioms of, say, Zermelo-Fraenkel set theory. Call this the "Fregean conception" of axioms. In the geometric case, primitive terms such as 'point', 'line', 'plane', 'coincident', 'between', 'congruent' are taken as determinate in meaning, so that axioms employing them have a determinate truth-value. For number theory, 'successor', 'plus', 'times', 'zero', etc. have definite meanings leading to true axioms (say the Dedekind-Peano axioms); and for set theory, of course, 'membership' is taken as understood, and the axioms framed in its terms true (or true of the real world of sets). In contrast, there are algebraic-structural axioms for groups, modules, rings, fields, etc., where now they are not even assertions, but rather *defining conditions* on types of structures of interest. The primitive terms are not thought of as already determinate in meaning but only as schematically playing certain roles as required by the "axioms". Call this the "Hilbertian conception". Any objects whatever interrelated in the ways required by the defining axioms constitute a structure of the relevant type, say, a group, or a module, ..., or a category. In the latter case, primitive terms such as 'object', 'morphism', 'domain', 'codomain', and 'composition' are not definite in meaning, but acquire meaning only in the context of a particular interpretation which satisfies the axioms.

Thus, "morphisms" need not be functions, and (so) "composition" need not be the usual composition of functions, etc.[1] A large part of the debate between Frege and Hilbert on foundations turned on their respective, very different understandings of "axioms" along precisely these lines.

It is worth noting, incidentally, that Dedekind [1888] presented his "axioms" for arithmetic explicitly as defining conditions, i.e. axioms in the Hilbertian sense, as part of his definition of a "*simply infinite system*", and not as assertions.

[1]Thus, "mophisms" may be realized as homotopy classes of maps (between topological spaces), as formal deductions of formulas in a logical system, as directed line segments in a diagram, etc.

One may, oversimplifying a bit, say that the tendency of modern mathematics toward a structuralist conception has been marked by the rise and proliferation of Hilbertian axiom systems (practically necessitated by the rise of non-Euclidean geometries), with relegation of Fregean axioms to a set-theoretic background usually only mentioned in passing in introductory remarks. Category theory surely has contributed to this trend; we now even have explorations of "Zermelo-Fraenkel algebras" (Joyal and Moerdijk [1995]).

This ambiguity over "axioms" is, of course, passed on to "theories" of algebraic structures, as in "group theory", "field theory", . . . , and, indeed, "category theory" and (with some qualifications to be discussed below) "topos theory" as well. On the one hand, there is the first-order theory (definition) of groups, or of categories or toposes; but, on the other, there is a body of substantive theorizing *about* such structures, which, while constantly appealing to the first-order definitional axioms, is intended as *assertory*, and takes place in an informal background whose primitive notions and assumptions usually require logical analysis and reconstruction to be identified. Standard practice refers (in passing) to a background set theory, as it is well known that that suffices for most purposes. But of course that cannot serve in the context of "categorical foundations" where autonomy from set theory is the name of the game.

So what is the background theory? It is not clear. And so we find ourselves uncertain when it comes to comparing categorical structuralism with other frameworks that have been proposed. Here are five fundamental questions that we would submit any such framework should address:

(1) What is the background logic? Is it classical? Is it modal? Is it higher order logic? If so, what is the status of relations as objects?

(2) What are the extra-logical primitives and what axioms–presumably assertory–govern them? Are 'collection', 'operation', 'category', 'functor', for example, on the list? Especially, what axioms of mathematical existence are assumed?

(3) Is indefinite extendability of mathematical structures recognized or is there commitment to absolutely maximal structures, e.g. of absolutely all sets, all groups, etc.?

(4) Are structures eliminated as objects, and, if not, what is their nature?

(5) What account, if any, is given of our reference and epistemic access to structures?

In the case of set-theoretic structuralism, it is fairly clear how to answer at least (1) - (4); similarly, in the case of Shapiro's [1997] *ante rem* structuralism, and he takes a stab at (5) as well. In the case of modal-structuralism, answers to (1) - (4) are also forthcoming. (As an eliminativist structuralism, questions of reference are replaced with questions of knowledge of possibilities, related to, but even more difficult than, questions of knowledge of consistency, for in central cases we are interested in *standard* structures.)[2] But when it comes to categorical structuralism, it isn't clear what to say even with regard to (1) - (4). At most, bearing on (3), one finds widespread opposition to the view that a fixed background of sets is the privileged arena of mathematics.[3]

11.2 McLarty's "Fregean" response

In a nutshell, McLarty claims that, while the algebraic-structuralist reading of CT axioms and general topos axioms is correct, nevertheless specific axioms for certain particular categories and toposes are intended as assertory. In particular, he singles out ETCS, the elementary theory of the category of sets, CCAF, the category of categories as a foundation, and SDG, synthetic differential geometry as a theory of the category of smooth spaces. The axioms of these systems are not to be read merely as defining types of structures but rather as assertions, true of existing parts of mathematical reality, much as the axioms of ZFC are normally understood. Indeed, in the case of ETCS, this could be understood as describing the very same subject matter as ZFC, although with the characteristic arrows machinery rather than a primitive set-membership relation.

It appears to me that CCAF, or, better, McLarty's own approach [1991] to axiomatizing a category of categories, is actually the most promising in relation to the above questions. Let us return to consider this below. First, let us take up the other two examples, ETCS and SDG.

Now, I would not wish to deny that ETCS provides an important part of a structuralist analysis of sets. Through its "arrows only" formulations and generalizations, it abstracts from a fixed set-membership relation and analyzes sets in their functional roles, "up to isomorphism", which is all that really matters for mathematics. What remains problematic, however, regarding McLarty's reading of ETCS (which he attributes to Mac Lane), is its apparent commitment to a fixed, presumably maximal, real-world universe of sets, "*the* category of sets". This just strikes me as a convenient fiction. First, there is the question of multiplicity of conceptions of sets, e.g. non-well-founded as well as well-founded, possibly choice-less as well as with choice, with or without Replacement, the various large cardinal extensions, and so forth. Presumably, all of these conceptions are mathematically legitimate, and it would be arbitrary to treat just one as ontologically privileged. But even if suitable qualifications of the "intended universe" are added to the (meta) description, the problem of indefinite extendability still looms. Whatever domain of sets we recognize can be transcended

[2] For detailed comparisons of these varieties of structuralism, see Hellman [2001].

[3] For further details, see Hellman [2003].

by the very operations that set theory seeks to codify, collecting, collecting everything "already collected", passing to collections of subcollections, iterating along available ordinals, etc. (This, incidentally, is entirely in accord with Mac Lane's expressed views [1986] on the open-endedness of Mathematics.) Set theoretic structuralism can be faulted precisely for failing to apply to set theory itself, especially in regard to the very multiplicity of universes of sets that it naturally engenders. Categorical structuralism promises to do better, but it is hard put to keep that promise if it falls back on a maximal universe of sets or, more generally, on an absolute notion of "large category".

When it comes to a realist interpretation of SDG, the problems are quite different but equally challenging. This is a non-classical theory of continua which can be developed independently of category theory, known as "smooth infinitesimal analysis" ("SIA"). (Topos theory has proved useful in providing models of SIA, but the essential analytic ideas do not depend on the topos machinery.) This is a theory intended as an alternative to classical, "punctiform" analysis; it introduces nilsquare (and nilpotent) infinitesimals, while at the same time limiting the class of functions of reals to continuous ones. A central axiom, the Kock-Lawvere axiom, stipulates that any function on the infinitesimals about 0 obeys the equation of a straight line. (The axiom is also called the "Principle of Microaffineness".)

This actually implies the restriction to continuous functions. And the constant slope of the "linelet" given by the axiom serves to define the derivative of a function. (The "linelet" can be translated and rotated, but not "bent".) To accommodate nilsquares, certain restrictions apply to the classical ordered field axioms for the reals: nil-squares do not have multiplicative inverses, nor are they ordered with respect to one another or with respect to 0. Indeed, one proves that "not every x is either $= 0$ or not $= 0$". Not only does SIA refrain from using the Law of Excluded Middle (LEM), it derives results that are formally inconsistent with it (similar in this respect to Brouwerian intuitionism but contrasting with Bishop constructivism, which, with LEM added, gives back classical analysis). But SIA, consisting of the (restricted) ordered field axioms, the KL axiom, and a certain "constancy principle", suffices for a remarkable development of calculus in which limit computations are replaced with fairly straightforward algebra, placing on an alternative, consistent and rigorous footing early pre-limit geometric methods in analysis and mechanics. (Cf. Bell [1998] for a nice survey of such results.)

Why is there a problem with thinking of this theory as an objective description of continuous functions or phenomena? After all, the charge by Russell and fellow classicists that infinitesimals lead to inconsistencies, while true of some naïve, informal practice, is demonstrably not true here, at least relative to the consistency of classical analysis. One simply must renounce LEM and, as already said, tolerate things like negations of generalizations of it, as just noted.

The difficulty comes when we attempt to explain why LEM fails, even though (under the realistic hypothesis we are entertaining) there really are nilsquares making up the "glue" of actual continua, the points of classical analysis now regarded as a useful but fictitious formal artifact of analytic methods. If there really are such "things", doesn't logical identity apply to them just as to everything else, regardless of our abilities to discriminate them from one another or from 0? The situation is really very different from that posed by intuitionistic analysis. There constructive meanings of the logical operators, disjunction, negation, the conditional, both existential and universal quantifiers, obviously do not sustain the formal LEM or related classically valid principles, e.g. quantifier conversions such as "not for every x $\phi(x)$" to "there exists x such that not $\phi(x)$". Apparent conflicts with classical analysis are only apparent due to these radically different meanings. But none of this is applicable in interpreting SIA, for constructive meanings do not seem appropriate to the subject. Nilsquares, for example, are not constructed at all; indeed, their existence cannot be asserted any more than it can be denied, on pain of contradiction. Rather one must settle for the double negation of existence. While this in itself might seem compatible with a constructive reading, the KL axiom itself seems, if anything, more dubious on such a reading. (What method do we have for finding the slopes of the linelets from a given constructive function on the nilsquares? Indeed, in what sense are we ever presented with such a function?) Lawvere and others have taken failure of LEM for nilsquares to express "non-discreteness", perhaps in analogy with familiar kinds of vagueness. But, once we are speaking of objects at all, however invisible or intangible, how can the predicate '$\_ = 0$' itself be vague? Though from a setting he never contemplated, one harks back to Quine: "No entity without identity!"

Indeed, with the tendency to speak of $\neq$ as "distinguishable" (Bell [1998]), it is natural to seek an interpretation of '$=$' in SIA as an equivalence relation broader than true identity, and this suggests trying to recover SIA in a *classical* interpretation. Such an interpretation has actually been carried out in detail (Giordano [2001]). Certain differences emerge: the class of functions treated is narrower than all continuous ones (a Lipschitz condition is invoked), but the KL axiom and much of the theory are recovered on a fully classical basis. Whether proponents of SIA and SDG will plead "change of subject" remains to be seen.

Turning to "category of categories", efforts towards axiomatization are at least grabbing the bull by the horns, laying down explicit assertory axioms on the mathematical existence of categories and providing a unified framework for a large body of informal work on categories and toposes (hence mathematics, generally). Three questions demand our attention: (1) What concepts are presupposed in such an axiomatization? (2) Are these such as to sustain the autonomy of CT vis-à-vis set theory or related background, or do they reveal a (possibly hidden) dependence thereon? (3) What is the scope of such a (meta) theory, in particular, what are the prospects for self-applicability and the idea of "*the* category of (absolutely) *all* categories"?

On (1) and (2), it is clear that these axioms (as in McLarty [1991]) are not employing the CT primitives ('object', 'morphism','domain', 'codomain', 'composition') schematically, as in the algebraic defining conditions, but with intended meanings presumably supporting at least plausible truth of the axioms. The objects are *categories*, the morphisms are *functors between categories*, etc. Commenting on this, Bell and I recently wrote: "Primitives such as 'category' and 'functor' must be taken as having definite, understood meanings, yet they are in practice treated algebraically or structurally, which leads one to consider interpretations of such axiom systems, i.e. their semantics. But such semantics, as of first-order theories generally, rests on the set concept: a model of a first-order theory is, after all, a set. The foundational status of first-order axiomatizations of the [better: a] metacategory of categories is thus still somewhat unclear." (Hellman and Bell [forthcoming])

In other words, when we speak of the "objects" and "arrows" of a metacategory of categories as *categories* and *functors*, respectively, what we really mean is "structures (or at least "interrelated things") satisfying the algebraic axioms of CT", i.e. we are using "satisfaction" which is normally understood set-theoretically. That is not to say that there are no alternative ways of understanding "satisfaction"; second-order logic or a surrogate such as the combination of mereology and (monadic) plural quantification of modal-structuralism would also suffice. But clearly there is some dependence on a background that explicates *satisfaction* of sentences by structures, and this background is not "category theory" itself, either as a schematic system of definitions or as a substantive theory of a metacategory of categories. But this need for a background theory explicating *satisfaction* was precisely the conclusion we came to in our [2003] paper, reinforcing the well-known critique of Feferman from [1977], which exposed a reliance on general notions of "collection" and "operation". It was precisely to demonstrate that this in itself does not leave CT structuralism dependent on a background set theory that I proffered a membership-free theory of large domains as an alternative. Although the reaction, "Thanks, but no thanks!", frankly did not entirely surprise me, it will also not be surprising if a perception of dependence on a background set theory persists.

As to the third question of scope, I think it is salutary that McLarty calls his system "a (meta) category of categories", rather than "*the* category of categories", which flies in the face of general extendability. No structuralist framework should pretend to "all-embracing completeness", in Zermelo's [1930] apt phrase. And we certainly had better avoid such things as "the category of exactly the non-self-applicable categories"! But it is, I believe, an open question just what instances of impredicative separation should be allowed.

To conclude this section, it seems a fair assessment to say that, while axioms for a (meta) category of categories do make some progress toward providing answers to some of our five questions put to the various versions of mathematical structuralism, we are left still well short of satisfactory, full answers, even to the first four.

11.3 Awodey's "Hilbertian" response

In contrast to the foregoing, this response takes as its point of departure an "anti-foundationalist" stance: mathematics should not be seen as based on a fixed universe of special objects, the elements of domains of structures, the relata of structural relations, as on the set-theoretic view. Instead, mathematics has a *schematic* character, which seems to mean two things: any theorem includes hypothetical conditions, which govern just what aspects of structure are relevant; and, in any case, the particular nature of individual objects is irrelevant. Moreover, whereas modal structuralism tries to get at this by open-ended modal quantification (which is not to be interpreted as ordinary quantification over a fixed background domain of *possibilia*, as Awodey seems to recognize), category theory itself provides a more direct expression, standing on its own without need of any further (assertory) background principles.[4] The central, general but flexible primitive notion is "morphism", capable of grounding talk of relations, operations, etc. (A summary list of relevant "arrows only" categorical concepts illustrates this.) A "top-down" metaphor, as opposed to "bottom-up", is used; it seems that it is sufficient simply to describe whatever ambient background structure we deem relevant to the mathematical purpose at hand, without needing to worry about any absolute claims of existence. (Clearly, this is reminiscent of Hilbert's view that sought to eliminate metaphysics from mathematics by, in effect, replacing absolute claims of existence with a combination of proofs from formal axioms, as defining conditions, together with a proof of formal consistency of the relevant axioms, although presumably, since we are in a post-Gödelian era, the latter demand is omitted.)

I see a dilemma in understanding all this. Either mathematics is adequately understood as just a complex network of deductive and conceptual interconnections, or it is not. On the first horn, what we are really presented with is a kind of formalism, in which theorems in conditional form, together with definitions, are all there is to mathematics, that is, we just give up on the notion of mathematical truth as anything beyond deductive logical validity. In this case, we

[4]It should be clear that modal-structuralism, although it does provide such background principles, is also quite explicitly "non-foundationalist" in Awodey's sense. To avoid confusion, I have preferred to speak of "structuralism frameworks" rather than "foundations", but I would certainly plead guilty to "foundational concerns", much in the spirit of Shapiro's *Foundations without Foundationalism*. Clearly, this is reflected in the questions (1) - (5) we have been putting to the various versions of structuralism.

really need not worry about primitives with meanings supporting basic axioms in the Fregean sense. Questions of truth (beyond first-order logical truth of conditionals) would arise only in certain limited cases, typically in applications of mathematics where we might be in a position to assert that the antecedent conditions are indeed fulfilled (e.g. that certain finite structures, say, are actually instantiated, or that even certain infinite ones are, say space or time or space-time as continua). Whether this is intended and is viable, after all (i.e. after all the criticisms that have been levelled against deductivism), remain to be determined.[5]

On the other horn of the dilemma, we take seriously the idea that 'morphism' is a primitive with definite, if multifaceted, meaning, giving genuine content to mathematics beyond mere inferential relations. But what is that meaning? And what is the content beyond inferential relations? Awodey himself [1996] has stressed the algebraic-structural character of the CT axioms, and, unlike McLarty, he does not appeal to any special topos axioms as assertions, nor does he appeal to a special category of categories. It seems clear, then, that the notion of morphism — which, unlike the notion of 'part/whole' employed in modal-structuralism and also exhibiting a kind of "schematic character", is a mathematical term of art, not a familiar one in ordinary English — depends on the context, viz. *on the category or categories presupposed or in which one is working*. As indicated at the outset, arrows (i.e. morphisms) need not be ordinary functions. They need only satisfy the conditions on "arrows" of the CT axioms or extensions thereof. Surely, this is what should be said in explaining "what morphisms are". Moreover, functors do more that ordinary functions, and they are centrally involved in "the usual language and methods of category theory" (Awodey [2004], p. 62). But then what we really have as primitives are "*satisfaction of axioms*" and "*functor between categories*"; i.e. we are presupposing "*category*" as a primitive as well. But this brings us right up against the same problem that confronted the previous view, namely that we are falling back on *prima facie* set-theoretic notions after all. The main difference seems to be that, whereas on the McLarty view we were at least being given axioms asserting the mathematical existence of various categories, here we are not even being given that. In any case, the CT "arrows only" explications of "relations", "operations", and so forth, are of no avail until we first understand "morphism" (i.e. "arrow"), "functor" and "category", i.e. until we already understand *satisfaction* or equivalent (second- or higher-order) notions. It would be plainly circular to appeal to "morphisms" to explain this!

[5] I have been taking as a ground rule for articulating structuralism that it should not collapse to formalism or deductivism. If CT structuralism is playing by different rules, that certainly should be made explicit.

11.4 Conclusion

The contrast between Fregean and Hilbertian axioms seems to present us with a stark choice. But really, unless we go back to formalism, mathematics requires both. For all the axiom systems of ordinary mathematics, for number theory, analysis, algebra, pure geometry, topology, and surely much of category and topos theory, i.e. for all commonly studied structures and spaces, not only is the Hilbertian conception appropriate, it is part and parcel of standard modern practice. But when we step back and contemplate fundamental and foundational issues — when we ask questions about what principles govern the mathematical existence of structures generally, or when we consider the closely related "unfinished business" of Hilbert's own program (as Shapiro puts it), the place of metamathematics, questions of absolute and relative formal consistency, questions of (informal) higher-order "consistency" or "coherence", relative interpretability, independence, etc.— then we are in the realm of outright claims, not mere hypotheticals as to what holds or would hold in any "structures" satisfying putative algebraic (meta) axioms of metamathematics. Rather we are seeking assertory axioms in the Fregean sense.[6] Thus, in connection with category theory, the advice of Berra [1998], "When you come to a fork in the road, take it!",[7] is quite apt, and the theory of large domains I sketched in my [2003] was one way of taking the advice (and the fork!). The alternative responses considered here, categories of categories (Fregean) or category theory as schematic mathematics (Hilbertian), lead us straight back to prima facie set-theoretic notions, only slightly beneath the surface, and so do not sustain category theory as providing an autonomous structuralist framework adequate to the needs of both mathematics and metamathematics.

References

Awodey, S., 1996, "Structure in Mathematics and Logic: A Categorical Perspective", *Philosophia Mathematica* 4, 209–237.

Awodey, S., 2004, "An Answer to Hellman's Question: 'Does Category Theory Provide a Framework for Mathematical Structuralism?' ", *Philosophia Mathematica* 12, 54–64.

Bell, J.L., 1998, *A Primer of Infinitesimal Analysis* (Cambridge: Cambridge University Press).

Berra, Y., 1998, *The Yogi Book: "I Really Didn't Say Everything I Said!"* (New York: Workman Publishing).

[6]This is highlighted in Shapiro [forthcoming]. As he reminds us, Hilbert clearly regarded the claims of proof theory as "contentful", closely related to statements of number theory (insofar, "mathematical" as well as "metamathematical"), and not conditional with respect to implicitly defined (meta) structures of metatheory.

[7]One road leads to Göttingen, the other to Jena.

Dedekind, R., 1888, *Was Sind und Was Sollen die Zahlen?* 3rd Edition (Braunschweig, Vieweg, 1911), tr. W. W. Beman, "The Nature and Meaning of Numbers" in *Essays on the Theory of Numbers* (New York: Dover, 1963).

Feferman, S., 1977, "Categorical Foundations and Foundations of Category Theory" , in R.E. Butts and J. Hintikka, eds., *Logic, Foundations of Mathematics, and Computability Theory* (Dordrecht, Reidel), 149–169.

Giordano, P., 2001, "Nilpotent Infinitesimals and Synthetic Differential Geometry in Classical Logic", in U. Berger, H. Osswald, and P. Schuster, eds., *Reuniting the Antipodes-Constructive and Nonstandard Views of the Continuum* (Dordrecht, Kluwer), 75–92.

Hellman, G., 2001, "Three Varieties of Mathematical Structuralism", *Philosophia Mathematica* 9, 184–211.

Hellman, G., 2003, "Does Category Theory Provide a Framework for Mathematical Structuralism?" *Philosophia Mathematica* 11, 129–157.

Hellman, G. and Bell, J.L. [forthcoming] "Pluralism and the Foundations of Mathematics", C.K. Waters, H. Longino, and S. Kellert, eds. *Scientific Pluralism*, Minnesota Studies in the Philosophy of Science, XIX (Minneapolis: University of Minnesota Press).

Joyal, A. and Moerdijk, I., 1995, *Algebraic Set Theory*, (Cambridge, Cambridge University Press).

Mac Lane, S., 1986, *Mathematics: Form and Function*, (New York, Springer Verlag).

McLarty, C., 2004, "Exploring Categorical Structuralism", *Philosophia Mathematica* 12, 37–53.

Shapiro, S., 1991, *Foundations without Foundationalism* (Oxford, Oxford University Press).

Shapiro, S., 1997, *Philosophy of Mathematics: Structure and Ontology*, Oxford, Oxford University Press.

Shapiro, S. [forthcoming] "Categories, Structures, and the Frege-Hilbert Controversy", *Philosophia Mathematica*.

Zermelo, E., 1930, "On Boundary Numbers and Set Domains: New Investigations in the Foundations of Set Theory", in W. Ewald, ed. *From Kant to Hilbert: Readings in the Foundations of Mathematics*, Oxford, Oxford University Press, 1996, 1208–1233, tr. M. Hallett, from the German original, "Über Grenzzahlen und Mengenbereiche: Neue Untersuchungen über die Grundlagen der Mengenlehre", *Fundamenta Mathematicae* 16, 29–47.

Chapter 12

CATEGORY THEORY AS A FRAMEWORK FOR AN *IN RE* INTERPRETATION OF MATHEMATICAL STRUCTURALISM

Elaine Landry
University of Calgary
elandry@ucalgary.ca

The aim of this paper is to present category theory as a framework for an *in re* interpretation of mathematical structuralism. The use of the term 'framework' is significant. On the one hand, it is used in distinction from the term 'foundation'. As such, what I propose is that we consider category theory as *a philosophical tool* that allows us to *organize* what we say about the shared structure of abstract kinds of mathematical systems.[1] On the other hand, the term 'framework' is used in the sense of Carnap [1956]. That is, category theory is taken as *a language*[2] used to frame *what we say* about the shared structure of abstract kinds of mathematical systems, as opposed to being a "background theory" which constitutes what a structure is.[3]

12.1 Foundation versus Framework

In this section, I consider what it means to say that category theory is a framework for mathematical structuralism, though not a foundation for mathematics. I will show, contra Feferman [1977] and Mayberry [1994], that the

[1]This in contrast to viewing category theory as a mathematical foundation that provides us with the "atoms" (of meaning or reference) of mathematics itself, e.g., that it tells us what, or whether, a structure is.
[2]See Landry [1999; 2001] for further elaboration of what is meant by taking category theory as a language.
[3]In this sense, the use of a category-theoretic linguistic frame is in contrast, to, for example, Shapiro's [1997] ontological, *ante rem*, reading of the concept of structure which uses "structure theory" to frame the claim that mathematical structures exist both over and above systems that exemplify them and independently of language.

J. van Benthem, G. Heinzmann, M. Rebuschi and H. Visser (eds.), The Age of Alternative Logics: Assessing Philosophy of Logic and Mathematics Today, 163–179.

reason category theory cannot provide a foundation for mathematics is not that it depends on set theory as either an ontological or conceptual base. Rather it is that category theory cannot be construed as *being about* either objects or structures *qua* (actually or possibly) existing things. Relying on the work of Lawvere [1966] and McLarty [1990], we will see the Feferman's criticisms miss their mark, and, moreover, we will see that category theory satisfies Mayberry's criterion of being a "foundational sea" to the same degree that set theory does. Yet, while category theory cannot provide a foundation for mathematics, it remains, as Bell [1981] notes, "foundationally significant".

12.1.1 Categories as "Structures"

Since Lawvere's work with the category of categories has provided much grist for the foundational mill, let us consider what he says of his aims in this regard:

> [i]n the mathematical development of recent decades one sees clearly the rise of the conviction that the relevant properties of mathematical objects are those which can be stated in terms of abstract structure rather than in terms of the elements which objects were thought to be made of. The question naturally arises whether we can give a foundation for mathematics which expresses wholeheartedly this conviction concerning what mathematics is about and in particular in which classes and membership in classes do not play any role... (Lawvere quoted in Feferman, [1977], pp. 149-150).

It is as an answer to this challenge, then, that Lawvere [1966] "formulated a (first-order) theory whose objects are conceived to be arbitrary categories and functors between them". (Feferman, [1977], p. 150). It is held, by Feferman (and Bell [1981]), that the problem with such an account is that when it comes to accounting for categories as themselves abstract "structures" and/or using categories to account for *abstract kinds* of "structures", one must appeal to notions which fall outside the range of category theory.[4] As Feferman explains:

> when explaining the general notion of structure and of particular kinds of structures such as groups, rings, categories, etc., we implicitly presume as understood the ideas of operation and collection; e.g., we say that a group consists of a collection of objects together with a binary operation satisfying such conditions ... when explaining the notion ... of functor for categories, etc., we must again understand the concept of operation ... (Feferman, [1977], p. 150).

[4]Specifically, Feferman claims that in either case, "[t]he logical and *psychological* priority if not primacy of the notion of operation and collection is ... evident" (Feferman, [1977], p. 150). And from this concludes that "[i]t is evidently begging the question to treat collections (and the operations between them) as a category which is supposed to be one of the objects of the universe of the theory to be formulated". (Feferman, [1977], p. 150.) Feferman's claim can be understood as follows: if we assume that mathematics is the study of abstract structure, then, insofar are categories themselves are structured (and, presumably, structured in terms of operation and collection), we need a general account of the very notion of structure itself.

That is, even if category theory can give a more general account of abstract kinds of structure than can set-theory,[5] we are still in need of a (meta) theory which makes "use of the *unstructured* notions of operation and collection to explain the structural notions to be studied". (Feferman, [1977], p. 150). To this end he provides a non-extensional type-free theory of operations and collections wherein "much of 'naïve' or 'unrestricted' category theory can be given an account..." (Feferman, [1977], p. 149).[6]

Now, if one had stopped one's inquiry here, one might be convinced, but as Feferman goes on to note, the schemes offer by Grothendieck Universes and the Gödel-Bernays theory of classes, readily offer the needed (meta) theory for category theory (though not in terms of operation and collection).[7] That is, Feferman himself concedes that

> there is no urgent or compelling reason to pursue foundations of unrestricted category theory, since the schemes... serve to secure all practical purposes... The aim in seeking a new foundation is mainly as a problem of logical interest motivated largely by aesthetic considerations (or rather by the inaesthetic character of the present solutions). (Feferman, [1977], p. 155)

To make his reasons compelling, then, Feferman needs to have demonstrated that the notions of operation and collection themselves are, in some sense, *constitutive* of the notion of structure, and he has not. Independently of Feferman offering-up these reasons, there are two possible, though not unrelated, responses to his claim that we yet need, to account for mathematics as the study of abstract structures, a non-extensional type-free theory of unstructured operations and collections. One that structuree' is not strictly a mathematical notion, and hence, such problems need not be resolved by providing a foundation for mathematics, but rather are best addressed by offering a philosophy for mathematics (see § 12.3). The second, though not unrelated, response is that a category *qua* a structured system is to be "algebraically" considered (see § 12.2). In either case, we note that category theory itself, i.e., without the

---

[5]Of the inadequacies of set theory as a foundation, Feferman says: "Since neither the realist (extensional) or the constructivist (intensional) point of view encompasses the other, there cannot be any present claim to *universal foundation* for mathematics..." (Feferman, [1977], p. 151.)

[6]While Feferman agrees with Mac Lane that work in elementary topos theory (ETS) shows the "formal" equivalence between ETS(Z) and ETS(ZF) and the theories of Z and ZF, respectively, he claims his point stands; because the "use of 'logical priority' refers not to the relative strength of formal theories but to the order of the definition of the concepts"... and "that the general concepts of operation and collection have logical priority with respect to structural notions (such as 'group', 'category', etc) because the latter are defined in terms of the former but are not conversely". (Feferman, [1977], p. 152.)

[7]Another scheme, offer by Bell [1986;1988], is to characterize (up to categorical equivalence) topoi as models of a higher-order, intuitionistically based, type theory; thus, allowing us to re-capture the sense in which set-theory and category theory are "formally" equivalent, i.e., by allowing for the specification of topoi as "local set theories".

background schemes, cannot provide a foundation any more than set theory: it cannot tell us what, or whether, structure is.[8]

12.1.2 The category of categories as a "Foundational Sea"

I now turn to consider Mayberry's claim that, because sets and their morphology are constitutive of the notion of structure, only set theory can provide a foundation for mathematics. My aim is to show that, while it can be agreed that "when we employ the axiomatic method we are dealing with structures",[9] it simply does not follow that "when we are dealing with mathematical structures, we are engaged in set theory". (Mayberry, [1990], p. 19) In particular, I will argue that there is no reason to hold that "each structure consists of a set or sets equipped with a morphology". (Mayberry, [1990], p. 19.)

Mayberry acknowledges that there are problems with his version of structuralism founded on an 'intuitive' set theory, viz., that it cannot be used to talk about the *large* categories,[10] for example, the category of *all* (small) groups. He further recognizes that

> to consider such categories seems a quite natural extension of ordinary structuralism, it appears to request the next level up in generality in which the notion under investigation is the notion of structure itself. (Mayberry, [1990], p. 35.)

His solution to this problem, however, is far from satisfying: it is to dismiss talk of such structures by simply denying that they are structures. He says

> [i]n fact, there can be no such structures, for the very notion of set is that of an extensional plurality limited in size, and the notion of set is constitutive of our notion of structure. (Mayberry, [1990], p. 35.)

The claim that the notion of set is that of an extensional plurality limited in size is both *ad hoc* and misleading: the only justification that Mayberry's privileging of 'intuitive' set theory has is that, given his claim that set is constitutive of our notion of structure, it makes his conclusion, that 'intuitive' set theory provides a foundation, follow. Consider, if, instead of defining a set intuitively as "an extensional plurality of determinate size, composed of definite

---

[8]Given set theory's inability to form the category of *all* structures of a given kind (groups, topological spaces, categories) and to form the category of all functors of any given category it cannot be used to 'foundationalize' category theory, and given category theory's inability to refer to *all* categories as 'objects' in the categories of categories, without making use of either Grothendieck Universes or a Gödel-Bernays theory of sets and classes, it cannot be seen as providing a foundation in and of itself. (See Feferman, [1977], pp. 154–155 for a brief but informative discussion of these issues.)

[9]That is, while it can be agreed that the aim of a structuralist foundation (or, more accurately, a structuralist philosophy) is to capture the belief that the subject matter of mathematics is structured systems and their morphology.

[10]Note, however, that it is not because it is large that the category of categories cannot be taken as a foundation. For a discussion of the various interpretations of large categories, (see McLarty, [1995], pp. 105–110).

property-distinguished objects" (Mayberry, [1990], p. 32) we define a category 'intuitively' as an object of *indeterminate* size,[11] composed of definite, functorially-distinguished objects. Then, following Lawvere, we could conclude that category theory provides a foundation for mathematics.[12]

While it seems clear, then, that neither set theory nor category theory can be a foundation in the sense of providing a *theory* which captures the idea that the subject matter of mathematics is structures and their morphology, it should also be clear that neither can it provide a foundation in the sense of providing "a sea in which structures swim".[13] Thus, while it is right to conclude that, on Feferman and Mayberry's "structuralist" criterion, category theory cannot provide a foundation for mathematics, this is not because it requires a prior notion of either operation or collection or 'intuitive' set theory. It is because, if it is to be counted as an "object language" for our talk of structures, it requires some prior, meta-theoretical, notion of structure that category theory itself cannot provide.[14]

12.2 Structures and structured systems

If we accept, then, that mathematics is the study of abstract structure, we must explain in what sense category theory provides the *philosophical tool* for organizing what we say about the shared structure of abstract kinds of mathematical systems. I begin first with Corry's [1996] historical investigation of the development of the 'algebraic' notion of structure. The aim here is to distinguish the set-theoretic path of the Bourbaki notion of structure from the algebraic path of the category-theoretic notion. Given this distinction, two observations can be made. The first, that the Bourbaki notion implicitly assumes an ontology out of which structures are made, i.e., assumes that types of structures are kinds of set-structured systems. The second, that this assumption leads to a reification of structure, i.e., leads to interpreting structures themselves as independently

[11] By 'indeterminate size' it is meant that we can define a category as large, either in the Gödel-Bernays sense, or in terms of Grothendieck Universes. That is, we do not have to restrict the size of a category by characterizing its objects and morphisms in terms of sets.

[12] To see this, in the following quote by Mayberry, simply replace 'set' with 'category' and 'universe of sets' with 'category of categories'. "The *fons et origo* of all confusion here is the view that set theory is just another axiomatic theory and that the universe of sets is just another mathematical structure ... The universe of sets is not a structure; it is the world that all mathematical structures inhabit, the sea in which they all swim." (Mayberry, [1990], p. 35.)

[13] And this fact cannot be altered by claiming that either stands along the shore of these issues since it is needed to provide a semantics for mathematics. As McLarty notes, "Mayberry ... has simply confused his own head with Lawvere's. [By claiming that "the idea of denying intuitive set theory its function in the semantics of the axiomatic method never entered Lawvere's head in his treatment of the categories of categories". (Mayberry, [1977]).] Lawvere believes 'intuitive' categories, and spaces, and other structures are just as real (or, more accurately, just as ideal) as 'intuitive' sets." (McLarty, [1990], p. 364.)

[14] For example, even though the category of categories can be used to talk about the shared structure of categories *qua* kinds of structured systems, it cannot be used to axiomatically define (all) categories *qua* structures.

existing things. In contrast to such set-theoretic and/or ontological readings of what structure is, I will use this history to point to a category-theoretic, *schematic*[15] *interpretation* of types of structured systems.

12.2.1 What *kind* structures number systems?

In the development of Abstract Algebra,[16] the use of kinds of "structures", as tool and/or unifying concepts, is evident. This development has its beginning in the various attempts at answering the question: "What structures number systems?". For Dedekind, the subject matter of "algebra"[17] may be considered in two different ways. On the one hand, we may consider the properties of number systems *qua* collections, wherein we overlook *the nature* of the elements involved. The *tools* which Dedekind used to talk about the algebraic structure of number systems, considered as such, were groups, ideals and modules. On the other hand, we may consider the properties of the elements of number systems and the interrelations among 'rational domains' contained in it. For Dedekind, the *unifying concept* for such an analysis was thought to be that of a field.[18]

Hilbert, continuing this "algebraic" analysis of number systems, maintained the distinction between properties of numbers systems (though not *qua* collections) and properties of the elements of number systems and their interrelations: he used invariants, ideas, rings, groups and fields as *tools* to talk about the latter. To talk about properties of number systems, he took a geometric turn, and considered them *qua* postulational systems. The *unifying concepts* for talking about number systems as such were the 'logical' (or meta-mathematical) properties of axiom systems themselves, namely, independence and consistency. In addition to this "algebraic" investigation was Hilbert's meta-mathematical analysis, which took axiom systems and their properties as objects of study in their own right. Thus, while we had, with Dedekind, that, in some sense, number systems themselves were the basis for algebraic analysis, the question at hand was "Could in his [Hilbert's] view the conceptual order be turned around

[15]I use the term 'schematic' in the sense of Goldfarb [2001].

[16]The reader is strongly encouraged to read Corry's [1996] insightful and informative account of this. While I stop short of fully accepting his account of the category theory's 'significance', I note here a debt to, and reliance on, his presentation of the 'facts' of the development of the notions of kinds and types of algebraic and mathematics structures.

[17]The term 'algebra' is placed in quotes since at the time this was not a well defined field. It may be characterized as the "theory of solving equations" (see Hasse, [1954], p. 11.)

[18]As Dedekind, himself, explains: "... I have attempted to introduce the reader to a higher domain, in which algebra and the theory of numbers interconnect in the most intimate matter ... I got convinced that studying the algebraic relationship of number is most conveniently based on a concept that is directly connected with the simplest arithmetic principles. I have originally used the term "rational domains". Which I later changed to "field". (*Werke*, p. 400)... The term [field] should denote here, in a similar fashion as in the natural sciences, in geometry, and in the social life of men, a system possessing a certain completeness, perfection and comprehensiveness, by mean of which it appears as a natural unity". (Dedekind, [1894], p. 452.)

so that the system of real numbers be dependent on the results of [the axiomatic analysis of] algebra rather than being the basis for it?" (Corry, [1996], p. 172)[19].

Appreciating the "foundational value" of the axiomatic method, Noether applied this shift in conceptual priority to Dedekind's subject matters. That is, to the properties of number systems *qua* collections (again, overlooking the nature of the elements) she proposed ideals, modules, groups and rings as *tools* for talking about their algebraic structure. Such tools, in light of Hilbert, where themselves now considered *qua* axiom systems. In a similar vein, for systems of abstract elements of any axiom system, the *unifying concept* was thought to be abstract rings, or the axiomatic presentation of rings themselves. Whereas Dedekind had considered properties of concrete elements of number systems and the field-theoretic interrelations between them as unifying, Noether considered the properties of abstract elements of abstract rings *qua* axiom systems as unifying. In this manner the unifying power is taken out of concrete number systems and put into *an abstract kind* of axiomatically presented structured system. As Corry explains:

> Noether's abstractly conceived concepts provide a natural framework in which conceptual priority may be given to the axiomatic definitions [of concepts] over the numerical systems considered as concrete mathematical entities. With Noether, then, the balance between the genetic and the axiomatic point of view begins to shift more consciously in favour of the latter. (Corry, [1996], p. 250)

These developments in the analysis of the algebraic structure of number systems gave rise to the independent branch of study of Abstract Algebra, wherein the focus of analysis was now the shared structure of the *abstract kinds* of algebraic systems (e.g., groups, rings fields) considered in themselves (typically considered qua axiom systems).[20] That is, those very tools and/or concepts that were once useful or unifying when talking about the algebraic structure of concrete, number, systems are now seen as systems of study in there own right.[21]

---

[19]Hilbert responded to such a query by distinguishing between the *genetic* and the *axiomatic* method, and, at least as regards the 'foundations' of mathematics, he held a preference for the latter: he says, "In spite of the high pedagogic value of the genetic method, the axiomatic method has the advantage of providing a conclusive exposition and full logical confidence to the contents of our knowledge." (Hilbert, [1900], p. 184) and "When we are engaged in investigating the foundations of a science, we must set up a system of axioms which contains an exact and complete description of the relations subsisting between the elementary ideas of the science. The axioms so set up are at the same time the definition of those elementary ideas, and no statement within the realm of the science whose foundation we are testing is held to be correct unless it can be derived from those axioms by means of a finite number of logical steps." (Hilbert, [1902], p. 447.)

[20]Exemplifying this shift is van der Waerden's *Modern Algebra*, in which "... different mathematical domains are considered as individual instances of algebraic structures, and therefore undergo similar treatments; they are abstractly defined, they are investigated by recurrently using a well-defined collection of key concepts, and a series of questions and standard techniques is applied to all of them." (Corry, [1996], p. 252.)

[21]As Hasse witnesses: "It is characteristic of the *modern* development of algebra that the tools specified about [i.e., groups and fields] have given rise to far-reaching autonomous theories which are more and more

12.2.2 What *type* structures abstract kinds of mathematical systems

Given this structural approach to abstract algebraic systems, the next question that arose was: What is the tool and/or unifying concept that allows us to talk about such *abstract kinds* of systems as instances of the same mathematical *type*? I begin first with Ore for whom the *type* which structures the various kinds of algebraic systems is the lattice. More specifically, what "structures" kinds of algebraic systems are the (union and cross-cut) properties of the lattice of certain subsystems of any given system. Here, then, is where we note both Hilbert's axiomatic influence and Noether's "set-theoretic"[22] influence. What is new, however, is that, in addition to overlooking the nature of the elements, we overlook too their *existence*. As Ore explains:

> In the discussion of the structure of algebraic domains, one is not primarily interested in the elements of these domains but in the relations of certain *distinguished sub-domains*... For all these systems there are defined *two operations* of union and cross-cut satisfying the ordinary axioms. This leads naturally to the introduction of new systems, which we shall call *structures*, having these two operations. The elements of the structure correspond isomorphically with respect to union and cross-cut to the distinguished subdomains of the original sub-domain while the elements of the original domain are completely eliminated in the structure. (Ore, [1935], p. 406.)

It is in this sense that the lattice-theoretic properties were taken as the unifying concepts for algebra; lattice theory, itself, was taken by Ore as *the formal tool* for providing a general structural account of the various kinds of algebraic systems, and, quite possibly, as having "foundational significance" insofar as it may further provide a structural account of the various kinds of mathematical systems as well.[23]

In contrast to Ore, for Bourbaki a type of structure is a system of elements that has a set-structure, that is, one overlooks the specific nature of the elements in favour of their algebraic, order or topological structure. As Shapiro notes:

---

replacing the basic problem of *classical* algebra ... Thus in the modern interpretation algebra is no longer merely the theory of solving equations, but the *theory of formal calculating* domains, as fields, groups, etc.: and its basic problem has now become that of obtaining an insight into the structure of such domains..." (Hasse, [1954], p. 11.)

[22]To explain the reading I give to Noether's use of the term 'set-theoretic', I point to Corry's telling remark that: "[t]he expression "purely set-theoretic considerations", in Noether's usage, does not refer to concepts nowadays related to the theory of sets (membership, power, etc.). It denotes arguments for proof in algebra, which do not rely on the properties of the operation defining the [system] under inspection, but rather properties of the inclusions and intersections of sub-[systems] of it. (Corry, [1996], p. 244) ... Such an approach would certainly correspond to the problem, mentioned by Alexandrov, of "axiomatizing the notion of a group from its partition into cosets as the fundamental concept" (Corry, [1996], p. 248.)

[23]As Corry notes: "At that opportunity [the 1936 International Congress of Mathematicians at Oslo] Ore claimed that the guidelines of his program, although originating with algebra, should not be limited to that domain alone, and he envisioned that they would be applied in additional fields of mathematics as well." (Corry, [1996], p. 276.)

> According to Bourbaki, there are three great types of structures, or "mother structures": algebraic structures, such as group, ring, field; order structures, such as partial order, linear order, and well order; and topological structures [which provides a formalization of the concepts of limit, neighbourhood and continuity]...(Shapiro, [1997], p. 176.)

Yet, as types of set-structured systems, one does not overlook the existence of elements: what structures the elements of kinds of mathematical systems into their respective types are the relations that hold between such systems *qua* set-theoretically presented axiom systems. Like Hilbert and Noether, Bourbaki's attention was focused on the axiomatic method. Unlike Hilbert, who focused on the logical properties of axiom systems, or Noether and Ore who focused on the properties of the inclusions and intersections of subsystems, Bourbaki used various set-theoretic types of structures *qua* axiom systems to unify what could be said of the various kinds of mathematically structured systems.

What remains open for discussion is whether, and in what sense, Bourbaki intended the theory of sets to be *constitutive* of the concept of structure, i.e., intended it as an answer to the question: "What is structure?". Whatever their intention might have been, the tension between the account of set theory as a *formal* language and the heuristic role of the formally, though implicitly, defined concept of structure was pulling at the seams of their 'algebraic' structuralism.[24] In any case, whether set-theory was intended to be used foundationally or heuristically, what appears to be true is that the efforts of Bourbaki were interpreted, both mathematically and philosophically, as providing a set-theoretically constitutive account of what structure is and, in so doing, shifted from the algebraic tradition's attempts to overlook *the nature of the elements* of kinds of mathematical systems in favour of abstractly characterizing their shared structure. As Bell explains:

> With the rise of abstract algebra... the attitude gradually emerged that the crucial characteristic of mathematical structure is not their internal constitution as set-theoretical entities but rather the relationship among them as embodied in the network of morphisms... However, although the account of mathematics they [Bourbaki] gave in their *Eléments* was manifestly structuralist in intention, actually they still defined structures as sets of a certain kind, thereby failing to make them truly independent of their 'internal constitution'. (Bell, [1981], p. 351.)

For the Bourbaki structuralist what unifies kinds of mathematical systems are types, and, more significantly, what appears to *make* these types *"powerful tools" for unification*, is the constitutive character of set theory.[25] In this manner,

[24] As they, themselves, note: "[t]he reader may have observed that the indications given here [of the concept of structure] are left rather vague; they are not intended to be other than heuristic, and indeed it seems scarcely possible to state general and precise definitions for structure outside the framework of formal mathematics". (Bourbaki, [1968], p. 347, footnote.), (Corry, [1996], p. 326.)

[25] Speaking to this "constitutive" reading, we note the following quotes of Bourbaki: "Each structure carries with it its own language, freighted with special intuitive references derived from the theories which the

types, or "structures", as set-structured systems are turn into "things". In contrast, the category-theoretic structuralist holds that what unifies kinds are types as cat-structured systems, yet, what makes these types tools for unification, is the *schematic* use of categories, in particular, and the organizational role of category theory, in general. Wherein, then, lies this distinction? It is that, nothing, in particular, is constitutive of what a category *is*. As Mac Lane explains:

> [i]n this description of a category, one can regard "object", "morphism", "domain", "codomain", and "composites" as *undefined terms or predicates*. (Mac Lane, [1968], p. 287, italics added.)

Like Bourbaki, we thus characterize the shared structure of abstract kinds of mathematical systems *qua* a type of structured system.[26] Yet unlike Bourbaki we need not take set, or, indeed, any particular kind of set, to be constitutive of what these types are themselves types of (though, of course, we might). Again, as Mac Lane explains:

> Bourbaki's concepts defined "mathematical structures" by taking an abstract set and appending to it an additional construct, in category theory there is no subordination of "mathematical structures" to sets, and this is the source of the supremacy of this theory over Bourbaki. (Mac Lane, [1980], p. 382.)

Moreover, in the spirit of Lawvere [1966], we can use Cat (or CAT) as the type used to talk about what structures these kinds of cat-structured systems, again, without having to appeal to set as constitutive of what this type is a type of. What we must note, however, is that, contra Lawvere, we, like our set-theoretic cousins, cannot use category theory as a *formal language*, or foundation. That is, we cannot use it to answer the question: "What is a mathematical structure *qua* a either a kind or type of category?". As Corry explains,

> [i]n no sense, however, has category theory provided, to this day, a definite, or even a provisionally satisfactory answer to the question of what is a "mathematical structure" .... Neither does category theory provide ultimate foundations for mathematics. (Corry, [1996], p. 389, italics added.)

---

axiomatic analysis ... has derived the structure... Mathematics has less than ever been reduced to a purely mechanical game of isolated formulas; more than ever does the intuition dominate the genesis of discoveries. But henceforth, it possesses the powerful tools furnished by the theory of the great type of structures; in a single view, it sweeps over immense domains, now unified by the axiomatic method ..." (Bourbaki, [1950], pp. 227–228) and further that "... whereas in the past it was thought that every branch of mathematics depended on its own particular intuitions which provided its concepts and primary truths, nowadays it is known to be possible, logically speaking, to derive practically the whole of mathematics from a single source, the theory of sets." (Bourbaki, [1968], p. 9.)

[26]For example, Set, Top, Group are types, i.e., kinds of cat-structured systems, that allow us to talk about the shared structure of abstract kinds of mathematical systems in terms of their being instances of the same type.

12.3 A schematic *in re* interpretation of mathematical structuralism

The final section of this paper brings together the above investigations to present a category-theoretically framed *in re* interpretation of philosophically positioned mathematical structuralism. The objective of this section is to show that it is in following the Bourbaki tradition too closely and, thereby, not appreciating the algebraic alternative, that philosophically interpreted mathematical structuralism has most failed us. Seen in this light, my aim is to first argue that category-theoretic analysis ought to be best seen as answering "What are the types that "structure" abstract kinds of structured systems?" (as opposed to speaking to the foundational/ontological claim that "structures" *are*) and, second, to separate these analyses from those which end with claims that types of structured systems, or "structures", are set-structured (or place-structured) "things".

12.3.1 Levels, interpretations and varieties of mathematical structuralism

Mathematical structuralism can be construed as the philosophical position that the subject matter of mathematics is structured systems and their morphology,[27] so that mathematical objects are nothing but "positions in structured systems" and mathematical theories aim to describe such objects and systems via their shared structure. At the level at which we consider *concrete kinds* of structured systems, *i.e.*, the level where 'system' means 'model', we have objects as positions in models and can use either isomorphisms or embeddings to talk about the shared structure of such kinds. For example, the theory of natural numbers aims to describe concrete systems of the natural-number structure, as characterized by the Peano axioms, so that its objects may be seen as von Neumann ordinals, Zermelo numerals, or any other object which shares the same structure, or morphology. If all systems that share this structure are isomorphic, we say that the natural number structure and its morphology determine its objects up to isomorphism. Analogous, then, to the shift in levels that one finds in the mathematical history of the development of the notion of algebraic structure, at the next level of philosophical analysis one finds the question: "What structures *abstract kinds* of structured systems"? In answer to this question, in the philosophical literature, one finds two interpretations of

[27]Note here that I have changed the slogan of structuralism from "mathematics is about structures and their morphology" to "mathematics is about structured systems and their morphology". This shift is intentional, it means to indicate that the aim of the structuralist is to account for the shared structure of mathematical systems in terms of kinds or types, as opposed to answering the question: "What is a structure?", or "What are the kinds or types that are constitutive of what a structure is?" This shift is further discussed in § (12.3.2).

mathematical structuralism: *ante rem and in re*. The latter is aligned with a realist view of structures insofar as it holds that "structures exist as legitimate objects of study in their own right. According to this view, a given structure exists independently of any system that exemplifies it ..." (Shapiro, [1996], pp. 149-150). *In re* structuralism, in contrast, is aligned with a nominalist view of structures insofar as it eliminates talk about structures in favour of talk about systems: "it does not countenance mathematical objects, or structures for that matter, as *bona fide* objects ... Talk of structure generally is convenient shorthand for talk about systems of objects". (Shapiro, [1996], p. 150.)

To further inform this debate, I rely on Aristotle's distinction between *prior in place* and *prior in definition*.[28] Against the *ante rem* structuralist, a category-theoretically framed *in re* interpretation of mathematical structuralism implies that there are no "structures", *qua* "things", over and above kinds of structured systems. As such, structures are not prior in place. Against the *in re* structuralist, categories *qua* schema are prior in definition insofar as they are needed, as an organizational tool (see Mac Lane [1992]), to talk about the shared structure of abstract kinds of structured systems as instances of the same type. Category theory, then, defines *what* a type of structured system is, but remains silent as to the claim *that* structure is.

Failing to heed Resnik's counsel (see Resnik, [1996], p. 96) that structuralism is not committed to asserting the existence of structures, yet, in response to this worry, three varieties of mathematical structuralism have been proposed, these are: the *set-theoretic*, the *sui generic*, and the *modal*.[29] In essence, these are suggested as "background theories" that allow us to talk about "structures" as either actually or possibly existing "things": they allow us to answer that either set-theory, structure-theory, or modal logic, provide the conditions for the actuality (or possibility) of a system being a "structure" of the appropriate kind.

12.3.2 The Bourbaki versus the "Algebraic" tradition

I now turn to my claim is that it is in following the Bourbaki tradition (which takes structures as set-structured "things") too closely and, thereby, not appreciating the algebraic alternative of mathematical structuralism that philosophically interpreted mathematical structuralism has most failed us. Witnessing this is Dummett's remark that:

[28] See the last two books, viz., M, N, of the Metaphysics (1076a5 – 1093b30), where Aristotle discusses mathematical objects and Ideas, and the manner in which these are prior in definition yet not, contra the Platonist, prior in place. See also *Metaphysics* Book V (1018b9–1019a14) where he discusses the various ways in which something can be correctly called prior to another.

[29] See (Hellman [2001]) for an excellent overview of these varieties and the problems associated with each.

> There is an unfortunate ambiguity in the standard use of the word 'structure', which is often applied to an algebraic or relational system - a set with certain operations or relations defined on it, perhaps with some designated elements; that is to say, a model considered independently of any theory which it satisfies. This terminology hinders a more abstract use of the word 'structure'; if, instead we use 'system' for the forgoing purpose, we may speak of two systems as having an identical structure, in this more abstract sense, just in case they are isomorphic. The dictum that mathematics is the study of structure is ambiguous between these two senses of 'structure'. If it is meant in the less abstract sense, the dictum is hardly disputable, since any model of a mathematical theory will be a structure in this sense. It is probably usually intended in accordance with the more abstract sense of 'structure'; in this case, it expresses a philosophical doctrine that may be labelled 'structuralism'. (Dummett, [1991], p. 295.)

While Dummett's analysis is, in some sense, helpful, it conflates two things: algebraic and set-theoretic accounts of types of structured systems, and concrete and abstract accounts of kinds of structured systems. Systems qua models can be used to account for the shared structure of a *concrete kind* of structured system, i.e., for the shared structure of the elements and/or properties of natural numbers *qua* set-structured systems. However, as we will see, algebraically read systems *qua* schematic types, as opposed to Bourbaki read "structures" *qua* set-theoretic types, may also be used to account for the shared structure of *abstract kinds* of structured systems. Instead, then, on focusing on the clarification of, and providing background theories for, the notion of structure as a "thing", I will focus on clarification of, and providing a framework for, the notion of a system as a *schema*. Thus, my aim as an algebraic structuralist is not the analysis of the constitutive character or modal status of "structures", but the analysis of the shared structure of abstract kinds of structured systems.[30]

I begin, then, with an *abstract* notion of a *system*, since, as we will see, this is where we find our corresponding notion of a *cat-structured* system. In its most general sense, a cat-structured system, then, has 'objects' and 'morphisms' as its abstract kinds which are structured by the category-theoretic axioms. So that, the schema for *a type* of structured system, *i.e.*, for a kind of mathematical system *qua* a category is

> ... anything satisfying these axioms. The objects need not have 'elements', nor need the morphisms be 'functions'... We do not really care what non-categorical

[30]We note, however, that Hellman [2002], does appreciate the distinction between the algebraic-schematic use of categories (what he calls the 'algebraico-structuralist perspective', p. 9), but his suggestion that the "problem of the 'home address' remains" (p. 8, p. 15), clearly indicates that he is stilling thinking of "structures" (be they categories of toposes) as 'things' requiring 'conditions for the possibility of existence'. In fact, however, if, on the algebraic approach, the aim of structuralism is to account for the shared structure of kinds of mathematical systems in term of schematic types, as opposed to answering "What is (or where is!) a structure?" then why should we be troubled by the fact that "[b]y themselves they [the category-theoretic axioms] assert nothing. They merely tell us what it is to be a structure of a certain kind" (p. 7) and thus are "unlike the axioms of set theory, [in that] its axioms are not assertory." (p. 7.)

> properties the objects and morphisms of a given category may have; that is to say, we view it 'abstractly' by restricting to the language of objects and morphisms, domains and codomains, composition, and identity morphisms. (Awodey [1996], p. 213.)

At once we see important differences: on the category-theoretic view, not only are there are no "objects" as either sets-with-structure (see Dummett, [1991], p. 295) or places-with-structure (see Shapiro, [1997, pgs. 73, 93]), there are no "structures" as either (equivalence types of) systems-with-structure or "the abstract form of a system, highlighting the interrelationships among the objects..." (Shapiro [1997], p. 74.) What this means is that the Bourbaki conception of a system (of a system whose "objects" are "positions in a set-structure",[31] or "places in a structure"[32]) is to be considered as a *kind* of structured system: it is not the archetype of either the concept 'system' or the concept 'structure'. A category, too, neither *constitutes* a privileged system or structure: it is a schematic type. It functions as a *philosophical tool* used to organize what we can say about the shared structure of the various abstract kinds of mathematically structured systems. The value, then, of this schematic notion of a cat-structured system is that it can be used to capture the shared structure of abstract kinds of structured systems, *independently* of its specific set-structure (independently of what its kinds are).[33]

We have shown, then, that if category theory is taken as the framework for what we say about the shared structure of abstract kinds of mathematical systems, then, we can account for a *schematic in re interpretation* of mathematical structuralism.[34] Against the *ante rem* structuralist, this category-theoretically framed *in re* interpretation of mathematical structuralism implies that there are no "structures'", *qua* "things" over and above kinds of structured systems.

---

[31]We can, however, present the underlying structure of a Bourbaki system, or equivalently present the *kind* of any set-structured system as a kind of cat-structured, by taking our objects to be sets and our morphisms to be functions. The result is the type of structured system called Set. But this does not mean that objects *are* sets and morphisms *are* functions, it means *in this type* of system propositions that talk about objects and morphisms can be interpreted as being about kinds of sets and functions.

[32]Shapiro's structure-theory itself is framed by ZF+ Coherence axiom.

[33]For example, in the kind of category called Top, we present the *topological-structure* by taking objects as kinds of topological spaces and morphisms as kinds of continuous mappings, independently of what those kinds are kinds of. As Awodey explains: "...suppose we have somehow specified a particular kind of structure in terms of objects and morphisms ... Then that category characterizes that kind of mathematical structure, independently of the initial means of specification. For example, the topology of a given space is determined by its continuous mappings to and from the other spaces, regardless of whether it was initially specified in terms of open sets, limit points, a closure operator, or whatever. The category Top thus serves the purpose of characterizing the notion of 'topological structure'." (Awodey [1996], p. 213.)

[34]Simply put, to talk about the shared structure of abstract kinds of mathematical systems in terms of kinds of cat-structured systems, there is no need for either set theory or structure theory or modal logic over and above category theory: a category acts as a schematic type that can be used to frame what we say about the shared structure of abstract kinds of mathematical systems, (in terms of types of cat-structured systems like Set, Group, or Top), *and* for kinds of cat-structured systems, (in terms of the types Cat or CAT). And, more significantly, it does so *without* our having to specify what these kinds are kinds of.

As such, categories as structures *are not* prior in place. Against the typical *in re* structuralist, however, categories as schema are prior in definition insofar as they are needed, as an organizational tool (Mac Lane [1992]), to talk about the shared structure of abstract kinds of structured systems as instances of the same type. Herein, then, lies the "foundational significance" (Bell, [1981]) of using category theory to frame an *in re* structuralist philosophy of mathematics: while the notion of a cat-structured system is privileged as a schema (is prior in definition) it is not reified as a constituting a structure (is not prior in place). Category theory, then, can act as *the other theoretical language* (see Carnap [1956]) because it permits us to *talk about* abstract kinds of structured systems *qua* cat-structured systems without our having to claim that category theory is either a "thing language" or that Cat (or CAT) is a "thing world". Thus, to be an algebraic *in re* structuralist about abstract kinds of mathematical systems, we need not provide a "background theory", that provides the conditions for the actuality (or possibility) of what, or whether, a category *qua* a structure is.

References

Alexandrov, P.S. and Hopf, H., [1935], *Topologie*, Springer, Berlin.

Awodey, S., [1996], "Structure in Mathematics and Logic: A Categorical Perspective", *Philosophia Mathematica*, (3), Volume 4, 209–237.

Bell, J.L., [1981], "Category Theory and the Foundations of Mathematics", *Brit. J. Phil. Sci.* 32, 349–358.

Bell, J.L., [1986], "From Absolute to Local Mathematics", *Synthese*, 69, 409–426.

Bell, J.L., [1988], *Toposes and local set theories*, Oxford University Press, Oxford.

Bourbaki, N., [1950], "The Architecture of Mathematics", *AMM* 67, 221–232.

Bourbaki, N., [1968], *Theory of Sets*, Hermann, Paris.

Carnap, R., [1956], "Empiricism, Semantics, and Ontology", in Benacerraf, P and Putnam, H., (eds.), [1991], *Philosophy of Mathematics*, (2nd ed.), Cambridge University Press, Cambridge.

Corry, L., [1996], *Modern Algebra and the Rise of Mathematical Structures*, Springer Verlag, New York.

Dedekind, R., [1930-1932], *Gesammelte mathematische Werke*, 3 vols. Fricke, R., Noether, E., and Ore., O., (eds.), Braunschweig. (Chelsea reprint, [1969], New York).

Dummett, M., [1991], *Frege Philosophy of Mathematics*, Harvard University Press, Massachusetts.

Eilenberg, S., and Mac Lane, S., [1945], "The general theory of natural equivalence", *Transactions of the American Mathematical Society*, 55, 231–94.

Feferman, S., [1977], "Categorical Foundations and Foundations of Category Theory", in *Foundations of Mathematics and Computability Theory*, Butts, R., and Hintikka, J., (eds.), Reidel Publishing Company, Dordrecht-Holland.

Goldfarb, W., [2001], "Frege's Conception of Logic", in, J. Floyd, J., and S. Shieh, S., (eds.), *Futures Past: Reflections on the History and Nature of Analytic Philosophy*, Harvard University Press, Massachusetts.

Hasse. H., [1954], *Higher Algebra*, Fredrick Ungar, New York. (English trans. of the 3rd ed. Of Hasse [1926] by Benac, T.J.).

Hellman, G., [2001], "Three Varieties of Mathematical Structuralism", *Philosophia Mathematica*, (3), Volume 9, 184–211.

Hellman, G., [2002], Does Category Theory Provide a Framework for Mathematical Structuralism", *Philosophia Mathematica*, (3), Volume 9, 184–211.

Hilbert, D., [1900], "Über den Zahlenbegriff", *JDMV* 8, 180–184.

Hilbert, D., [1902], "Mathematical Problems", *BAMS* 8, 437–479 (English trans. Of Hilbert [1901], by Newson, M.W.).

Landry, E., [1999], "Category Theory: The Language of Mathematics", *Philosophy of Science* 66 (Proceedings), S14–S27.

Landry, E., [2001], "Logicism, Structuralism and Objectivity", *Topoi*: Special Issue – Mathematical Practice, Volume 20, 79–95.

Lawvere, F.W., [1966], "The Category of Categories as a Foundation of Mathematics", Proc. Conference Categorical Algebra (LaJolla 1965), Springer Verlag, New York, 1–20.

Mac Lane, S., [1968], "Foundations of Mathematics: Category Theory", in *Contemporary Philosophy: A Survey*, Klibansky, R., (ed.), Firenza, la Nuova Italia Editrice, 286–294.

Mac Lane, S., [1980], "The Genesis of Mathematical Structures", *Cahiers Topol. Geom. Diff.* 21, 353–365.

Mac Lane, S., [1992], "The Protean Character of Mathematics", in *The Space of Mathematics*, Echeverra, J., Ibarra, A., and Mormann, J., (eds.), de Gruyter, New York, 3–12.

Mayberry, J., [1994], "What is Required of a Foundation for Mathematics?", *Philosophia Mathematica 3*, Volume 2, Special Issue, "Categories in the Foundations of Mathematics and Language", Bell, J.L., (ed.), 16–35.

McLarty, C., [1990], "The Uses and Abuses of the History of Topos Theory", *Brit. J. Phil. Sci.*, 41, 351–375.

McLarty, C., [1995], *Elementary Categories, Elementary Toposes*, Clarendon Press, Oxford.

Noether, E., [1921], "Idealtheorie in Ringbereichen", *MA* 83, 24–66.

Noether, E., [1926], "Abstrakter Aufbau der Idealtheorie in alebraischen Zahlun Funktionskörper", *MA* 96, 26.61.

Ore, O., [1935], "On the foundations of Abstract Algebra, I", *AM* 36, 406–437.

Resnik, M.D., [1996],"Structural Relativity", *Philosophia Mathematica*, (3), Volume 4, Special Issue, "Mathematical Structuralism", Shapiro, S., (ed.), 83–99.

Shapiro, S., [1996], "Space, Number and Structure: A Tale of Two Debates", *Philosophia Mathematica, 3*, Volume 4, Special Issue, "Mathematical Structuralism", Shapiro, S., (ed.), 148–173.

Shapiro, S., [1997], *Philosophy of Mathematics: Structure and Ontology*, Oxford University Press, Oxford.

Waerden, B.L. van der, [1930], *Modern Algebra, 2 vols*, Springer, Berlin.

Chapter 13

CATEGORIES, SETS AND THE NATURE OF MATHEMATICAL ENTITIES

Jean-Pierre Marquis
University of Montreal
Jean-Pierre.Marquis@umontreal.ca

13.1 Our claim

It is claimed that category theory cannot provide an adequate foundation for mathematics. The main reasons seems to be the following:

1 Category theory cannot provide an adequate foundation for mathematics for *epistemological* reasons, i.e. it presupposes other, more simple, concepts for its understanding;

2 Category theory, perhaps useful in certain areas of mathematics, for instance in algebraic topology, homological algebra, algebraic geometry, homotopical algebra, K-theory, theoretical computer science or even mathematical physics, cannot provide a comparable picture of mathematics as set theory does. First, there is an *informal* set theory that provides a framework for mathematics. What this informal set theory amounts to is not entirely clear, but it seems to play an important role. Second, there is a well-known and well-understood *universe*, namely the cumulative hierarchy, and a well-known and well-understood theory written in a well-known and well-understood formal language, namely ZF (of NBG) written in first-order logic. Thus, the objection goes, category theory does not fulfill some obvious philosophical and metamathematical requirements one might expect or ask from a foundational framework.

In this paper, we want to address these issues in the following manner. I want to argue that:

J. van Benthem, G. Heinzmann, M. Rebuschi and H. Visser (eds.), The Age of Alternative Logics: Assessing Philosophy of Logic and Mathematics Today, 181–192.

1 Category theory, as it already is, is based on a conception of mathematical object which is, from an ontological point of view, radically *different* from the conception underlying set theory; this fact has numerous consequences, one of which is that the epistemological argument against category theory is ill-founded and therefore can be discarded;

2 Although many category theorists believe that category theory is fine as it is, even for foundational purposes[1] — a view that I will not examine here, for it would take us away from our main concern — an alternative picture is being developed, mostly by the logician Michael Makkai at McGill and upon which I will rely heavily, a picture that comprises a universe of mathematics based on a *different* conception of sets, radically different from the cumulative hierarchy, although there is a hierarchy of a different nature, and a formal language in which a theory of that universe can be presented and developed. In a nutshell, the universe, technically the universe of weak ω-categories, is highly heterogeneous in the sense that there are various *kinds* of entities and the variety of these kinds is reflected by the variety of criteria of identity for them. The formal language is an extension of first-order logic, namely it is first-order logic with dependent sorts, FOLDS, which in this context takes the form of a diagrammatic language. We should add immediately that this picture extends radically the nature of mathematical objects presented in the first step of the argument. When we get to this stage of the presentation, I submit that, not only do we have answers to the main objections to a categorical framework, but we can see clearly that the views involved are based on radically different conceptions of mathematical objects. At that point, we can evaluate the situation both from a technical point of view, i.e. what are the technical benefits and the drawbacks of each view, from a philosophical point of view, i.e. which view, if any, is philosophically justified, in particular, which view represents best the way mathematicians work and think about mathematical objects.

13.2 The nature of mathematical entities

Let us start with the nature of mathematical entities in general and with a rough and classical distinction that will simply set the stage for the picture we want to develop. We essentially follow Lowe 1998 for the basic distinctions. We need to distinguish between abstract and concrete entities, on the one hand, and universals and particulars on the other hand. For our purpose, it is not necessary to specify a criterion of demarcation between abstract and concrete entities. We

[1] Probably one of the best illustration of that position can be found in Taylor's fascinating book. See Taylor 1999.

simply assume that such a distinction can be made, e.g. concrete entities can change whereas abstract entities cannot. We assume that a universal is an entity that can be instantiated by entities which themselves are not instantiable, the latter being of course particulars. Given these distinctions, an entity can be a concrete particular, a concrete universal, an abstract particular or an abstract universal.

Our focus here is between the last two possibilities. For we claim that the current conception of sets makes them abstract particulars whereas for objects defined within categories, mathematical entities are abstract universals.[2] This, we claim, is true of category theory as it is.

Sets, as they are generally conceived and as they are represented in ZFC or NBG, are indisputably abstract particulars. We assume that they are abstract. The fact that they are particulars is established by looking at the criterion of identity for sets in these theories, namely the axiom of extensionality. As is well known, a set is completely determined by its elements and two sets are identical if and only if they have the same elements. Thus a specific set cannot be instantiated by another entity and is therefore a particular. It could be claim that sets are particulars *because* there is a unique criterion of identity for them. A theory in which some entities are universal in the previous sense has to have at least two different criteria of identity: one for the universals themselves and one for the particulars that instantiate these universals. In category theory as it is, we find many different criteria of identity. Three are well-known and common to standard category theory. Others arise in more complex situations. In the universe of categories, there is a whole spectrum of criteria of identity: at one end of the spectrum, we have criteria of identity for particulars, and at the other end, we have a hierarchy of criteria of identity for universals, all related to one another in a systematic manner. What is philosophically interesting, is that the traditional distinction between universal and particular is in some ways inadequate.

13.3 Criteria of identity in a categorical context

A brief look at the axioms of a category should be enough to convince anyone that there is an implicit criterion of identity at work for morphisms. Indeed,

---

[2]Our choice of terminology is radically different from Ellerman 1987 where a similar proposal is made. However, Ellerman argues that category theory is a theory of *concrete* universals whereas set theory is a theory of *abstract* universals. Needless to say, the difference lies in the way the abstract/concrete distinction is articulated. Thus, whereas Ellerman argues that both concepts of category theory and set theory are universals and that the difference lies in the fact that the former is concrete whereas the latter is abstract, we believe that both are abstract and the former are universals and the latter are particulars. See Marquis 2000 for some critical remarks on Ellerman 1987. We should point out, however, that the terminology is not clarified in Marquis 2000. We should also point out that our terminology is different from the one found in Makkai 1998 and 1999 where Makkai argues that the concept of collection implicit in a categorical framework is abstract. Once more, the terminology is justified if it is made in a certain way. However, our overall point of view owes a great deal to Makkai's technical work.

we have that, for instance, f(gh) = (fg)h, the associativity for morphisms, is an *identity*. Morphisms are treated as particulars from the very beginning. Recall that a morphism f: X $\rightarrow$ Y is an *isomorphism* if there is a morphism g: Y $\rightarrow$ X such that fg = 1_Y and gf = 1_X. Two objects X and Y are said to be *isomorphic* if there is an isomorphism between them (at least one). Thus a *specific* pair of maps is used to identify objects as being isomorphic. One could of course immediately stop at this point and reflect on the necessity that these morphisms be particulars. It seems reasonable to ask for a certain type of morphisms that should satisfy certain conditions, as is clear already in the case of a homotopy category. We will come back to this point in due course.

A second criterion of identity appears when we consider how objects are *defined* in a category. Consider the simple and well-known case of the definition of a product for two objects X and Y in a category C. A product for X and Y in a category C is an object P of C *together* with morphisms π_X : P $\rightarrow$ X and π_Y : P $\rightarrow$ Y such that for any object Q of C together with morphisms f : Q $\rightarrow$ X and g : Q $\rightarrow$ Y, there is a *unique* morphism h : Q $\rightarrow$ P such that π_Xh = f and π_Yh = g. The important point here is that this definition characterizes P up to a unique isomorphism. This means that for any object Q *isomorphic* to P in C, if P is a product of X and Y, then Q is a product of X and Y and, moreover, if Q and P are both products of X and Y, then they are isomorphic. Thus, this definition does not give us a *particular* object as a product, nor does it characterize P by stipulating what its elements should be. It specifies under what conditions an object is an instance of the universal *product*, when a given object is a *token* of the product *type*. The criterion of identity in this case is given by the unique isomorphism existing between two tokens of the product of two objects of the category. From a global point of view, the criterion of identity is given by the underlying groupoid of the category C.

This is typical of the way objects are defined in category theory. Mathematical entities and their properties *in* a category C are only given up to isomorphism. It is also true of other entities, e.g. adjoint functors. Category theory specifies what are the abstract universals of mathematics and to know the abstract universals of a domain is to know the fundamental features of that domain. For instance, once one has shown that a given category C has finite products, more generally finite limits, then one knows about various constructions and results that hold in C. A more important example is provided by the notion of abelian category, given by the existence of certain limits (and colimits) and what are called "exactness conditions" (the terminology comes from homological algebra). Another example is provided by the categorical definition of the natural numbers: they too are presented as abstract universals.[3] There is no

[3]Lowe 1998 also argues, but on different grounds, that the natural numbers are abstract universals. For a categorical analysis of the natural numbers, see for instance McLarty 1993.

doubt that the use of category theory in certain contexts, e.g. algebraic topology, homological algebra or algebraic geometry, reveals what is *fundamental* in these domains.

Now, collections or sets are a special kind of mathematical entities. Can they be thought of as entities like any other entity in a category? Already in the mid-seventies, Lawvere had suggested a way to think of sets in this context. Unfortunately for us, he called them "abstract sets", using the term "abstract" in the different sense from what we have assumed here.[4] Here is how he describes these "abstract sets":

An abstract set X has elements each of which has no internal structure whatsoever; X has no internal structure except for equality and inequality of pairs of elements, and has no external properties save its cardinality. (Lawvere, 1976, 119.)

The first sentence could be reformulated by saying that the elements of an abstract set are "atoms" or faceless points.[5] Nonetheless, there is an internal criterion of identity for the elements of each sets: we can tell, given two elements of a given set whether they are the same or they are different. Thus, a set in this sense comes equipped with a criterion of identity for its elements. There is no *global* criterion of identity for elements: one cannot ask, given two arbitrary objects ("elements"), whether they are the same or not. The criterion of identity is always relative to a given abstract set. Furthermore, there is no global relation of elementhood: one cannot ask, for any object x and any set A, whether x is an element of A or not. Finally, one cannot ask, given two sets X and Y, whether X = Y or not. Sets in the above sense are *isomorphic* or not. This is what the last sentence of the quote means: their only external property is their cardinality, i.e. two sets are "identical" when they are isomorphic. However, as Lawvere remarks, these sets are "more refined (less abstract) than a cardinal number in that it does have elements while a cardinal number does not." (Lawvere, 1976, 119.) It is as if such a set would be a *representative* of a cardinal number, or I would like to say a *token* of a cardinal number, but seen as a token of that cardinal number, i.e. with any specific property erased. Notice that this is in general *how* we look at tokens as tokens of a type: we ignore all specific properties of

[4]We are using "abstract" in an ontological sense, whereas Lawvere clearly has an epistemological notion in mind, something which is not uncommon among mathematicians. In the paper we are referring to, Lawvere says that his notion of sets is "less abstract" than the notion of cardinality. Clearly, one entity cannot be more (or less) abstract than another entity in the ontological sense. However, in the epistemological sense, one can have different levels of abstraction, assuredly a common phenomenon within mathematics. This is an issue we will explore elsewhere. See Marquis 2002.

[5]It could certainly be argued that Cantor was developing a conception of sets along these lines. For the process of double abstraction underling Cantor's conception yields a "form", not a particular entity. This is especially clear when one looks at order-*types*. Ordinals are tokens of a type for Cantor, they are not abstract particulars *à la* Von Neumann. Lawvere has himself recently developed this idea along a different line by looking at abstract sets as *Kardinalen*. See Lawvere 1994.

the token and see only the properties that exemplify the type. In Aristotelian terminology, one would say that such a set is a collection *qua* collection.

Should these sets be treated as particulars or as universals? Since we do not have the standard axiom of extensionality — we cannot compare elements that belong to different sets, it seems that we cannot treat them as particulars. One could argue, though, as follows: since these sets can be the domain or the codomain of morphisms, in particular, they are the domain and the codomain of their own identity morphism. Since the latter *are* assumed to be particulars, the sets have to be particulars too. But this argument fails for two reasons. First, it fails because the criterion of identity for sets is given by isomorphisms. Thus, in particular, any automorphism is acceptable, i.e. a set can be identical with itself in more than one way. This might sound odd, but we are perfectly at ease with this idea for *geometrical* objects. Second, it fails because in the universe we will be considering, even morphisms won't be treated as particulars. The identities are replaced by isomorphisms systematically.

I suggest that we call these sets "transcendental sets" since they are purely the *form* of sets.[6] Another possibility would be to call them "perfect sets" or again, following a suggestion also made by Lawvere, "pure sets", but I favor the previous terminology. We have already underlined the fact that the *totality* of these transcendental sets *cannot* constitute a set. (For there is no set-theoretical criterion of identity for them.) As Lawvere has already observed, these transcendental sets can support mappings, the latter notion being taken as a primitive notion. Composition of mappings can be defined and it clearly satisfies the usual axioms of a category. Thus, the *totality* of transcendental sets constitute a category. It is a different *kind* of entity. As Lawvere has argued in his paper, it is reasonable to say that the universe of these transcendental sets form a topos. In fact, in a topos, any object can be considered to be a transcendental set.

The first conclusion we can immediately draw is that a coherent conception of sets can be developed in a categorical context; this conception is different from the conception inherent to traditional set theory; thus, at the very least we can say that there are various conceptions of sets (of course, there is also the *naïve* conception, but to claim that, say ZFC, with the cumulative hierarchy, is the correct formalization of that conception is a problematic claim).

We could stop here and start arguing for the foundational relevance of category theory as it is. We could give various technical results obtained within toposes or about toposes and expose their foundational significance and importance. We could also articulate a view in which mathematics is done in toposes. The main claim, I guess, would be the following: given any piece

[6]I am using the term "transcendental" based on an analogy with Kant's usage. It has nothing to do with the expression "transcendental numbers".

of mathematics M, it is possible to find a topos E in which the concepts and theorems of M can be defined and proved. This in itself is an interesting position that deserves to be examined carefully. But we will rather move on to a different, emerging, position, a position that has an intrinsic beauty and that can be presented as an alternative to the set theoretical picture.

13.4 Systems of categories

Let us come back to categories. Given the importance of isomorphisms *in* categories, one would expect that the notion of isomorphism would provide the criterion of identity *for* categories. However, this is not the case.[7] A criterion of identity for categories is given by the notion of *equivalence* of categories: two categories C and D are *equivalent* if there are functors F: $C \to D$ and G: $D \to C$ such that the composite FG is isomorphic to 1_D and GF is isomorphic to 1_C. Notice that the identity between the composites FG and GF and the respective identity functors are replaced by isomorphisms. We can immediately conclude two important facts from this situation: a category of categories, no matter what it turns out to be, *cannot* simply be a category. For, as we have seen, in a category, the criterion of identity is given by the isomorphisms and since categories are not individuated by isomorphisms, a category of categories will have to be something else. Second, transcendental sets and categories have *different* criteria of identity; for transcendental sets, it is given by isomorphisms, for categories it is (at least at this level) given by equivalences; hence, categories *cannot* be said to be structured sets, in the same sense, say, that one can say that groups are structured sets. This is now crucial: categories in our universe cannot be said to be (structured) sets.

Before we move on to the universe as a whole, let us briefly consider how the criteria of identity build up in a categorical context.

Often in category theory, the objects of investigation are functors F: $C \to D$, G: $A \to B$. One has to determine the correct criterion of identity for such functors. It turns out that the right notion can be presented as follows: two functors F: $C \to D$ and G: $A \to B$ are *equivalent* if there are equivalence of categories E_1: $C \to A$ and E_2: $D \to B$ and an isomorphism η: $\mathrm{GE}_1 \to \mathrm{E}_2\mathrm{F}$. This situation can be represented by the following diagram:

$$\begin{array}{ccc} C & \xrightarrow{F} & D \\ E_1 \downarrow & {}^{\eta} \nearrow & \downarrow E_2 \\ A & \xrightarrow[G]{} & B \end{array}$$

[7] As far as I can tell, this was first discovered, or at the very least emphasized, by Grothendieck in his Tohoku paper, published in 1957. However, it might have been discovered by Yoneda some times before. This is an open problem.

We see that this is more involved: we use the notion of equivalence of categories together with the notion of isomorphism of parallel functors.

We could go on like this and introduce the criterion of identity for fibrations, bicategories, etc. In each case, the criterion of identity would be more involved. The thrust should however be clear: the identity of objects in a categorical context is *derived* from that context, i.e. the underlying category in each case. Furthermore, there is a hierarchy of criteria of identity that seems to be endless. This is the first sense in which the categorical universe is heterogeneous: the criterion of identity for objects in a category is not the same as the criterion of identity for categories, which in turn, is not the same as the criterion of identity for functors, which in turn, is not the same as the criterion of identity for fibrations, etc. We cannot refrain at this moment to quote from Lowe:

The idea that one can "introduce" a kind of objects simply by laying down an identity criterion for them really inverts the proper order of explanation. As Locke clearly understood, one must first have a clear conception of what kind of objects one is dealing with in order to extract a criterion of identity for them from that conception. (...) So, rather than "abstract" a kind of objects from a criterion of identity, one must in general "*ex*tract" a criterion of identity from a metaphysically defensible conception of a given kind of objects. (Lowe, 1995, 517.)

In categorical practice, the kind of objects one has to deal with are very often clear from the context. One then determines the proper criterion of identity for the objects of that kind.[8] This is strikingly different from the prevalent situation in set theory. A general theory of identity reflecting the order of presentation we have just given has been proposed by Michael Makkai in the form of FOLDS. It is the very purpose of that formal framework to be able to formulate in a precise and rigorous fashion, for various kinds of mathematical entities, corresponding criteria of identity. We will come back to FOLDS later.

This informal hierarchy of criteria of identity already indicate the heterogeneity of the universe. At the bottom of the universe, we find the transcendental sets with their criterion of identity: isomorphisms. Thus, their totality does not constitute a set. In fact, they form various categories. These categories, in turn, can be collected into totalities: what are these totalities? They certainly cannot be categories. There is more structure involved. They are at least what are called strict 2-categories or 2-categories. Then again, 2-categories form totalities and these totalities, to be described accurately, require more structure: they form weak 3-categories. In order to see this more clearly, let us look at equivalences more carefully.

[8]Indeed, in many textbooks, various notions are defined, examined and developed and the criterion of identity for these objects is not even mentioned, *e.g.*, fibrations. See for instance Borceux 1994, Jacobs 1999.

An equivalence is given by a pair of functors F: $C \rightarrow D$, G: $D \rightarrow C$ and natural isomorphisms α : FG $\rightarrow 1_D$ and β : GF $\rightarrow 1_C$. The fact that α and β are isomorphisms means that the identities $\alpha\alpha^{-1} = 1_{1_D}$ and $\alpha^{-1}\alpha = 1_{\mathrm{FG}}$ hold. In other words, we have reintroduced particulars at the last stage. To be entirely consistent with the underlying conception of object we are assuming, these identities should be replaced by isomorphisms (of the right type, i.e. satisfying certain conditions.) A complete description of this situation is given by what are called bicategories, or, more commonly nowadays, weak 2-categories. The next step is provided by tricategories or weak 3-categories.[9] The latter notion takes six pages to be defined and 13 pages are required to present the various conditions that have to be satisfied by the various levels. Fortunately, a general and very compelling picture is emerging. I want to insist on the fact that it is technically and philosophically compelling. The general picture of the resulting universe is given by what are called "higher dimensional categories" or weak ω-categories. (See Leinster 2002 for a review of different definitions.)

Here is an extremely simplistic sketch of the universe of weak n-categories. 0-categories are transcendental sets. One and the same set can be the same as itself in various ways; i.e. it can have various automorphisms. More generally, two sets can be the same in different ways. Each and every isomorphism between them stipulates how they are the same and we can keep track of these various identities. Moreover, 0-categories, i.e. sets, are linked to one another by morphisms and these morphisms compose in the obvious way. Let us call morphisms between 0-categories 1-morphisms. There is a motivation behind the terminology, for 0-categories can be represented as points and 1-morphisms as directed lines between points. It would be tempting to say that 1-morphisms satisfy various identities, e.g. associativity, and that they form a category. But as we have seen, that would amount as treating them as abstract *particulars*. Hence, instead of having identities between 1-morphisms, we require that isomorphisms exist between them (with extra conditions). This implies that 2-morphisms between 1-morphisms have to be introduced and that they provide a criterion of identity for 1-morphisms. When this is done, it can be seen that the collection of 1-morphisms form a weak 2-category. How about 2-morphisms? Clearly, once more, we have to go up the ladder and introduce 3-morphisms. These will stipulate how 2-morphisms behave, how they compose and under what conditions they are identical. The general pattern should now be obvious: to connect and identify n-morphisms, (n+1)-morphisms are required. Notice that it is possible to stop at any n and stipulate that at that

[9]It can be shown that any weak 2-category is equivalent, in a specific sense, to a strict 2-category. Thus, it would appear that weak 2-categories, those we are describing here and that are relevant to this discussion, are dispensable. However, this is no longer true for 3-categories. In the latter case, there are weak 3-categories which cannot be replaced by strict 3-categories.

point, equalities between n-morphisms exist but that for all $j < n$, identities are given by (j+1)-morphisms.

The general picture is therefore this. The collection of 0-categories forms a 1-category. If we were to stop at this stage, it would mean that we take equalities between 1-morphisms and that the latter are treated as abstract particulars. But we can consider the collection of 1-categories and this is a 2-category. Again, if we were to stop at this point, it would mean that we consider equalities between 2-morphisms but that 1-morphisms, that is 1-categories, are now treated as abstract universals. Thus, for each n, there is an (n+1)-category of all n-categories. Of course, one can consider the ω-category of n-categories for all n: this amounts to defining n-categories for all n simultaneously. The ω-category of weak n-categories is the alternative picture to the cumulative hierarchy of sets.

The technical problems involved in the study of higher-dimensional category theory are daunting. We have not even mentioned the simplest obstacle. We refer the reader to the literature. (See Baez 1997, Baez & Dolan 1998a, Baez & Dolan 1998b, Batanin 1998, Makkai 1998, Makkai 1999 and Leinster 2002.) The point I wanted to make is purely conceptual. I am deliberately ignoring the practical motivations underlying actual research into higher dimensional categories, although they are probably more important than the conceptual ones within the research community, for they go from computer science to topological quantum field theory via homotopy theory. What I *do* want to emphasize are the following points:

1 Although we have started with a simple opposition between universals and particulars, the final picture forces us to think about this opposition with care. In the original picture, we had abstract universals and abstract particulars. Now, we seem to be forced to think about the realm of abstract universals in a more elaborate way: within abstract universals, there is a complex structure of relationships between kinds of universals. A simple case of a similar hierarchy can be given: start with a *specific* metric space X, given as a particular. Consider its group of automorphisms. As such, the latter group is also a particular. However, as a group, it is a token of a type: a group whose elements are unidentified but with an isomorphic structure. In turn, this group can be seen as a one object category, that is an object in the universe of categories. At this stage, we are back in the foregoing picture. Now, the elements of the original groups are automorphisms of the one object category and they can be related to one another either by equalities, in which case we treat them as particulars, or they can be related by morphisms of higher order logic, in which case they are universals. There is a general ontological picture emerging from this analysis that will force us to look more carefully at the nature of universals.

2 The argument against category theory usually rests on the way categories are presented, i.e. as classes or sets with a certain structure. Thus, it is assumed that category theory has to be presented or understood in a set-theoretical framework. As we have seen in the foregoing section, this misses the fundamental aspect of categories; if the nature of categories is revealed by its criterion of identity, then we can say that categories are *not* structured sets.

3 It might very well be that the concept of a collection as an abstract universal rests on our *understanding* of abstract particulars. One cannot but think of representation theory of groups where interplay between an abstract universal, e.g. an abstract group, and abstract particulars, e.g. its representations, is crucial. However, it by no means implies that a coherent conception of such collections cannot be developed and depends for its development upon a specific choice of abstract particulars.

The requirements of a foundation for mathematics might vary, depending upon one's conception of the foundational enterprise.[10] We do believe that category theory is such that it can answer any requirement one might expect from a foundational framework. But one has to look at it properly and see how and in what sense it is universal. John Bell, in his paper on category theory and the foundations of mathematics, claimed that "far from being in opposition to set theory, [category theory] ultimately enables the set concept to achieve a new universality." (Bell, 1981, 358) Bell could not be closer to the point: sets are not particulars in a categorical framework, they *are* universals and they are the first universals in a complex and rich hierarchy that ought to be foundationally appealing.

References

Baez, J., 1997, "An Introduction to n-Categories", in 7th *Conference on Category Theory and Computer Science*, E. Moggi & G. Rosolini (eds.), SNCS, vol. 1290, Berlin: Springer-Verlag, 1–33.

Baez, J. & Dolan, J., 1998a, "Higher-Dimensional Algebra III. N-Categories and the Algebra of Opetopes", *Advances in Mathematics*, 135, 145–206.

Baez, J. & Dolan, J., 1998b, "Categorification", in *Higher Category Theory*, E. Getzer & M. Kapranov (eds.), Contemporary Mathematics, vol. 230, Providence: AMS, 1–36.

Batanin, M., 1998, "Monoidal Globular Categories as a Natural Environment for the Theory of Weak n-Categories", *Advances in Mathematics*, 136, 39–103.

[10]See, for instance, Mayberry 1994 and Marquis 1995.

Bell, J., 1981, "Category Theory and the Foundations of Mathematics", *British Journal for the Philosophy of Science*, 32, 349–358.

Borceux, F., 1994, *Handbook of Categorical Algebra 2: categories and structures*, Cambridge University Press.

Ellerman, D. P., 1988, "Category Theory and Concrete Universals", *Erkenntnis*, 28, 409-429.

Jacobs, B., 1999, *Categorical Logic and Type Theory*, New York: Elsevier.

Lawvere, W., 1976, "Variable quantities and variable structures in topoi", *Algebra, topology, and category theory,* New York, Academic Press, 101–131.

Lawvere, W., 1994, "Cohesive Toposes and Cantor's '*lauter Einsen*'", *Philosophia Mathematica*, (3), vol. 2, 5–15.

Leinster, T., 2002, "A Survey of definitions of n-category", *Theory and Applications of Categories*, vol. 10, 1–70.

Lowe, E.J.,1995, "The Metaphysics of Abstract Objects", *Journal of Philosophy*, 92(10), 509–524.

Lowe, E.J., 1998, *The Possibility of Metaphysics*, Oxford: Clarendon Press.

Makkai, M., 1998, "Towards a Categorical Foundation of Mathematics", *Logic Colloquium '95(Haifa)*, LNL 11, Berlin: Springer, 153–190.

Makkai, M., 1999, "On Structuralism in Mathematics", in *Language, Logic, and Concepts, Essays in Memory of John Macnamara*, R. Jackendoff, P. Bloom, K. Wynn, eds., Cambridge: MIT Press, 43–66.

Marquis, J.-P., 1995, "Category Theory and the Foundations of Mathematics : Philosophical Excavations", *Synthese*, 103, 3, 421–447.

Marquis, J.-P., 2000, "Three kinds of Universals in Mathematics?", in *Logical Consequence: Rival Approaches and New Studies in Exact Philosophy: Logic, Mathematics and Science*, Vol. II, B. Brown & J . Woods, eds., Oxford: Hermes, 191–212.

Mayberry, J., 1994, "What is Required of a Foundation for Mathematics?", *Philosophia Mathematica,* (3), vol. 2, 16–35.

McLarty, C., 1993, "Numbers Can be Just What They Have to", *Noûs*, 27, 487–498.

IV

INDEPENDENCE, EVALUATION GAMES AND IMPERFECT INFORMATION

Chapter 14

TRUTH, NEGATION AND OTHER BASIC NOTIONS OF LOGIC

Jaakko Hintikka
Boston University
hintikka@bu.edu

14.1 What is the logic of ordinary language?

According to a story, Albert Einstein was once asked how he had come upon his strange revolutionary ideas. He replied: "By asking the questions that children are discouraged to ask." If we want to follow Einstein's strategy in the philosophy of logic, we are thus led to ask such questions as we ourselves discourage our own introductory logic students to ask. But what are such questions? One of them might very well be: What is the logic of our ordinary language? It is convenient to us logic teachers to pretend initially that it is the logic we are teaching, in other words, that the notation of the usual first-order logic is nothing but a streamlined version of ordinary English. In older textbooks this claim is sometimes made explicitly. If pressed, we might appeal to Chomsky (e.g. 1986) whose Ersatz logical forms alias LFs differ only inessentially from the logical forms of ordinary first-order formulas. Yet such appeals should evoke pangs of intellectual conscience, for our actual Sprachlogik differs in several disturbing ways from the received ("Frege-Russell") first-order logic. I have shown (in Hintikka 1997) that even one of the most general notions of formal logic, the notion of scope, is not a primitive notion but one which can be applied to natural language only indirect ways. It can also be shown that the logic of natural-language conditional sentences can only be captured by going way beyond ordinary first-order logic. But even apart from such theoretical differences, there are lots of ordinary language sentences whose logic is not captured by their *prima facie* translations into first-order logical notation. For instance, consider the sentence

J. van Benthem, G. Heinzmann, M. Rebuschi and H. Visser (eds.), The Age of Alternative Logics: Assessing Philosophy of Logic and Mathematics Today, 195–219.

(1) For this problem, there is a person such that if he or she can solve it, anyone can.

This seems to have the logical form (if we look away from the initial demonstrative identification of the problem)

(2) $(\exists x)(A[x] \supset (\forall y)A[y])$.

But (2) turns out to be a logical truth whereas (1) is not naturally taken to be one. Likewise, C.S. Peirce already noted that the following pair of sentences clearly have a different meaning even though our logic tells us that they do not (*Collected Papers* 4.546):

(3) Someone is such that he will commit suicide if he fails in business.

(4) Someone is such that he will commit suicide if everyone fails in business.

Or, rather, Peirce noted that the following *prima facie* translations (5)–(6) of (3)–(4) into the notation of first-order logic are logically equivalent, in spite of the difference in meaning between (3)–(4):

(5) $(\exists x)(F[x] \supset S[x])$

(6) $(\exists x)((\forall y)F[y] \supset S[x])$

Likewise, the reasoning that leads to some of the best-known paradoxes seems to be impeccable reasoning, in spite of giving rise to paradoxical conclusions. The sorites paradox is a case in point. If we abbreviate "a has n hairs" as H(a,n), then the inductive inference leading to the paradox appears to be unobjectionable. It could be taken to be of the form

(7) $(H(a,0) \supset B(a))\&(\forall x)((H(a,x) \supset B(a)) \supset (H(a,x+1) \supset B(a))$
ergo $(\forall x)((H(a,x) \supset B(a))$

where $B(a)$ says that a is bald. In a simpler form, the structure of (7) can be taken to be

(8) $S[0](\forall x)(S[x] \supset S[x+1])$, ergo $(\forall x)S[x]$.

which looks like a perfectly valid instance of mathematical induction. Each of these anomalies might look insignificant, but their cumulative impact ought to be a clue that shows that the logic of our ordinary discourse is far from being adequately understood.

14.2 What is truth?

Another Einsteinian question is surely what precisely is meant by truth in first-order logic. We teach students all about truth-functions and truth-values, but are not likely to give an adequate answer when a student inquires what the mysterious notion of truth is. Maybe we tell our students to look up a

Tarski-type truth-definition from an advanced text (or from Tarski 1935), trying to suppress the guilty awareness of how a student will not find any real insight into the notion of truth from Tarski-type conditions on valuations and on infinite sequences. A student's puzzlement over such complications is part and parcel of those philosophers' dissatisfaction who claim that there is nothing in Tarski-type truth definitions that show that they are definitions of *truth*. The striking fact here nevertheless is that in the case of first-order logic (the logic of quantification) there is an accurate general answer easily available (see Hintikka 1998 and 2001). What is more, this answer is nothing more and nothing less than an explication of our natural pretheoretical notion of truth for quantificational sentences. In order to see what this answer is, consider a sentence of the form

(9) $(\forall x)(\exists y)F[x,y]$

When is (9) true? Obviously if and only if for any given value of x it is in principle possible to find a "witness individual" y depending on x such that $F[x,y]$. And this colloquial location "it is possible to find, given x" is in the eyes of a mathematician nothing but an euphemism for the existence of a function $f(x)$ which produces a suitable witness individual as its value for any given argument x, in other words a function f such that the following is true:

(10) $(\forall x)F[x,f(x)]$

The generalization of this observation is that a first-order sentence S is true if and only if there exists a full array of its Skolem functions. But what are the Skolem functions of S? In order to recognize them, let us assume that S is in a negation normal form. What this means is that its propositional constants are &, $\vee$ and $\sim$ and that all negation signs precede immediately atomic formulas or identities. Then the Skolem form of S is obtained y_1 by replacing each existentially quantified subformula $(\exists x)F[x]$ of S by $F[f(y_1,y_2,\ldots)]$ and prefixing the entire sentence with $(\exists f)$. Here f is a new function variable, different for different existential subformulas, and $(\forall y_1)$, $(\forall y_2)$, ... are all the universal quantifiers in S on which the quantifier $(\exists x)$ depends on in S. The truth-making choices of the values of the function variable f are the Skolem functions of S. And what these functions do is to produce the witness individuals (usually dependent on other such individuals), which according to our pretheoretical conception show the truth of S. Thus what we have here is a straightforward generalization of the truth-condition for (9), identified above. For some purposes, the notion of Skolem functions can - and must - be extended to relate also to the propositional connectives of S. Assuming still that S is in a negation normal form, this means replacing each disjunction $(S_1 \vee S_2)$ that occurs as a subformula of S by

(11) $(S_1\&g(y_1,y_2,\ldots)=0)\vee(S_2\&g(y_1,y_2,\ldots)\neq 0)$.

At the same time, the entire sentence is prefixed by $(\exists g)$. In (11), g is a new function variable, different form all the f's and different for different disjunctions and $(\forall y_1)$, $(\forall y_2)$... are all the universal quantifiers on which the disjunction in question depends on in S. Furthermore, 0 can be any designated member of the domain. In the special case of a sentence of the form $(\exists x)\, F[x]$ its sole Skolem function reduces to a constant individual. This individual serves as the "witness individual" which according to our pretheoretical conception vouchsafes the truth of the sentences in question. The general case becomes as obvious as this paradigm case as soon as we realize that in general the requisite witness individuals that show the truth of the sentence depend on other witness individuals, in mathematicians' jargon, are functions of them. These functions are precisely the Skolem functions of S. Hence appropriate witness individuals exist for S if and only if there exists an array of all the Skolem functions of S. Then and only then is S true (see Hintikka 2001). Some philosophers have played with the notion of a truth-maker. As far as quantificational languages are concerned, there is only one kind of truth-makers, and they are Skolem functions.

14.3 Compositionality and the meaning of quantifiers

This definition of truth is so perspicuous, and so obviously but an explication of our very own notion of truth, that one could legitimately expect that is has been acknowledged and exhaustively discussed by philosophers of our time. It boggles one's mind that this has not happened. Philosophers and logicians have discussed Tarski-type truth-definitions *ad nauseam*, notwithstanding the fact that a much simpler and much more natural truth-definition is readily available for them. Why this neglect? The reason is not that the truth definitions for first-order sentences which turn on the existence of Skolem functions cannot be formulated in the same language, for nor can a Tarski-type truth definition. The real reasons are different. One of them is Tarski's tacit insistence that truth definition must be compositional. Tarski did not spell out this requirement, but a closer examination reveals his commitment to it. (Such an examination is found in Hintikka and Sandu 1999.) In contrast, game-theoretical truth definitions of the kind explained violate prima facie the requirement of compositionality. This is shown by the definition of the Skolem form of a first-order sentence given above. In it, the selection of the arguments $y_1, y_2, \ldots$ of the new function constant do not depend only on the subformula $F[x]$, but also on which the outside universal quantifiers are on which $(\exists x)$ depends in the sentence in question. (As should be obvious, the principle of compositionality amounts essentially to the assumption of semantical context-independence.) As a consequence, the eminently natural game-theoretical definition of truth cannot be implemented without violating compositionality. In view of the popularity of compositionality among linguists and logicians, it seems likely that in different direct and indirect ways a commitment to compositionality is one of the main factors that have conspired to suppress the Skolem-function definition of truth from philosophers' attention.

Commitment to compositionality is connected with another oversight of the majority of contemporary philosophers. It is the belief that the semantics of quantifiers is exhausted by the idea that quantifiers "range over" a certain class of values. If so, the truth of a universally quantified sentence $(\forall x)\, F[x]$ reduces to the truth of all its substitution-instances F[b], where b is a member of the domain; and likewise for existentially quantified sentences. From this idea of quantificational truth it is only a short trip to Tarski-type truth definitions, which are conditioned by Tarski's requirement as to what an acceptable truth-definition must be like. What is wrong with the exclusiveness of the "ranging over" idea is that it does not address at all the other component of the semantics of quantifiers. This other component is the representation of dependencies and independencies between actual real life variables by means of the dependencies and independencies of the quantifiers to which the variables in question are bound. Such dependencies are expressed in so many words by Skolem functions. Their role in the game-theoretical truth predicate shows how the dependence relations between different variables are taken care of in the game-theoretical characterization of truth. Only when these dependence relations are looked away from will a truth-definition in terms of substitution-instances or, for that matter, a Tarski-type truth-definition, appear natural.

Another important reason for neglect of Skolem-type truth definition that logicians and philosophers have been suspicious of second-order logic and tried to stick to the first-order level. Such a goal seems to be guiding already Tarski. Now remaining on a first-order level might be a commendable aim, but it has not been implemented in the right way. Instead of second-order logic, philosophers have preferred to it set theory practiced on the first-order level. We are all familiar with Quine's misplaced quip about higher-order logic being set theory in sheep's clothing. It is turning out, Quine notwithstanding, that it is axiomatic set theory, not higher-order logic, that is the big bad wolf here. Outside philosophical fairy tales, the cold sober fact is that first-order axiomatizations of set theory cannot do an adequate job in their foundational role of capturing set-theoretical truths. Their failure is discussed in (Hintikka 2004(a)).

This failure of the only viable-looking rival means that there are no valid objections to defining truth in terms of the existence of Skolem functions. Such definitions may not be the last word on our concept of truth but they are an eminently useful first word. Their plausibility can be further enhanced by dramatizing the production of witness individuals by Skolem functions as steps in certain explicitly definable search games. The definition of these games, known as semantical games, makes the role of Skolem functions eminently intuitive. Full arrays of Skolem functions for a sentence S are precisely the winning strategies for the verifier in the semantical game $G(S)$ correlated with S and starting with S. Thus the truth of S can be defined as the existence of a winning strategy for the verifier in the game $G(S)$.

In a critical philosophical perspective, however, such a use of game-theoretical concepts is nevertheless merely a dramatization of the basic insight into the role of Skolem functions as implementing our natural notion of truth. I shall nevertheless call the definition of truth for quantificational sentences in terms of the existence of Skolem functions the game-theoretical truth definition. The game-theoretical framework is in any case useful in several respects. For one thing, it shows that the game-theoretical truth definition in not subject to criticism from an intuitionistic or from a constructivistic viewpoint. If we want for some reason or other to restrict ourselves to a constructivistic notion of truth, it can be done simply by restricting the values of Skolem function quantifiers to constructive functions (whatever they are or may be). Likewise, an intuitionistic notion of truth can be captured by restricting Skolem functions to known ones. I think that we can leave the question as to which functions are known for Brouwer and his followers to decide. One major advantage of such game-theoretical truth-definition was already noted. They allow variation in a way that captures different nonclassical conceptions of truth in a natural way. This makes it possible to compare competing logics with each other in an informed way.

14.4 IF logic as the natural basic logic

Even more importantly, when we start thinking in game-theoretical terms, we can at once see that there are lots of perfectly natural semantical games that do not correspond to any sentences of the received first-order logic. In other words, certain second-order sentences behave just like truth-conditions for nonexistent first-order sentences in terms of perfectly well-defined semantical games. For instance, the second-order sentence

(12) $(\exists f)(\exists g)(\forall x)(\forall y)F[x, f(x), y, g(x, y)]$

is the truth-condition of the sentence

(13) $(\forall x)(\exists z)(\forall y)(\exists u)F[x, z, y, u]$.

In the correlated game, the verifier is searching for a truth-making value of z on the basis of his or her (or its, if the player is a computer) knowledge of a given value of x, and searching for a value of u on the basis of his, her or its knowledge of the values of x and y. Likewise, the second-order sentence

(14) $(\exists f)(\exists g)(\forall x)(\forall y)F[x, f(x), y, g(y)]$

asserts the existence of a winning strategy in a similar game whose only novelty is that the search for the second "witness individual" is carried out with the verifier's knowledge limited to the value of y. Such games are perfectly well defined, and the second-order (14) is related to them precisely the same way as (12) is related to the semantical game played with (13). Once we see this, we can see that we can formulate first-order sentences related to (14) in the same way as (12) is related to (13), as soon as we relax our notation so as to allow a quantifier (Q_2y) to be independent of another quantifier, say (Q_1x), even though it occurs its syntactical scope. This can be done by writing it (Q_2y/Q_1x). Then the first-order counterpart to (14) is expressible as

(15) $(\forall x)(\exists z)(\forall y)(\exists u/\forall x)F[x, z, y, u]$

which is equivalent to

(16) $(\forall x)(\forall y)(\exists z/\forall y)(\exists u/\forall x)F[x, z, y, u]$.

Then the semantical game correlated with (15) is such that (14) expresses the existence of a winning strategy for the verifier in it.

This can obviously be generalized. The result is what has been called independence-friendly (IF) logic. For its theory, the reader is referred to Hintikka (2002(b)). IF logic is our natural basic logic. It is richer in its expressive capacities than the received first-order logic, which can be thought of as the slash-free fragment of IF first-order logic. (But cf. below.) In it, several crucial notions can be expressed that were not expressible in the received old first-order logic. For instance, the equicardinality of two sets, say α and β, can be expressed on the first-order level as follows:

(17) $(\forall x)(\forall z)(\exists y/\forall z)(\exists u/\forall x)$
$((x \in \alpha \supset y \in \beta) \& (z \in \beta \supset u \in \alpha) \& ((y = z) \leftrightarrow (x = u)))$

This may be compared with the second-order sentence serving the same purpose:

(18) $(\exists f)(\exists g)(\forall x)(\forall z)$
$((x \in \alpha \supset f(x) \in \beta) \& (z \in\supset g(z) \in \alpha) \& ((f(x) = z) \leftrightarrow (x = g(z))))$

The equivalence of (17) and (18) illustrates a most remarkable thing about IF first-order logic: It is tantamount to the $\sum_1^1$ (sigma one-one) fragment of second-order logic. It is easily shown that each sentence of this fragment has an equivalent IF first-order sentence. Conversely, each IF first-order sentence is equivalent to its on game-theoretical truth condition, which is a $\sum_1^1$-sentence. It can also be seen that the truth conditions of different IF first-order sentences can be integrated into a $\sum_1^1$-truth predicate. (It is assumed here that we are dealing with an IF first-order language strong enough to express its own syntax.) Since that predicate has an equivalent in the corresponding IF first-order language, which can admit of a truth predicate definable in the same language (see Hintikka 1998 and 2001). Thus the notion of truth and its definability are put to a radically new light by the simple step of allowing quantifiers to be independent of each other. In every other respects, we can preserve all the classical semantical rules, as they must be formulated in game-theoretical terms.

14.5 Two negations

But the notion of truth is not the only one which now has to be re-examined. Another one is the notion of negation. With respect to it, we are in for an intriguing surprise. The concept of negation that results from perfectly "classical" semantical rules where independence is allowed does not obey the law of excluded middle. Is IF logic therefore "nonclassical"? The truth is that there is no obvious definition of "classical" that we could appeal here to decide the issue, unless we resort to the quaint old sense of the word as referring to what is taught in classrooms. Since the negation $\sim$ used in IF logic obeys the most classical semantical rules imaginable and yet violates *tertium non datur*, the right conclusion to be drawn here that the law of excluded middle is not part and parcel of "classical" logic.

This strong negation $\sim$ has to be distinguished from the familiar contradictory negation $\neg$. The same distinctions must be extended to conditionals and equivalences. A conditional "If A, then B" may have the logical force of either $(\sim A \vee B)$ or $(\neg A \vee B)$, and an equivalence can mean either $(A\&B) \vee (\sim A\& \sim B)$ or $(A\&B) \vee (\neg A\&\neg B)$.

Here we are witnessing yet another apparently trivial question which nevertheless leads to surprising new perspectives. What has been found out is that there is a strong (dual) negation implicit in all our use of the basic logical notions. It is the negation that naturally goes together with the game-theoretical concept of truth which was seen to be but an implementation of our pretheoretical notion of truth. Such a strong negation must thus be tacitly present also in the logic of ordinary language. This strong negation is in a game-theoretical perspective even more fundamental than contradictory negation. But if so, how is the contradictory negation to be handled in our explicit logic? How come that the negation that is present in natural language in the sense of having syntactical markers for it is the contradictory one? How is the contradictory negation to be interpreted semantically in GTS? And what is there to be said in the light of the distinction about the conditionals of ordinary language?

Three possibilities can be investigated here separately. The first is suggested by the naturalness of the game-theoretical truth condition also when applied to natural language. It suggests that appearances notwithstanding it is the game-theoretical semantics that governs also the semantics of natural language, including the behavior of negation and conditionals in them. But how can this make much difference? When no slashes are present, the difference between $\sim$ and $\neg$ should not make any difference. Indeed, the received first-order logic can apparently be identified with the slash-free fragment of IF first-order logic. So how can the distinction between $\sim$ and $\neg$ make any difference for slash-free sentences like (1)–(6) or for inferences like (8)? An answer here is that the distinction between $\sim$ and $\neg$ makes no difference for slash-free formulas only if it is assumed that atomic sentences obey the law of excluded middle. If they do not, there is a difference after all. Among other things, the same sentences are no longer logically true. And it is in fact easy to ascertain that then (2) is no longer logically true, (5) and (6) no longer logically equivalent and (8) no longer a valid inference. (The "then" here means of course taking $(A \supset B)$ to mean $(\sim A \vee B)$.)

Thus independence-friendly logic offers an interesting general perspective on the different mini-paradoxes of first-order logic. They can be dissolved if we assume that the conditionals of natural language are of the form $(\sim A \vee B)$ rather than $(\neg A \vee B)$, that is, that they are IF conditionals (as we will call them) rather than traditional ones. This dissolution strongly suggests that the logic of ordinary language is primarily independence-friendly logic rather than Frege-Russell one. This result is especially interesting philosophically in the case of (7)-(8), that is, in the case of sororities paradox. There exists a large and inconclusive literature on this paradox and on its variants. It is often surmised that the paradox should not arise in connection with predicates like "bald" which are unsharp, that is, whose attribution to a particular case need not always be either true or false. (This is sometimes expressed by speaking of "truth-value gaps".) However, no simple way of implementing this idea is found in the literature. Now we can see that the failure of *tertium non datur* for the predicate in question is after all that is needed to disarm the paradox, assuming that the basic logic of natural language is IF first-order logic. This assumption is in fact strengthened by its success in disarming the prima facie paradoxes of first-order logic. This observation can be generalized. The IF first-order logic that has been examined promises to be a far more natural logic of unsharp concepts than the so-called fuzzy logic of Lofti Zadeh. (see e.g. Zadeh and Yager 1991.) Of perhaps what can be said here is that the logic of natural language we are in effect already using can serve as a "fuzzy logic" better than its trade name variant without any additional assumptions or constructions.

Another well-known paradox is likewise disarmed by IF first-order logic. It is the liar paradox. When we use IF logic in a theory of elementary arithmetic, we can of course formulate a truth predicate $W[x]$ for it in the same arithmetic. Hence by means of the diagonal lemma we can formulate the Gödel-type sentence

(19) $\sim W[\mathrm{g}]$

whose Gödel number is g and which says that the sentence with the Gödel number g is false. (In (19) g is the numeral that represents g.) The sentence (19) is true if false, and false if true. Hence it must be neither true nor false, which is perfectly possible in IF logic. No contradiction is hence forthcoming.

But now it might at first seem that the extended IF first-order logic must run afoul of the so-called strong liar paradox. In elementary arithmetic using IF logic we can formulate a truth predicate, that is a predicate $W[x]$ that applies to the Gödel number $x = g(S)$ of an arithmetical sentence S if and only if S is true. Why cannot we apply the diagonal argument to the contradictory negation of $W[x]$ so as to obtain a sentence that so to speak says "I am not (i.e. contradictorily not) true"? The answer is that one cannot prefix $\neg$ to an open formula like $W[x]$, only to closed sentences. Hence the crucial liar sentence (Gödel-type self-referential sentence) is in this case ill formed. Again, no contradiction is in the offing.

Thus we have found an excellent first approximation to the logic of natural language. It is not the "ordinary" (i.e. received) first-order logic but the slash-free part of IF first-order logic, with the tacit provision that the predicate constants may be unsharp, that is, may fail to obey the law of excluded middle. The only negation used in this logic is the dual negation $\sim$. This nevertheless makes a difference only when the given predicate constants fail to conform to the *tertium non datur*.

Thus we have found exceedingly simple solutions to some of the oldest and most intriguing puzzles of the entire canon of logic. These solutions might at first seem too good meaning too simple to be true. Now I firmly believe that these solutions are definitive ones, but I also believe that further discussion is needed to back them up and to put them into perspective. But in order not to trivialize the issues that discussion must not pertain to the details of the paradoxes or to the purported lines of reasoning that lead into them. The solutions I have explained depend essentially on only one assumption. This assumption is that the natural, preferred logic of ordinary language is IF logic. Hence the further discussion that is needed here should pertain to the status of IF logic as compared with alternatives to it, especially when it comes to the treatment of negation. The rest of this paper is accordingly devoted to certain extensions of IF logic that might seem to have a claim to be our genuine *Sprachlogik*.

Indeed, it is unmistakable that the contradictory negation is needed in the semantics of natural languages. Hence we have to develop an explicit formal logic that will involve $\neg$ and not only $\sim$. A minimal step in that direction is to introduce $\neg$ by a fiat. The result is what has been called the extended IF logic. Studying it is the second one of the three lines of thought mentioned above. However, since there cannot be any game rules for $\neg$, the semantics of extended IF logic will have to be introduced by the bland metalinguistic stipulation that $\neg S$ is true if and only if S is not true. And here the italized not must itself be a contradictory negation. Hence the semantics of $\neg$ can along these lines be specified only by relying on the same notion in a metalanguage.

By the same token, the contradictory negation can in the extended IF logic occur only sentence-initially. For if it occurred otherwise, its semantics would presuppose a game rule for it. With proper care, it is possible to relax this requirement somewhat, however, as long as does not occur within the syntactical scope of any quantifier. If we now assume that the logic of natural language is like the extended IF logic, a number of phenomena in natural language become explainable. Some of them are mentioned in Hintikka (2002(a)), especially the fact that contradictory negation is in natural languages a barrier to anaphora.

The extended IF first-order logic is an interesting logic in its own right. It is obviously equivalent to the $(\sum_1^1 \cup \prod_1^1)$ fragment of second order logic. It might at first sight seem rather similar to the unextended IF first-order logic. On a closer examination, however, the differences are seen to be profound. Most of the "nice" metatheorems that hold for IF first-order logic are no longer valid in the extended IF logic, such as compactness, upwards Skolem-Löwenheim theorem, and the separation theorem. We will return to this matter, but it can already now be seen that the extension in question is important.

This is connected with the expressive richness of the extended IF first-order logic. In order to see this richness, consider an attempt to reconstruct the entire simple theory of types on the first-order level, construing it as a many-sorted first-order logic with different sorts. The structure of types is easy to specify on the first-order level. The only thing that cannot be expected by ordinary first-order logic is the requirement that for each arbitrary class of n-tuples of entities of a certain type there exists the embodiment of that class on the next higher type (order) level. This requirement can obviously be implemented by means of sentences of the extended IF first-order logic. This logic is therefore in a sense as rich as the entire theory of all finite types, and hence capable of codifying most traditional mathematics.

14.6 Extending IF logic with the help of tertium non datur

However, it seems clear that the extended IF logic cannot be the last word here. On the one hand, it is unsatisfactory simply to introduce $\neg$ by a fiat, without giving any account of the actual rules by means of which its semantics is determined. On the other hand, it can easily be seen that the logic of natural language is richer than even the extended IF first-order logic, in that what is unmistakably a contradictory negation can occur within the scope of quantifiers. The most obvious case in point is offered by negative quantifiers like no. If someone says

(20) nobody has the winning lottery ticket

it does not mean that everybody has something else. It simply means that it is not the case that someone has the winning ticket. Such a sentence therefore has the logical form

(21) $(\forall x)\neg W[x]$

Since W[x] is allowed here to be an IF formula which is not necessarily true or false for different substitution values for x, (21) is not necessarily equivalent to

(22) $\neg(\exists x)W[x]$

Now the semantics of (21) is not determined by the game-theoretical rules of IF first-order logic. It was just seen that it is not determined by the semantics of the extended IF first-order logic, either. How, then, can a natural semantics be defined for sentences like (21)? An eminently natural answer is available here. It can be approached from two different directions. The general question concerns the interpretation of sentences S_0 where $\neg$ is allowed to occur within the scope of quantifiers. One way of doing so is by considering a hierarchy of semantical games. The first begins with S_0 and comes to an end either with an atomic sentence or with a sentence (closed sentence) of the form $\neg S_1$. (This sentence usually is a substitution-instance of a subformula of S_0.) The truth-value of $\neg S_1$ is either true or false, and it is determined by the facts of the subordinate game $G(S_1)$ which can then be handled in the same way. In other words, $\neg S_1$ is deemed true for the purposes of $G(S_0)$ if and only if there exists no winning strategy for the verifier in $G(S_1)$, otherwise false.

For instance, a play of the game with the sentence (21), i.e. of the game $G((\forall x)\neg W[x])$, will stop after a choice of an individual (say b) by the falsifier. This endpoint sentence is of the form $\neg W[b]$. It is true if and only if there exists no winning strategy for the verifier in the game $G(W[b])$, otherwise false.

It is immediately seen that on this interpretation (21) is true if and only if all the sentences of the form $\neg W[b]$ are true, where b is a constant representing some member of the domain. This assigns a meaning to (21) game-theoretically if $W[x]$ does not contain $\neg$, for in $\neg W[b]$ the contradictory negation is then sentence-initial (i.e. prefixed to a closed formula). Otherwise we are dealing with a clause in a recursive truth-definition.

What all this amounts to is that the interpretation that extends our semantics to nested contradictory negations is a kind of substitutional interpretation quantifier. It will be called in the following the substitutional interpretation quantifier, but without thereby prejudicing its precise relation to what has in the past been called the substitutional interpretation quantifier of quantifiers. In the simplest cases it does coincide with substitutional interpretations in the received sense. In such cases, the truth of a universally quantified sentence is tantamount to the truth of all its substitution-instance, and the truth of an existentially quantified one with the truth of at least one of its substitution-instances. Restricting one's attention to such simple cases has led some philosophers to the conclusion that there is no deep difference between substitutional and objectual interpretations of quantifiers. (For this kind of view, see Kripke 1976.) This is nevertheless a mistake belied by what is found in IF logic. Even in the absence of contradictory negation, a strict inside-out (recursive) definition of truth in a substitutional sense is impossible in the presence of irreducible independence. This is especially blatant in the case of mutually dependent quantifiers. Their logic can be considered a strict counter-example to the substitutional interpretation quantifier of quantifiers.

In any case, in the presence of independence and dependence indicators there will have to become restrictions on the occurrence of $\neg$. The main such restriction is that no quantifier outside the scope of a given occurrence of $\neg$ can depend on a quantifier inside its scope. More generally, the scopes of different occurrences of $\neg$ must be nested, that is, they must form a tree structure.

14.7 Elementary versus non elementary logics

The logic definable in this way will be called the fully extended IF first-order logic. It calls for a number of explanations and comments.

First, it might be tempting to consider the fully extended first-order logic as the natural logic of ordinary language. This temptation is perhaps strengthened by the belief that something like the substitutional interpretation quantifier of quantifiers constitutes their natural semantics. There is perhaps a true element to this temptation. However, it is not the whole story. For one thing, the usual substitutional interpretation quantifier of quantifiers relies on the assumption that the semantics of quantifiers is exhausted by the "ranging over" idea. This interpretation hence cannot do justice to the role of quantifiers as expressing relations of dependence and independence between variables. It is therefore only a part of the story. Indeed, it is a secondary part, for the solutions to the mini-paradoxes of first-order logic outlined above strongly suggests that our basic logic operates like the unextended IF logic and not like its full version. The substitutional component in the truth definition for the full IF logic is thus an additional ingredient over and above the game-theoretical conception of truth codified by the existence of Skolem functions. The naturalness of this game-theoretical truth definition is the best testimony against relying too much on the "ranging over" idea alone. Indeed, we can now begin to appreciate the reasons for the complexity of the semantics of natural languages. This semantics is a mixture of different ingredients, where the basic game-theoretical conception of truth is supplemented by essentially different substitutional ideas.

The specious plausibility of the substitutional interpretation quantifier of quantifiers may perhaps be partly dispelled by asking what has to be known in order to understand a quantificational sentence or such an interpretation. The most important part of the answer is that the domain of individuals (aka the universe of discourse) has to be known. This rules out all uses of quantifiers where their range is open-ended. Such an open-endedness does not make it impossible to play semantical games and to understand statements as to what can or cannot happen in them. Accordingly, a quantificational language can be understood and used even when the language users do not know precisely what the domain is. Hence game-theoretical semantics of quantifiers is more widely applicable than a substitutional one.

If it strikes you as outlandish to apply a quantificational language without a sharply defined universe of discourse, you can contemplate Aristotle's syllogistic logic. No idea of a sharply delineated range of quantifiers was presupposed there, and existential force was not carried primarily by the particular "quantifier" but by the predicate term. Or think of higher-order quantifiers in our own logic. What precisely is their range? Believers in the standard interpretation in Henkin's sense will give you one answer, believers in nonstandard interpretation another one. And in this case the idea of a quantifier "ranging over" its values does not help us very much.

These remarks are not calculated to show that the substitutional interpretation quantifier is not possible or that it is not interesting. However, they suggest strongly that it is not the whole story of our pretheoretical understanding of quantifiers.

This point can be elaborated further. In a sense semantical games are what I shall call concrete processes. They are playable by actual humans. Indeed, Charles S. Peirce already envisaged them as games between a human "proponent" and an equally human "interpreter" (e.g. *Collected Papers* 3.479-482, 5.542, see Hilpinen 1982). (For most applications, it is nevertheless more natural to think of them as games between a human agent ("knower") and nature in the familiar game-theoretical sense of "games against nature".) Each play of these games which is connected with a finite sentence consists of a finite number of concrete moves. The players need not be initially familiar with all the members of the domain in which the game is played. One can in fact develop an instructive fallibilist epistemology starting from the assumption that the only information an inquirer receives about the world are the outcomes of the semantical games the inquirer plays against nature with different strategies. All this is possible even when the given domain is infinite. The crucial point is that the domain never plays any role in these activities as a closed totality.

It need not be assumed that every member of the domain has a name. It suffices to require that when a player of a semantical game chooses an individual from the domain, the players can give it a name and amplify their language by adjoining the newly coined name to it. Only a finite number of such extensions is needed in any play of a semantical game. Admittedly, for a truth definition we need the totality of the strategies that the inquirer has at his, her or its disposal. But if one's logical conscience is sensitive, one can restrict this strategy set to strategies that are constructive, known, computable or otherwise in conformity with one's principles of logical morality. (Or should I say, one's moral logic?) Apart from that qualification, the basic features of IF logic should be acceptable to everyone. That everyone includes Wittgenstein, for whom (as was pointed out) basic semantical games must be humanly playable. In the spirit of virtual history, I cannot help wondering how much greater progress the philosophy of logic would have experienced if Wittgenstein had realized that semantical games are the true logical home of our basic logical concepts. (Wittgenstein associated an importance to the activities of seeking and finding, but he related them to the notion of object rather than to the nature of quantifiers. see Hintikka, forthcoming (c).)

The game-theoretical truth definition and unextended IF logic should likewise be acceptable to intuitionists, at least if we allow them to restrict the verifier's strategies to known ones. No infinite operations are involved in playing the semantical games that are the logical home of unextended IF first-order logic.

In contrast, the substitutional interpretation quantifier (and its objectual counterparts) involve the given domain as a complete totality. For instance, the substitutional truth condition (21) requires that $W[b]$ is true, for each and every name b of a member of the domain. If different variables range over different classes of values, all the totalities of such values have to be considered as completed totalities. It has to be assumed, importantly, that all the individuals (members of the domain) have names. Thus substitutional truth conditions are infinitistic and nonconstructive in a way the game-theoretical truth predicate is not. With a side-glance at Hilbert, it may be said that the classes of values of the different first-order quantifiers are the only true "ideal objects" we need in mathematics.

What has been found has major repercussions for the traditional philosophy of mathematics. For one thing, it is seen that the main source of trouble is not the infinity of the domain of numbers. Semantical games can be played on infinite models and not only on finite ones. It makes sense to speak for instance of seeking and finding a rational number of a certain kind. Conversely, IF logic is one of the few rivals of the ordinary first-order logic that affects also the theory of finite models for first-order formulas. For such reasons, I am not calling the approach favored here finitistic. The infinity of the domain is not the main issue.

However, in contrast Brouwer seems to have been right in blaming a large part of the interpretational problems in the foundations on the unrestricted use of the tertium non datur principle in mathematics. For the substitutional interpretation quantifier with all its infinitistic burdens becomes unavoidable only when we have to interpret contradictory negations occurring within the scope of quantifiers.

This point must be pushed deeper, however. As I have argued elsewhere, intuitionistic logic should be thought of, not as the logic of mathematical truths per se, but as a logic of our knowledge of mathematical (and logical) entities. Now the game-theoretical truth definition shows that the crucial entities in the logic of mathematics are Skolem functions. Therefore we can hope to interpret sentences in a logical language intuitionistically only as long as their semantics can be formulated by reference to Skolem functions. (Of course we may have to restrict them to known functions.) Now it is precisely when we begin to use contradictory negation in arbitrary positions that we cannot any longer interpret our sentences by reference to Skolem functions.

However, Brouwer's insights have not been implemented in an instructive way in the earlier discussion. The best known way of attempting to do so has been to set up an "intuitionistic logic" to replace ordinary first-order logic. (see e.g. Heyting 1956.) But this is barking up the wrong logic. Of course ordinary first-order logic has to be generalized so as to become a fragment of IF first-order logic. But understood as such a fragment, there is nothing wrong with ordinary first-order logic.

On the other hand, even when an explicit intuitionistic logic is formulated, it does not do the job of IF first-order logic, either. It does not capture adequately the epistemic element in Brouwer's thinking, and more importantly it does not deal any better than ordinary first-order logic with the representation of dependence and independence relations between variables, either.

This failure is not automatically avoided by the introduction of the substitutional interpretation quantifier of quantifiers, either. Even though the substitutional interpretation quantifier is objectionable to an intuitionist, it does not make any difference in ordinary first-order logic. Hence Brouwer's objections to classical logic can be met by using a game-theoretical interpretation rather than a substitutional one. It is only when independence indicators are present that the two interpretations, the game-theoretical and the substitutional one, differ from each other. It is therefore only then that the tertium non datur principle begins to depend on distinctly infinitistic element into the logic of classical mathematics. It does so because the tertium non datur principle can only be backed up by the substitutional interpretation quantifier. But this is a much deeper issue than what can be handled by tinkering with the inference rules of ordinary first-order logic. Hence the real target of intuitionistic criticism ought to be the substitutional interpretation quantifier of quantifiers rather than the inference rules of first-order logic.

This point is somewhat obscured by the fact that prima facie failures of the law of excluded middle can also be caused by the epistemic element in intuitionistic logic. A failure of the law of excluded middle is hence merely a symptom of trouble. The need of a substitutional interpretation quantifier is the trouble.

The infinitistic character of substitutional first-order logic also makes it a poor candidate for the role of the true logic of ordinary language and ordinary discourse. Here I can in fact appeal to many of the standard finitistic arguments once one of the persisting mistakes in this area is eliminated. This mistake confuses the infinity of the domain with the infinity of the operations we need to carry out to apply our logic and our languages to it . This mistake is undoubtedly due to the more general mistake of assuming that the semantics of quantifiers must be explained by reference to their "ranging over" all the members of the domain. What matters is the question whether infinite operations are needed to apply our logic, not what the cardinality of the domain is to which it is being applied. As long as our logic is purely game-theoretical, its application can be thought of as being implemented by plays of semantical games. Such plays involve a finite number of moves that in principle can be carried out by human players. Hence the possible infinity of the domain does not matter.

Thus there is no obstacle to thinking of IF logic as the natural logic of our colloquial language. But the application of a substitutionally interpreted quantificational language in an infinite domain presupposes infinite operations. If so, it cannot very well be the logic of ordinary discourse for in our ordinary thinking we cannot even in principle rely on the assumption that infinite operations actually are carried out. IF logic is the natural logic of natural language, and thereby supports the solutions of the paradoxes discussed above.

14.8 Fully extended IF logic is equivalent to second-order logic

We nevertheless have to take the substitutional interpretation quantifier seriously in general logical theory. There are in fact interesting further insights to be reached here concerning first-order logics that rely on substitutional interpretation quantifier. We have defined a hierarchy of *first-order* sentences with an increasingly complex structure of nested *contradictory negations*. How is it related to the *quantificational* hierarchy ($\sum_n^1 - \prod_n^1$ hierarchy) of *second-order* sentences? The surprising answer turns out to be: the two hierarchies are equivalent.

The validity of this result is in fact fairly easy to see. It is well known that (and how) a $\sum_1^1$-sentence can be reduced to an IF first-order sentence. Furthermore, a $\prod_1^1$-sentence is equivalent to a sentence of the form $\neg S$, where S is an IF first-order sentence without contradictory negations. By the same token, if each $\sum_n^1$-sentence S is equivalent to a sentence of the full IF first-order logic with $n-1$ layers of contradictory negations, then clearly each $\prod_n^1$ sentence has an equivalent translation of the form $\neg S$.

Furthermore, consider a $\sum_n^1$ sentence. By definition, it has the form of a string of existential second-order quantifiers followed by a $\prod_{n-1}^1$ formula. Now these second-order existential quantifiers can be replaced by first-order independent quantifiers in the same way as in showing that each $\sum_1^1$ sentence is equivalent to an IF first-order sentence.

To illustrate this step, assume that the given $\sum_n^1$ sentence is

(23) $(\exists f)F[f]$

where f is a zero-argument function variable and $F[f]$ is a $\prod_{n-1}^{1}$ formula. It is assumed that $F[f]$ is in the negation normal form. Then (23) is clearly equivalent to the following sentence:

(24) $(\forall x)(\forall y)(\exists z/\forall y)(\exists u/\forall x)((x = y) \supset (z = u))\ \&\ F^*[x, y, z, u])$

Here x, y, z, u are new variables not occurring in $F[f]$ and F^* is obtained from $F[f]$ as follows:

(i) Every occurrence of a subformula of the form $(f(w) = v)$ is replaced by $((x = w) \supset (z = v))$ and likewise for subformulas of the forms

$$(v = f(w))$$

$$(f(w) = a)\ (a = f(w)), \text{etc}$$

(ii) Every occurrence of an atomic subformula of the form $A(f(w))$ is replaced by $((x = w) \supset (A(z)))$, and likewise for other kinds of atomic subformulas containing an argument of the form $f(w)$ or $f(a)$. Nested functions are handled in the same way as in the $\sum_1^1$ case. Predicates can be handled by means of their characteristic functions.

Thus the entire second-order logic turns out to be equivalent to the substitutional first-order logic. We shall call this result the *negation reduction* of second-order logic to the first-order level. (An essentially equivalent result is proved in Väänänen 2001.) It throws light on several questions in the foundations of logic and mathematics. One of them is the relation of first-order logic to higher-order logics. We may in fact think of this problem as one of the Einsteinian questions mentioned in the beginning of this paper, that is, questions that are so subtle that they appear trivial. The obvious-looking way of answering this question by saying that what distinguishes the two is the ontological status of the entities which our quantifiers range over in the two kinds of logic. In first-order logic, they are particulars (individuals); in the second-order logic, the values of quantified variables can also be sets or relations of particulars or functions from particulars to particulars. What can be simpler that this? Admittedly, in first-order axiomatic set theory sets are admitted as values of first-order variables. However, such a set theory is turning out to be a disaster area when considered as a foundational project. (see Hintikka, 2004(a).) Again, the peculiarities of higher-order logic come to play only when a standard interpretation in Henkin's sense is imposed on ranges of higher-order variables. If that is not done, higher-order logic can in effect be dealt with as if it were a many-sorted first-order logic. Apart from such qualifications, the distinction between first-order logic and higher-order logic seems to be exhaustively characterized by reference to the categorical (ontological) status of the values of the variables of quantification.

The result we have reached shows that the first-order vs. higher-order distinction is in reality more complicated than that. (This point has been emphasized aptly by Jouko Väänänen (2001).) The reduction to the first-order level marks a definite gain in conceptual clarity. Any philosophical nominalist will rejoice at this reduction. Among other things, the reduction shows that foundationally speaking we do not have to worry in second-order logic about the thorny question of the existence or nonexistence of different kinds of higher-order entities, such as sets. All we are trafficking in are different kinds of structures of particular objects. This satisfies one of the major desiderata of Hilbert (1996, p. 1121) who blamed the entire *Grundlagenkrise* on the use of higher-order entities by Frege, Dedekind and Cantor.

Hilbert's worry about mathematicians' reliance on higher-order quantification has not received the attention it deserves. Apparently the problem of the existence of higher-order objects has been tacitly transformed to the technical-looking question as to what existence assumptions to make in axiomatic set theory. The reappearance of serious problems in the foundations of set theory shows that this attempted transformation does not help us. The negation reduction shows that substitutional first-order logic satisfies Hilbert's wishes, even though it involves serious other problems.

In a different direction, the negation reduction vindicates the status of second-order logic as genuine logic. Since the existence of higher-order entities like sets plays no role in it, we do not need any set theory to back it up. This reinforces our reversal of Quine's quip. To deal with sets as if they were particular objects is to admit dangerous higher-order conceptualizations in sheep's clothing. (How very dangerous they are is shown in Hintikka, 2004(a).) In contrast, the prima facie dangerous second-order quantifiers turn out to be reducible in a sense to the sheepish first-order level.

14.9 Logic and mathematical reasoning

In a foundational perspective results like the negation reduction aid and abet mightily the cause of logicism. Admittedly, some of the earlier formulations of the tenets of logicism are now inapplicable, including those that claim that all the axioms we need in mathematics are theorems of a suitable axiomatization of logic. There is no such thing as a complete axiomatization of even the (unextended) IF first-order logic. In other words, our basic logic is semantically incomplete. Hence it makes no sense to speak of an axiomatic reduction of mathematics to logic. The real question is whether all the different conceptualizations and all the different modes of reasoning used in mathematics can be reconstructed by means of logic.

It is usually thought and said that all modes of reasoning needed in mathematics can be represented by means of second-order logic. If so, the negation reduction theorem shows that they can in at least one natural sense be reduced to modes of logical reasoning, which is precisely what logicists are supposed to claim.

This is perhaps not the last word on the subject, however. On the one hand, the substitutional first-order logic which is the target of the reduction is not unproblematic philosophically. On the other hand, it is not obvious that literally all assumptions that can be considered in mathematics can in fact be captured by means of second-order logic. For instance, it is not immediately clear that maximality assumptions like Hilbert's Axiom of Completeness (Hilbert 1903) can be so formulated. However, even apart from such qualifications, the results reached here, especially when they are combined with the realization of the failure of first-order axiomatic set theory to capture set-theoretical truths (see Hintikka 2004(a)), show impressively the fundamental role of logic in mathematical reasoning.

At the same time, the negation reduction raises our awareness of what separates concrete unproblematic reasoning from questionable one. What makes the difference was seen not to be the finitude of the domain. Now it is seen not to lie (Hilbert notwithstanding) in the first-order character of unproblematic reasoning, either, for all second-order reasoning and hence virtually all mathematical reasoning can in principle be conducted on first-order level. The crucial step is to allow contradictory negations into the scopes of quantifiers or, to use logicians' jargon, to "quantify into" a context governed by a contradictory negation.

The negation reduction theorem is of interest also from the vantage point of hierarchy theory. (For it see Addison 1961, and forthcoming). In this theory, different quantifier hierarchies are studied comparatively. Now by utilizing the notion of informational independence (independence of a quantifier on another one) it can be shown that certain important quantifier hierarchies are equivalent to hierarchies of contradictory negation. This seems to open a possibility of extending the scope of the entire hierarchy theory.

There has been in the literature some discussion of the question whether IF first-order logic is perhaps "really" (part of) higher-order logic. The results reached here show that the entire question is ill formulated. By the ontological criterion, IF first-order logic is unproblematically first-order, for all values of bound variables in its semantics are individuals (particular members of the domain). But if we do not go by this criterion, the notions of "first-order" and "second-order" have to be redefined. In the light of the results reached here, it might be maintained with a greater plausibility that the entire second-order logic is "in reality" but full IF first-order logic in a different notation. It seems that those philosophers who claim that IF logic is "in reality" second-order logic are tacitly requiring that the semantics of genuine first-order logic must be compositional and must rely exclusively on the "ranging over" idea. If so, the upshot of the line of though carried out here is to show how hopelessly restrictive such a conception of first-order is. Such a truncated first-order logic will not cut much ice even as the supposed logic of ordinary discourse.

It has to be admitted, however, that the borderline between first-order logic and second-order logic is much less sharp than first meets an untrained eye. This interplay of the two logics is manifested in the role of Skolem functions in the theory of first-order logic. It is also natural to generalize the rule of existential instantiation so as to allow the introduction of new function constants and not only new individual constants. (The instantiating "witness individuals" may depend on other individuals.) Moreover, first-order formulas can entail second-order formulas, even existential ones. This is exemplified by the fact that each first-order sentence logically implies its own Skolem form, as (9) implies

(25) $(\exists f)(\forall x)F[x, f(x)]$

Similar crossings are not found where two parts of logic are truly separated. For instance, no positive epistemic conclusion is implied by non-epistemic premises, and similarly for other parts of logic.

14.10 A prescriptive postscript

But what do all these results have to do with the title notion of this volume, the notion of alternative logic? The answer depends on what this singularly ill-defined term is taken to mean. In its most superficial sense, an alternative logic is any logic different from the received logic which is usually taken to be the "classical" first-order logic. What has been found in this paper (and in its predecessors) shows that in this sense the term "alternative logic" is an oxymoron. This is shown once and for all by the independence-friendly (IF logic) examined in this paper. IF logic is not an alternative to the received first-order logic. Rather, IF logic replaces the received "classical" first-order logic and accommodates it as a special case. The received first-order logic turns out to be a result of logicians' failure to acknowledge the important limitations that restrict the expressive power of the "classical" first-order logic. These limitations also show that the received first-order logic does not deserve this honorific appellation, unless the term "classical" is taken in one of its earlier senses as "what is taught in class-rooms".

In a somewhat less insipid sense an alternative logic is often taken to be a logic whose system of axioms and inference rules is different from the "classical" one. But what is a valid rule of inference? It is in its usual sense a rule that preserves truth. (It is already a symptom of an invidious confusion that in the literature the preservation of truth is not always distinguished from the preservation of logical truth.) Admittedly, in some cases what is to be preserved is merely probable truth or truthlikeness. However, those variations do not make an essential difference to the line of thought pursued here. If a putative rule of inference does not satisfy such a preservation requirement, it can scarcely serve any realistic purpose in the applications of logic and hence should not be called rule of inference. But if so, if truth-preservation is a condition sine qua non of a rule of inference, the validity of rules of inference has to be studied in a semantical theory of the language in which the inferences are couched, for the notion of truth belongs to semantics (model theory). More fully expressed, the study of inferences involving certain logical notions must turn on the role these notions play in the way reality is represented in language. Now what is in this perspective the semantical task of quantifiers, those central notions of our basic logic? It is usually thought that the semantical function of quantifiers is exhausted by their variables' "ranging over" a class of values. This idea is among other places epitomized by Frege's unfortunate idea that quantifiers are higher-order predicates whose task is to express whether lower-order predicates are empty or not.

In reality, this "ranging over" is only a part of the real job description of quantifiers. The other part of what quantifiers do is to express through their formal dependence or independence of each other the real-life dependence or independence of their respective variables. Once this is realized, it is easily seen that the received Frege-Russell notation does not allow the representation of all possible patterns of dependence and independence between variables. It hence fails to do full justice to the meaning of quantifiers. What IF logic does is to eliminate this shortcoming. Unlike the "classical" first-order logic, it fulfills the whole task of quantifier logic and not only a part of it.

Since this is precisely the task that any general logic of quantifiers has to accomplish, there cannot be any genuine alternatives to IF logic, either. What look like such alternatives, principally intuitionistic logic and constructivistic logic, can be construed as resulting from restricting the modes of dependence between variables, perhaps to knowable ones or constructive ones. If this is taken to be a sufficient reason to label them "alternative logics", there is no need to object, as long as their character as variants of IF logic is acknowledged.

Thus the verdict is clear as far as independence-friendly logic is concerned. IF logic is not a logic of certain kinds of games. It is a study of Skolem functions of quantifier sentences. These functions receive their significance from the fact that they codify relations of dependence between different variables. Rightly understood, IF logic is not alternative to any other logic, nor does it have any genuine alternatives.

This leaves most of the so-called alternative logics still unexplained. It is impossible to do justice to all of them here, but it might be in order to try to indicate what is interesting about them. As an example, what is known as the theory of circumscription will do. An inference from certain premises to a conclusion by circumscription relies, over and above the information that the premises convey, on a tacit assumption to the effect that the premises provide all the relevant information. This is a contingent assumption, not a logical or even necessarily a common sense truth. As any puzzle fan knows, often the solution of a puzzle requires precisely a violation of the sufficiency presumption in that it requires the presence of a factor not foreshadowed in the given information.

Now at first such inference might not seem to require any new principles of reasoning. Indeed, reasoning from partly tacit premises is one of the oldest topics of logical theory. It has an established name, viz. enthymemic reasoning. Why should circumscriptive reasoning nevertheless require a special alternative logic? Perhaps it does not. What is peculiar and interesting about it is that there does not seem to be any way of expressing the tacit sufficiency premise in the language in which the circumscriptive inferences are carried out.

The theory of circumscriptive inference is therefore an attempt to elicit information from a premise which is not only unspoken but unspeakable in the language used, by introducing special rules of inference. This is an intriguing enterprise, independently of whether it is deemed fully successful or not, but it need not involve a logic alternative to our old ones. It is a branch of the theory of enthymemic inference, viz. the branch which studies inferences in which the tacit premise is not expressible in the language in which the reasoning is carried out.

This can be generalized to several other highly interesting "logics". Probabilistic inductive logic depends on assumptions concerning the orderliness of one's universe of discourse. In the simplest case, such an assumption is codified in the constant of Carnap's λ-continuum. The choice of a value of λ codifies such a regularity or irregularity assumption, even when it is not expressed in the form of a proposition in the explicit language of inductive reasoning.

In mathematics, we find fascinating assumptions that are not expressible, or at least not easily expressible, by means of the usual mathematical and logical concepts in the mathematical notation itself. They are assumptions of extremality (maximality and minimality). Hilbert's struggles with his "Axiom of Completeness" in geometry vividly illustrate this problem. Extremality assumptions have not given rise to a new logic except perhaps in Hintikka (1993). They can nevertheless play the same role as the tacit premises of circumscriptive logic or inductive logic. What is common to all these "alternative logics" is that they are methods of eliciting consequences of certain tacit assumptions. They are not theories of inference in general; they are chapters of a theory of enthymemic reasoning. It is not even clear that they involve in the last analysis any peculiar modes of logical inference.

In the light of these observations, perhaps what you should do next time when you are tempted to use the expression "alternative logic", is to look for an alternative locution.

References

Addison, John W. (1960). "The theory of hierarchies", in E. Nagel, A. Tarski, and P. Suppes, editors, *Logic, Methology and Philosophy of Science, Proceedings of the 1960 International Congress*, Stanford University Press, Stanford, 26–37.

Addison, John W., (forthcoming). "Tarski's theory of definability: Common themes in descriptive set theory, recursive function theory, classical pure logic, and finite-universe logic".

Chomsky, Noam, (1986). *Knowledge of Language*, Praegen, New York.

Henkin, Leon (1950). Completeness in the theory of types. *Journal of Symbolic Logic* vol. 15, 81–91.

Heyting, A. (1956). *Intuitionism: An Introduction*, North-Holland, Amsterdam.

Hilbert, David, (1903). *Grundlagen der Geometrie*, Zweite Auflage, Leipzig. (First ed., 1899. The axiom of completeness was added in the second edition.)

Hilbert, David, (1996: original 1922), "The new grounding of mathematics", translation from German, in William B. Ewald, editor, *From Kant to Hilbert 1–2*, Clarendon Press, Oxford, 1115–1134.

Hilpinen, Risto, (1982). "On C.S. Peirce's theory of propositions", in Eugene Freeman, editor, *The Relevance of C.S. Peirce*, The Hegeler Institute, La Salle, 264–270.

Hintikka, Jaakko, (1993). "New foundations for mathematical theories", in J. Väänänen and J. Oikkonen, editors, *Logic Colloquium 90: Lecture Notes in Logic*, no. 2, Springer, Berlin, 122–144. [ASL summer meeting in Helsinki.]

Hintikka, Jaakko, (1996). *The Principles of Mathematics Revisited*, Cambridge University Press, Cambridge.

Hintikka, Jaakko, (1997). "No scope for scope", *Linguistics and Philosophy* vol. 20, 515–544.

Hintikka, Jaakko, (1998). "Truth-definitions, Skolem functions, and axiomatic set theory", *Bulletin of Symbolic Logic*, vol. 4, 303–337.

Hintikka, Jaakko, (2001). "Post-Tarskian Truth", *Synthese*, vol. 126, 17–36.

Hintikka, Jaakko, (2002a). "Negation in logic and in natural language", *Linguistics and Philosophy*, vol. 25, 585–600.

Hintikka, Jaakko, (2002b). "Hyperclassical logic (aka IF logic) and its implications for logical theory", *Bulletin of Symbolic Logic*, vol. 8, 404–423.

Hintikka, Jaakko, (2004a). "Independence-friendly logic and axiomatic set theory", *Annals of Pure and Applied Logic*, vol. 126, 313–333.

Hintikka, Jaakko, (2004b). "On the epistemology of game-theoretical semantics", in J. Hintikka, T. Czarnecki, K. Kijania-Placek and A. Rogszczak, editors, *Philosophy and Logic: In Search of the Polish Tradition: Essays in Honour of Jan Wolenski on the Occasion of his 60th Birthday*, Synthese Library 323, Kluwer Academic Publishers, Dordrecht, 57–66.

Hintikka, Jaakko, (forthcoming), "The crash of the philosophy of the *Tractatus*".
Hintikka, Jaakko and Jack Kulas, (1983). *The Game of Language*, D. Reidel, Dordrecht.
Hintikka, Jaakko, and Gabriel Sandu, (1996). "Game-theoretical semantics", in J. van Benthem and Alice ter Meulen, editors, *Handbook of Logic and Language*, Elsevier, Amsterdam, 361–410.
Hintikka, Jaakko, and Gabriel Sandu, (1999). "Tarski's guilty secret: compositionality", in J. Wolenski and E. Köhler, editors, *Alfred Tarski and the Vienna Circle*, Kluwer Academic Publishers, Dordrecht, 217–230.
Kripke, Saul, (1976). "Is there a problem about substitutional quantification?", in G. Evans and J. McDovell, editors, *Truth and Meaning*, Oxford University Press, New York, 325–419.
Peirce, Charles S., (1931-1966). *Collected Papers of Charles Sanders Peirce*, 8 vols., ed. By C. Hartshorne, P. Weiss, and A.W. Burks, Harvard University Press, Cambridge.
Quine, W.V., (1970). *Philosophy of Logic*, Prentice Hall, Englewood.
Tarski, Alfred, (1935). "Der Wahrheitsbegriff in den formalisierten Sprachen", *Studia Philosophica*, vol. 1, 261–405.
Väänänen, Jouko, (2001). "Second-order logic and foundations of mathematics", *Bulletin of Symbolic Logic*, vol. 7, 504–520.
Zadeh, Lofti, and R.R. Yager, editors, 1991, *An Introduction to Fuzzy Logic: Applications in Intelligent System*, Kluwer Academic Publishers, Dordrecht.

Chapter 15

SIGNALLING IN IF GAMES: A TRICKY BUSINESS

Theo M.V. Janssen
University of Amsterdam
T.M.V.Janssen@uva.nl

Francien Dechesne
Tilburg University &
Technische Universiteit Eindhoven
f.dechesne@tue.nl

15.1 Introduction

IF logic is introduced by J. Hintikka and advocated in a number of publications; the main ones are [Hintikka, 1996] and [Hintikka and Sandu, 1997]. The difference with predicate logic concerns the interdependency of quantifiers. In predicate logic quantifiers may depend on the quantifiers in whose scope they occur, e.g. in the quantifier sequence $\forall x \exists y \psi$ the value chosen for y may depend on x. So scope indicates possible dependencies. It is not possible in predicate logic to express that a quantifier should be independent of another one. In IF logic it is possible to express that an existential quantifier is independent of a preceding universal quantifier (IF abbreviates 'Independence Friendly'); the issue of (in)dependence also applies to disjunctions.

In IF logic existential quantifiers (and disjunctions) are by definition independent of other existential quantifiers. A natural generalization of IF would be that these can at choice be dependent or independent of existential quantifiers. We indicate this generalization by IFG (IF Generalized). The generalizations that are investigated by [Hodges, 1997] and [Caicedo and Krynicki, 1999] go even further and allow for independency with universal quantifiers and conjunctions.

J. van Benthem, G. Heinzmann, M. Rebuschi and H. Visser (eds.), The Age of Alternative Logics: Assessing Philosophy of Logic and Mathematics Today, 221–241.

IF and IFG are logics with unexpected properties. [Hodges, 1997] has a section called *Deathtraps*, and says (p. 546): 'the idea "what a player is allowed to know", though it has strong intuitive content, can be very misleading'. [Caicedo and Krynicki, 1999] state (p. 22): 'One has to be careful about wrong extrapolations from classical semantics.' [Janssen, 2002] concludes (p. 375): 'The examples given in the previous sections show that strategies are on several points in conflict with intuitions on independence'.

The cause of the unexpected properties lies in a phenomenon which is called 'signalling', and which will be explained in the next sections. Although [Caicedo and Krynicki, 1999] have warned the reader, they fell in the trap of wrong extrapolations. Several of their theorems are incorrect, and this is, in our opinion, not due to some accidental oversight, but because the deathtraps of signalling are not well known.

In this paper we will present several tricky examples of signalling. The results of [Caicedo and Krynicki, 1999] and the fundamental claim by Hintikka that IF is a conservative extension of predicate logic all are incorrect due to signalling. Probably they can reformulated in a weaker sense ([Caicedo *et al.*, to appear]).

15.2 What is signalling?

In the context of game theoretical semantics for IF, 'signalling' is the phenomenon that the value of a variable one is supposed not to know, is available through the value of another variable. Below we present the earliest example of this phenomenon, but first we explain informally the game theoretical interpretation of IF and IFG. Formal definitions will be given in section 15.5.

The interpretation of a formula proceeds by a game between two players; $\forall$belard and $\exists$loise. $\forall$belard aims to refute the formula, $\exists$loise to confirm it. We suppose the formula to be in negation normal form. In that case $\forall$belard makes the choices for $\forall$ and $\wedge$, and $\exists$loise for $\exists$ and $\vee$. If a choice is to be made independent of the values of certain variables, that is indicated by mentioning those variables after a slash that is attached to the quantifier or connective. For instance, in $\exists x_{/y}$ the x has to be chosen independent of y, and in $\vee_{/x}$ a disjunct has to be chosen independent of x. Again, $\exists$loise has to make the choices in these cases, and $\forall$belard for the other ones. A formula is defined to be true if $\exists$loise has a winning strategy (and false if $\forall$belard has a winning strategy, but this will hardly play any role in the paper).

Time for an example. We start with a classical formula:

$$\forall x \exists y[x = y] \tag{15.1}$$

The game proceeds as follows. First $\forall$belard chooses a value for x, and next $\exists$loise chooses one for y. If she chooses the same value as $\forall$belard she wins this play. As a matter of fact, that method always is successful; it is a winning strategy, and therefore the formula is true.

An IFG formula that resembles (15.1) is:

$$\forall x\, \exists y_{/x}[x = y] \tag{15.2}$$

Here $\exists$loise has to make her choice independently of x. Maybe she by accident, but she does not have a winning strategy for the game. Hence the formula is not true. It is not false either, because also $\forall$belard has no winning strategy.

Consider now

$$\forall x \exists z\, \exists y_{/x}[x = y] \tag{15.3}$$

A vacuous quantifier is inserted, and in classical predicate logic that would make no difference; its truth value remains unchanged. But in IFG it makes a difference. The winning strategy for $\exists$loise is to choose z to be equal to x, which is allowed because no independence restrictions are put on z. Next she chooses y to be equal to z, which is allowed because she does not ask for the value of x. The result is that x equals y. So by proceeding in this way, $\exists$loise has a winning strategy. The strategy recognizes the value of z as a *signal* for the value of x, and uses that signal to choose the value of y. For original IF logic this example of signalling is not possible (because existential quantifiers are there by definition independent of each other), but we will meet several other examples concerning IF logic where signalling is used.

Example (15.1) was the first example of signalling that was discovered [Hodges, 1997] p. 548. The example of a wrong extrapolation from classical logic given by [Caicedo and Krynicki, 1999] p. 22 has a completely different appearance, but in fact it is a reformulation of (15.1), so it is based upon signalling. [Janssen, 2002] gives many other examples of signalling, and so does the present paper.

15.3 Signalling is needed

Before we consider cases where signalling causes a problem, it first will be illustrated that signalling is an essential component of the semantics of IF and IFG.

Consider (15.4), interpreted in a model with 2 elements: $\{0, 1\}$.

$$\forall x[x = 1 \vee x \neq 1] \tag{15.4}$$

This formula is classically true, and it is so in IF and IFG semantics: $\exists$loise can make her choice based upon the value of x, and has as strategy: if $x = 1$ then L else R.

A related example is:

$$\exists u[u = 1 \vee u \neq 1] \tag{15.5}$$

This formula is true in classical logic, and it is true in IF. But the winning strategy is *not*, as you might expect, to choose for $\vee$ the left disjunct in case $u = 1$ and the right disjunct otherwise. This is not allowed: the choices of $\exists$loise are in IF by definition independent of her own earlier choices. This independence is analogous to the formation of the Skolem form for classical predicate logic: the Skolem function for an existential quantifier has as arguments only variables that are universally quantified. For example: $\forall x \exists y \exists z[x < y < z]$ has Skolem form $\exists f \exists g \forall x[x < f(x) < g(x)]$ The fact that z $(= g(x))$ is greater than y $(= f(x))$, is accounted for by first recalculating internally in g what $f(x)$ is; one might say $g(x) = g'(x, f(x))$.

As explained, when playing (15.5) it is not allowed to recalculate the value of u for the decision on $\vee$. What can than be a winning strategy? The solution is to use constant strategies: for $\exists u$ the value 0, and for $\vee$ always choose R.

In IFG the same problem arises for (15.6) where it is made explicit that the value of u may not be used for making a choice for the disjunction. In (15.6) the same strategy is winning as for (15.5).

$$\exists u[u = 1 \vee_{/u} u \neq 1] \tag{15.6}$$

We return to IF, and make the example more complicated. Consider:

$$\forall x \exists u[u = x \wedge [u = 1 \vee u \neq 1]] \tag{15.7}$$

This formula is classically true, but for IF semantics the situation seems difficult. The constant strategies from the previous example will not work. $\exists$loise cannot take a strategy that always yields the same value for u because it must satisfy $u = x$, and the strategy for $\vee$ must vary with the value of u. The solution is to use the value of x as signal for the value of u. $\exists$loise wins by choosing u equal to x, and making for $\vee$ the choice L if $x = 1$ and R otherwise.

Example (15.7) is an important example because it illustrates the need for signalling in IF. The language of IF logic is an extension of the language of predicate logic, and it is claimed to be a conservative extension [Hintikka, 1996] p. 65. Without signalling $\exists$loise has no winning strategy in example (15.7). *Without* signalling, (15.7) would be a formula without slashes that is not true in IF, whereas classically it is true; it would be a counterexample to the claim of conservative extension. This shows: signalling is needed.

This story can directly be transferred to IFG. The IF example (15.7) is reformulated in IFG as (15.8), where of course the same strategy is winning.

$$\forall x \exists u[u = x \wedge [u = 1 \vee_{/u} u \neq 1]] \tag{15.8}$$

When [Hodges, 1997] proposes to switch to IFG, he says (p. 22): 'Obviously this won't diminish the expressive power of the language.' But in order to allow a reconstruction of IF in IFG, we must have the same possibilities for signalling in IFG (for (15.8)) as in IF (for (15.7)).

So, signalling has effects that cannot easily be missed: giving it up would cause considerable changes in the semantics. An alternative semantics, in which signalling cannot occur, is given in [Janssen, 2002] but at the cost of loosing some of the expressive power. However, the argument given above for signalling is not the end of the story: in sections 15.4.4 and 15.7.4 we will show that even *with* signalling there are problems with the claim of conservative extension.

15.4 Counterexamples

In the literature one finds some theorems and claims which are incorrect due to the fact that the possibilities of signalling are not well known. In this section we present the counterexamples informally; in section 15.7 we will prove the results in a more technical way. For instance, in this section we do not quote the original versions of the theorems, and furthermore we will not give complete proofs here. Although we will prove that certain formulas are true, by providing winning strategies for ∃loise, we will not prove that certain formulas are *not* true. The reason is that negative proofs are more difficult to obtain: the collection of possible strategies is rather unwieldy. Instead we will (following [Caicedo and Krynicki, 1999]) first define (in section 15.5) an alternative for the game interpretation which they also use for formulating their theorems: an interpretation which uses sets of valuations.

15.4.1 Renaming of bound variables

Claim 1. (cf. Lemma 3.1a [Caicedo and Krynicki, 1999] p. 26)
Let Qx be $\forall x$ or $\exists x$ where the quantifiers may be with or without a slash. If ϕ is a sentence in which $Qx[\psi(x)]$ occurs as subformula, and z does not occur (free or bound) in ψ, then the subformula $Qx[\psi(x)]$ may be replaced by $Qz[\psi(z)]$ without changing the truth value.

We will show that this claim is *incorrect*. A counterexample is given by the following two formulas, where (15.10) is obtained from (15.9) by replacing s by y. They are played on a model with 2 elements: $\{0, 1\}$, and it does not matter whether you see them as IFG or IF formulas.

$$\forall x \forall y \forall z[x \neq y \vee \exists s\, \exists u_{/x}[u = x \wedge s = z]] \tag{15.9}$$

$$\forall x \forall y \forall z[x \neq y \vee \exists y\, \exists u_{/x}[u = x \wedge y = z]] \tag{15.10}$$

In (15.9) the winning strategy is as follows. We let $f_\vee \equiv$ *if* $x \neq y$ *then* L *else* R; $f_{\exists s} \equiv s := z$ and $f_{\exists u_{/x}} \equiv u := y$. This strategy is winning because y signals the value of x to $\exists u_{/x}$. In (15.10) the corresponding strategy is $f_{\exists y} \equiv y := z$. That strategy is not winning because when it comes to the choice of u, y is equal to z and not to x: the $\exists y$ blocks the signal from $\forall y$.

We expect that a general version of the renaming theorem is not possible. No matter how we restrict the choice of the new variable: there might always be a context in which this new variable blocks a signal from outside. So we always run the risk of changing the truth value in some context if we rename the variable. In order to obtain a kind of renaming theorem, we have to change the notion of equivalence. Instead of the absolute notion 'equivalent in all contexts', a stricter notion seems to be required.

15.4.2 Prenex normal form

In predicate logic the prenex normal form theorem describes how quantifiers can be shifted to the front of a formula, e.g. $\forall x[\phi] \vee \psi$ can be replaced by $\forall x[\phi \vee \psi]$, under the condition that x does not occur free in ψ. In IF and IFG these two formulas are not necessarily equivalent because quantifiers in ψ become dependent on $\forall x$. Therefore [Caicedo and Krynicki, 1999] present a rephrasing.

Claim 2. (cf. Lemma 3.1.c,d [Caicedo and Krynicki, 1999] p. 26)
Let Qx be $\forall x$ or $\exists x$ where the quantifiers may be with or without a slash and let $\phi_{/x}$ denote the result of adding to all quantifiers in ϕ the independence condition $_{/x}$. Then any subformula of the form $[Qx[\psi] \vee \theta]$ can be replaced by $Qx[\psi \vee \theta_{/x}]$ without changing the truth value.

This quantifier extraction claim is *incorrect* because it may interfere with signalling. There are two ways in which this can happen.

The first counterexample will show that in $Qx[\psi \vee \theta_{/x}]$ the x can be used to send signals to $\theta_{/x}$. Indeed, the formulation of the claim given above, was intended to prevent this, but as we have seen in section 15.3, also decisions on disjunctions may depend on signals. The second counterexample is based upon the possibility that in $Qx[\psi \vee \theta_{/x}]$ the quantifier may block signals from outside to $\theta_{/x}$.

Counterexample 1

This first example is based upon a situation where a formula that is not true, becomes true by quantifier extraction. It is an IFG example; the situation for IF formulas still has to be investigated further. Consider:

$$\forall z[\forall x[x \neq z] \vee \exists u_{/z}[u = z \vee_{/z} u \neq z]] \tag{15.11}$$

It will be clear that no z satisfies the left disjunct. In the right disjunct there is for $f_{\exists u_{/z}}$ no argument available, so it is a constant strategy. Furthermore $f_{\vee_{/z}}$ is not allowed to depend on z, but it may depend on u. Since $f_{\exists u_{/z}}$ is constant, this means that $f_{\vee_{/z}}$ is constant: always L or always R. For at least one value of z such a choice will not be winning, hence ∃loise has no winning strategy; so, formula (15.11) is not true.

The result of the quantifier extraction transformation in (15.11) is:

$$\forall z \forall x[x \neq z \vee \exists u_{/z,x}[u = z \vee_{/z} u \neq z]] \tag{15.12}$$

This formula is true as the following strategy shows. Let $f_\vee \equiv$ *if* $x \neq z$ *then* L *else* R. So either the left disjunct is satisfied, or the right disjunct has to be satisfied in a situation where $x = z$. For u we take an arbitrary (but fixed) value. For $\vee_{/z}$ the value of x can be used as a signal: $f_{\vee_{/z}} \equiv$ *if* $u = x$ *then* L *else* R. So if the players arrive at the subformula $u = z$, this one is satisfied because $u = x$ and $x = z$, and if they arrive at the subformula $u \neq z$, that one is satisfied because $u \neq x$ and $x = z$. So these choice functions for the disjunctions form a winning strategy for ∃loise.

Note that in variant (15.13) of (15.12) for $f_{\vee_{/z,x}}$ no information about x can be used. So (15.13) is not true, just as the sentence in which the quantifier extraction has not yet taken place, viz. (15.11).

$$\forall z \forall x[x \neq z \vee \exists u_{/z,x}[u = z \vee_{/z,x} u \neq z]] \tag{15.13}$$

This counterexample suggests that in claim 2 not only the quantifiers, but also the connectives have to be slashed.

Counterexample 2

This example, again an IFG example, shows that by giving ∀ wide scope a signal can be blocked. The counterexample is obtained from:

$$\forall y \exists u[\forall u[u \neq u] \vee \exists x_{/y}[x = y]] \tag{15.14}$$

For this formula the winning strategy is: $f_{\exists u} \equiv u := y$, $f_\vee \equiv R$, and let $f_{\exists x_{/y}} \equiv x := u$. Since $u = y$ it follows that $x = y$. Hence (15.14) is true in our model.

Quantifier extraction changes formula (15.14) into

$$\forall y \exists u \forall u[[u \neq u] \vee \exists x_{/y,u}[x = y]] \tag{15.15}$$

The strategy $x := u$ given above for $\exists x_{/y}$ is not allowed for $\exists x_{/y,u}$, and the formula is not true. The quantifier $\forall u$ blocks the signal $u = y$ from outside to $\exists x_{/y,u}[x = y]$.

We think that a general version of the quantifier extraction lemma (as in claim 2) is not possible. One can always construct a context in which a moved quantifier blocks a signal. Therefore we suggest investigating the idea that the normal form theorem has to be restricted to situations where this cannot arise.

15.4.3 Slashed disjunction elimination

It is claimed that slashed disjunction can be eliminated. In Hintikka's work this follows from his translation procedure from IF logic to second order logic and back (cf. [Hintikka, 1996] p. 52 and p. 62–63). [Caicedo and Krynicki, 1999] give the result within IFG.

Claim 3. (cf. Lemma 3.2, [Caicedo and Krynicki, 1999] p. 25)
$\phi \vee_{/Y} \psi \equiv_{\mathsf{G}} \exists u_{/Y} \exists s_{/Y,u}[[[u = s \wedge \phi] \vee [u \neq s \wedge \psi]] \vee \exists! u[u = u \wedge [\phi \vee \psi]]]$

The last part of the formula in the claim deals with the situation that there is one element in the model. In our counterexample we assume that the model has at least three elements, so we may neglect that part.

Consider now:

$$\forall y \forall t[\exists x_{/t}[x = t] \vee_{/y} \exists x_{/t}[x = t]] \tag{15.16}$$

The strategies for the two occurrences of $\exists x_{/t}$ may depend on y, but not on u. $\exists$loise may follow for the left occurrence of $\exists x_{/t}$ another strategy than for the right one. Because $\forall$belard can choose from at least 3 different values for t, there will always be one occurrence $\exists x_{/t}$ that $\exists$loise has to satisfy for two or more different values of t. This she cannot do, so the sentence is not true.

The proposed elimination rule changes (15.16) into (15.17), where we omitted the last part (since it is false in our model).

$$\forall y \forall t\, \exists u_{/y}\, \exists s_{/u,y}[[u = s \wedge \exists x_{/t}\, x = t] \vee [u \neq s \wedge \exists x_{/t}\, x = t]] \tag{15.17}$$

$\exists$loise has a winning strategy for this sentence: $f_{\exists u_{/y}} \equiv u := t$, $f_{\exists s_{/y,u}} \equiv s := t$, $f_{\vee} \equiv L$, and $f_{\exists x_{/t}} \equiv x := s$. This strategy never violates the independence restrictions, and it guarantees $\exists$loise to win. Note that the y plays no essential role in the strategies.

We conclude that also the rule for the elimination of slashed disjunctions has to formulated in some restricted way.

15.4.4 Conservative extension

Claim 4. ([Hintikka, 1996] p. 65.) *Technically speaking IF first-order logic is a conservative extension of ordinary first-order logic*

Also this claim is undermined by the tricky properties of signalling, as will be explained below.

A variant of example (15.7) from section 15.3 is the IFG example:

$$\forall x \exists u[u = x \wedge [u = 1 \vee_{/u} u \neq 1]] \tag{15.18}$$

This example is true: the x could be used as a signal for the value of u at the disjunction. An extended version of (15.18) is:

$$\forall x \forall y \exists u\, [u = x \wedge \forall s[s = y \vee [u = 1 \vee_{/u} u \neq 1]]] \tag{15.19}$$

This formula is again true: for the first $\vee$ choose R, and for the second $\vee$ use x as a signal for the value of u. Next we replace s by x.

$$\forall x \forall y \exists u\, [u = x \wedge \forall x[x = y \vee [u = 1 \vee_{/u} u \neq 1]]] \tag{15.20}$$

Here the signal is blocked, and consequently the formula is not true.

Next we write (15.20) in the language of IF logic, what means that the independence of $\vee$ from $\exists u$ is not expressed. Then we get:

$$\forall x \forall y \exists u \; [u = x \wedge \forall x[x = y \vee [u = 1 \vee u \neq 1]]] \qquad (15.21)$$

Above we have argued that $\exists$loise has no winning strategy for (15.20), hence not for (15.21); a proof will be given in Section 15.7.4. So, according to the IF interpretation, (15.21) is not true. At the same time it is a formula from classical predicate logic, and classically it is true. Hence IF-logic is *not* a conservative extension of predicate logic.

15.5 Definitions

15.5.1 The language

Definition 5. *The language of IFG-logic is defined as follows:*

1. The language has **variables**. Typical examples are x, y, z, u, s, and t. In general discussions variables range over a domain A of model $\mathcal{A}$, but in the examples the domain is $\{0, 1\}$ or $\{0, 1, 2\}$.
2. The language has **constants**. In general discussions typical constants are a and b. In the examples 0 and 1 are used.
3. A **term** is a variable or a constant We choose our fragment not to contain any function symbols.
4. The **relation symbols** are $R_1, R_2, \ldots$; each with a fixed arity. In the examples the binary relation symbols $=, \neq, <$, and $\leq$ are used.
5. If $t_1, \ldots, t_n$ are terms, and n is the arity of R, then $R(t_1, \ldots, t_n)$ and $\neg R(t_1, \ldots, t_n)$ are **formulas**.
6. If ψ and θ are formulas, z is a variable, and Y a set of variables, then also the following expressions are **formulas**: $\psi \wedge \theta$, $\psi \vee \theta$, $\psi \vee_{/Y} \theta$, $\forall z \psi$, $\exists z \psi$. If $z \notin Y$ then $\exists z_{/Y}\, \psi$ is a formula.

After a slash we will omit brackets of set denotations, and write $\exists y_{/x}$ and $\exists y_{/x,u}$.

Definition 6. *$FV(\phi)$ is the set of* **free variables** *in ϕ. It consists of those variables in ϕ which do not occur in ϕ bound. $FV(\phi)$ includes variables in Y's occurring in $\vee_{/Y}$ and $\exists x_{/Y}$ as far as they are unbound in ϕ.*

A comparison of our definition of the logic with the literature gives rise to the following remarks:

1. We assume all formulas to be in negation normal form, as is done in almost all publications of Hintikka and in [Väänänen, 2002]. In some other publications about variants of IF-logic negation may occur freely (e.g. [Hodges, 1997] and [Caicedo and Krynicki, 1999]). We do not need it, however, for

our discussion of signalling. A happy consequence is that there is no role switch between ∀belard and ∃loise, what makes the discussion easier to understand.

2. In Hintikka's version choices depend only on moves of the opponent, but can be made independent of them by slashing those moves away. We allow also (in)dependency between own choices. Therefore also existentially quantified variables may arise after a slash.
3. We do not consider strategies for ∀belard, as they are not necessary to make our points.

15.5.2 Playing

The main ingredient of a game is a formula from IFG logic. The aim of the game is to determine the truth of the formula in a model $\mathcal{A}$. The two players have different aims: ∃loise tries to confirm the truth and ∀belard to refute it. We are, however, not so much interested whether ∃loise accidently wins (or looses), but whether she has a winning strategy for the game, because that is the criterium whether the formula is true or not. Therefore we will describe IFG on two levels: the level of actual playing the game, where the two players move, and win or lose, and the level of a set of sets of plays where the players may have a winning strategies.

We first describe how a game is played: which player has to move in a given position, what are his/her possible moves, and what is the effect of the move. In the course of the game the players will encounter subformulas like $\psi \vee_{/Y} \theta$ or $\exists x_{/Y}\, \psi$. The subscript indicates that the choice of the move has to be made independent of the variables in Y. This is a restriction on the motivation for the choice, but not on the choice itself. Therefore in the description of playing it makes no difference whether $_{/Y}$ occurs as subscript or not. Its role will be defined when we consider strategies in section 15.5.3.

A valuation describes at least the values of the free variables in ϕ; an alternative name would be finite assignment. For valuations we mainly use v and w.

Definition 7. *A* **valuation** *v for a formula ϕ in a model $\mathcal{A}$ is a function $v \in \mathrm{dom}(\mathcal{A})^X$ where $FV(\phi) \subseteq X$.*

Definition 8. *We use the following notations concerning valuations:*

$\{x\colon a\}$	the valuation that assigns a to x ($a \in \mathrm{dom}(\mathcal{A})$)
$v * \{x\colon a\}$	the valuation obtained from v by changing the value assigned to x into a if v was defined for x, or by extending the domain of v such that it now assigns a to x.
ϵ	(*the empty valuation*) the valuation that is defined for no variable at all

$v \sim_Y w$ (v *is an* Y*-variant of* w) valuations v and w are defined for the same variables, the values they assign may differ for the variables in Y, but are the same outside Y

Definition 9. *A* **play** *is a triple* $\langle \mathcal{A}, \phi, v\rangle$ *where* ϕ *is a formula from IFG-logic,* $\mathcal{A}$ *a model and* v *a valuation where* $FV(\phi) \subseteq \text{dom}(v)$*. A* **position** *is a pair* $\langle \psi, w\rangle$*, where* ψ *is a subformula of* ϕ *and* w *a valuation where* $FV(\phi) \subseteq \text{dom}(w)$*. A* **move** *is a transition from a position to a position. The possible moves are determined by the form of* ϕ*. We distinguish the following cases:*

1. $\langle \phi, v\rangle \equiv \langle \psi \wedge \theta, v\rangle$
   ∀belard chooses L or R. If he chooses L, then the play is continued from position $\langle \psi, v\rangle$, otherwise from position $\langle \theta, v\rangle$.
2. $\langle \phi, v\rangle \equiv \langle \psi \vee \theta, v\rangle$ or $\langle \phi, v\rangle \equiv \langle \psi \vee_{/Y} \theta, v\rangle$
   ∃loise chooses L or R. If she chooses L, the play is continued from position $\langle \psi, v\rangle$ otherwise from position $\langle \theta, v\rangle$. For the role of $_{/Y}$ see section (15.5.3) below.
3. $\langle \phi, v\rangle \equiv \langle \forall x \psi, v\rangle$
   ∀belard chooses a value for x, say a, and the play proceeds from position $\langle \psi, v * \{x\colon a\}\rangle$.
4. $\langle \phi, v\rangle \equiv \langle \exists x \psi, v\rangle$ or $\langle \phi, v\rangle \equiv \langle \exists x_{/Y}\, \psi, v\rangle$
   ∃loise chooses a value for x, say b. Then the play is continued from position $\langle \psi, v * \{x\colon b\}\rangle$.
5. $\langle \phi, v\rangle \equiv \langle R(t_1, \ldots, t_n), v\rangle$
   Here the play ends. Let $a_i = v(t_i)$ if t_i is a variable, en $a_i = t_i^{\mathcal{A}}$ if t_i is a constant. If $(a_1, \ldots, a_n) \in R^{\mathcal{A}}$ then ∃loise has won the play, otherwise she has lost.
6. $\langle \phi, v\rangle \equiv \langle \neg R(t_1, \ldots, t_n), v\rangle$
   Here the pay ends. Let a_i be defined as in the previous clause. If $(a_1, \ldots, a_n)$ $\notin R^{\mathcal{A}}$ then ∃loise has won the instance of the game, otherwise she has lost.

15.5.3 The game

We now switch to the level where strategies can be defined. We consider a game as set of plays ([Hodges, 1997] calls this level a 'contest').

Definition 10. *A game is a triple* $\langle \mathcal{A}, \phi, V\rangle$*, where* $\mathcal{A}$ *is a model with domain* A*,* ϕ *a formula of IFG-logic and* V *a collection of valuations such that there is a set* X *of variables with* $FV(\phi) \subseteq X$ *and* $V \subseteq A^X$*.*

A choice function f_ϕ is a function that describes which choices ∃loise may make, depending on the values previously chosen for the variables. The requirement that a choice does not depend on the variables in a set Y is formalized by requiring that f_ϕ yields the same choice for different values assigned to the variables in Y.

Definition 11. *The possible* **choice functions** f_ϕ *for* ∃*loise at position* $\langle\phi, v\rangle$ *in a game* $\langle\mathcal{A}, \eta, V\rangle$ *are defined by the following cases:*

$\phi \equiv \psi \vee \theta$	$f_{\psi\vee\theta}\colon V \to \{L, R\}$
$\phi \equiv \psi \vee_{/Y} \theta$	$f_{\psi\vee_{/Y}\theta}\colon V \to \{L, R\}$ such that from $v_1 \sim_Y v_2$ follows that $f_{\psi\vee_{/Y}\theta}(v_1) = f_{\psi\vee_{/Y}\theta}(v_2)$.
$\phi \equiv \exists x\psi$	$f_{\exists x\psi}\colon V \to A$, where A is the domain of $\mathcal{A}$.
$\phi \equiv \exists x_{/Y}\, \psi$	$f_{\exists x_{/Y}\,\psi}\colon V \to A$, where A is the domain of interpretation and $v_1 \sim_Y v_2$ implies $f_{\exists x_{/Y}\,\psi}(v_1) = f_{\exists x_{/Y}\,\psi}(v_2)$.

We say that a choice function f *is* **independent of** Y *on* V *if for all* $v, w \in V$ *from* $v \sim_Y w$ *follows* $f(v) = f(w)$.

Definition 12. *A* **strategy** F_ϕ *for a game* $\langle\mathcal{A}, \phi, V\rangle$ *is a collection choice functions* $\{f_\psi\}_{\psi\in Sub(\phi)}$ *which for each subformula* ψ *where* ∃*loise has to make a choice, provides a choice function* f_ψ. *Different occurrences of* ψ *in* ϕ *have their own choice function.*

Definition 13. *A strategy* F_ϕ *is called a* **winning strategy** *in game* $\langle\mathcal{A}, \phi, V\rangle$ *if playing in accordance with that strategy guarantees* ∃*loise to win in all possible plays* $\langle\mathcal{A}, \phi, v\rangle$, $v \in V$, *of the game. Notation:* $\mathcal{A} \models_{\mathsf{G}} \phi[V, F_\phi]$.

Definition 14. *Sentence* ϕ *is called* **game-true**, *shortly 'true', if there exists a winning strategy* F_ϕ *such that* $\mathcal{A} \models_{\mathsf{G}} \phi[\{\epsilon\}, F_\phi]$.

15.6 Valuations

In this section we will present an alternative definition for IFG that resembles the classical definition of satisfiability using valuations. One of the reasons is that several theorems from [Caicedo and Krynicki, 1999] are formulated with such a definition. There is a difference however between the use of valuations for IF (and IFG), and in the traditional approach to predicate logic. Whereas classically a formula is interpreted with respect to a single valuation, for IF and IFG this will be done with respect to a set of valuations. Therefore several notions from definition 8 are lifted to the level of sets.

Definitions 15. *The following notations concern sets of valuations:*

$\{xy\colon aa, bb\}$	(*is an example of the explicit notation we use for a set of valuations*) the set of valuations that assign to x and y identical values from $\{a, b\}$
$V * \{x\colon a\}$	$\{v * \{x : a\} \mid v \in V\}$
$V * \{x\colon A\}$	$\{v * \{x : a\} \mid v \in V, a \in A\}$

$V \sim_Y W$ (V *is an Y-variant of W*) for each $v \in V$ there is a $w \in W$ such that $v \sim_Y w$ and for each $w \in W$ there is a $v \in V$ such that $w \sim_Y v$.

In order to express a counterpart of independence in the realm of valuations, we borrow the notion of Y-saturatedness from [Caicedo and Krynicki, 1999]. That a choice function is independent of Y will correspond with having a certain partition of its domain into Y-saturated sets (see definition 20).

Definition 16. *A subset W of V is called Y***-saturated in** *V if for all $w, v \in V$ from $w \sim_Y v$ and $w \in W$ follows that also $v \in W$.*

Lemma 17. *Let V_1 and V_2 be Y-saturated subsets of V. Then $V_1 \cup V_2$, $V_1 \cap V_2$, and $V_1 \setminus V_2$ are Y-saturated subsets in V.*

Lemma 18. *The equivalence classes in V of the relation $\sim_Y$ are Y-saturated.*

Definition 19. *The partition $V/_{\sim_Y}$ of V into $\sim_Y$ equivalence classes is called the Y***-saturated partition** *of V.*

We are now prepared for the definition of the interpretation in a model $\mathcal{A}$ of a formula ϕ with respect to set a of valuations V; notated as $\mathcal{A} \models_{\mathsf{V}} \phi[V]$. Note that the subscript $_{\mathsf{V}}$ is fixed part of the notation, whereas for V any denotation for a set of valuations can be used.

Definition 20. *Let $\mathcal{A}$ be a model with domain A, ϕ a formula, X a set of variables for which $FV(\phi) \subseteq X$, and $V \subseteq A^X$. Then ϕ* **is true under** *V* **in** *$\mathcal{A}$, notated $\mathcal{A} \models_{\mathsf{V}} \phi[V]$, iff:*

1. For atomic ϕ
   $\mathcal{A} \models_{\mathsf{V}} R(t_1, \ldots, t_n)[V]$ iff for all $v \in V$ we have $(a_1, \ldots, a_n) \in R^{\mathcal{A}}$, where $a_i = v(t_i)$ if t_i is a variable, and $a_i = t_i^{\mathcal{A}}$ if t_i is a constant.
   $\mathcal{A} \models_{\mathsf{V}} \neg R(t_1, \ldots, t_n)[V]$ iff for no $v \in V$ we have $(a_1, \ldots, a_n) \in R^{\mathcal{A}}$, where a_i as just defined.
2. $\mathcal{A} \models_{\mathsf{V}} [\psi \wedge \theta][V]$ iff $\mathcal{A} \models_{\mathsf{V}} \psi[V]$ and $\mathcal{A} \models_{\mathsf{V}} \theta[V]$.
3. $\mathcal{A} \models_{\mathsf{V}} [\psi \vee \theta][V]$ iff $V = V_1 \cup V_2$ for some V_1 and V_2, such that $\mathcal{A} \models_{\mathsf{V}} \psi[V_1]$ and $\mathcal{A} \models_{\mathsf{V}} \theta[V_2]$.
4. $\mathcal{A} \models_{\mathsf{V}} [\psi \vee_{/Y} \theta][V]$ iff $V = V_1 \cup V_2$ for some V_1 and V_2, such that V_1 and V_2 are Y-saturated in V and $\mathcal{A} \models_{\mathsf{V}} \psi[V_1]$ and $\mathcal{A} \models_{\mathsf{V}} \theta[V_2]$.
5. $\mathcal{A} \models_{\mathsf{V}} \forall x \phi[V]$ iff $\mathcal{A} \models_{\mathsf{V}} \phi[V * \{x \colon A\}]$.
6. $\mathcal{A} \models_{\mathsf{V}} \exists x \psi[V]$ iff there is a $W \sim_x (V * \{x \colon A\})$ such that $\mathcal{A} \models_{\mathsf{V}} \psi[W]$
7. $\mathcal{A} \models_{\mathsf{V}} \exists x_{/Y} \psi[V]$ iff $V = \cup_i V_i$, where each V_i is Y-saturated in V and for each i there is an a_i such that $\mathcal{A} \models_{\mathsf{V}} \psi[\cup_i (V_i * \{x \colon a_i\})]$.

It might be helpful to compare the clause 7 with clause 4: the slashed existentional quantifier is seen as a slashed disjunction for the different values x may take. Clause 6 could be expressed in an analogous way.

Note that if we let $V = \emptyset$, this inductive definition of satisfaction yields, for any IFG-formula, $\mathcal{A} \models_{\mathsf{V}} \phi[\emptyset]$. This might look anomalous, but it is actually necessary for the situation with disjunction, where the empty sets of valuations can occur if V is split into V and $\emptyset$. Be aware that this is different from saying that formulas are always satisfied by the singleton set $A^{\emptyset} = \{\epsilon\}$: this is not the case. In fact, satisfaction with respect to $A^{\emptyset}$ is only *defined* for formulas with no free variables, i.e. sentences.

Definition 21. *For ϕ a sentence, we say that ϕ is* **valuation true** *iff $\mathcal{A} \models_{\mathsf{V}} \phi[\{\epsilon\}]$. Notation: $\mathcal{A} \models_{\mathsf{V}} \phi$.*

Definition 20 differs for clause $\exists x_{/Y}$ essentially from the definition in [Caicedo and Krynicki, 1999]. Their definition has a typo that turns $_{/Y}$ into a vacuous addition, and, after the obvious correction, gives rise to a difference between the game interpretation and the valuation interpretation in certain cases. We will not go into details.

Next we establish the equivalence of the game interpretation and the valuation interpretation.

Theorem 22. *A sentence ϕ is game true iff ϕ is valuation true.*

Proof.
We will show that for any formula ϕ and for all $V \subseteq A^X$ with $FV(\phi) \subseteq X$ the statements (i) and (ii) are equivalent.
(i) $\mathcal{A} \models_{\mathsf{V}} \phi[V]$
(ii) there is a winning strategy F_ϕ such that $\mathcal{A} \models_{\mathsf{G}} \phi[V, F_\phi]$
In particular this shows for sentences ϕ: $\mathcal{A} \models_{\mathsf{V}} \phi[\{\epsilon\}]$ *iff* there is a winning strategy F_ϕ such that $\mathcal{A} \models_{\mathsf{G}} \phi[\{\epsilon\}, F_\phi]$, which proves the theorem.

Proof (⇒).
We only consider the clauses where a choice function for ∃loise has to be designed.

3. $\mathcal{A} \models_{\mathsf{V}} \psi \vee \theta[V]$
   By definition 20 there are V_1 and V_2 such that $\mathcal{A} \models_{\mathsf{V}} \psi[V_1]$ and $\mathcal{A} \models_{\mathsf{V}} \theta[V_2]$. Then, by induction hypothesis, there are winning strategies F_ψ and F_θ such that $\mathcal{A} \models_{\mathsf{G}} \psi[V_1, F_\psi]$ and $\mathcal{A} \models_{\mathsf{G}} \theta[V_2, F_\theta]$. Define $f_{\psi\vee\theta}(v) = L$ if $v \in V_1$, and R otherwise. Let $F_\phi = \{f_{\psi\vee\theta}\} \cup F_\psi \cup F_\theta$. Then $\mathcal{A} \models_{\mathsf{G}} \phi[V, F_\phi]$.
4. $\mathcal{A} \models_{\mathsf{V}} \psi \vee_{/Y} \theta[V]$
   Follow the construction from the previous case. Since V_1 and V_2 are Y-saturated in V, so is $V_1 \setminus V_2$, hence $f_{\psi\vee_{/Y}\theta}$ indeed is independent of Y on V.

6. $\mathcal{A} \models_{\mathsf{V}} \exists x\psi[V]$
   According to clause 6 of definition 20 there is a $W \sim_x (V * \{x\colon A\})$ such that $\mathcal{A} \models_{\mathsf{G}} \psi[W]$. By induction hypothesis there is a winning strategy F_ψ such that $\mathcal{A} \models_{\mathsf{G}} [W, F_\psi]$. For each $v \in (V * \{x\colon A\})$ choose a $w_v \in W$ such that $w_v \sim_x v$. Define $f_{\exists x\psi}(v) = w_v(x)$. Then $\mathcal{A} \models_{\mathsf{G}} \exists x\psi[V, \{f_{\exists x\psi}\} \cup F_\psi]$

7. $\mathcal{A} \models_{\mathsf{V}} \exists x_{/Y}\, \psi[V]$
   Let $\{W_j\}$ be the Y-saturated partition of V (see def. 19). Let V_i and a_i be as in clause 7 of definition 20. For each W_j choose a V_i such that $W_j \subseteq V_i$ and define $b_j = a_i$. Since $\mathcal{A} \models_{\mathsf{V}} \psi[V_i * \{x\colon a_i\}]$ we have $\mathcal{A} \models_{\mathsf{V}} \psi[W_j * \{x\colon b_j\}]$, and since the W_j are pairwise disjunct $\mathcal{A} \models_{\mathsf{V}} \psi[\cup_j (W_j * \{x\colon b_j\})]$. By induction hypothesis there is a winning strategy F_ψ such that $\mathcal{A} \models_{\mathsf{G}} \psi[\cup_j (W_j * \{x\colon b_j\}), F_\psi]$. Define $f_{\exists x_{/Y}\, \psi}(v) = b_j$ if $v \in W_j$. This defines a function because the sets W_j are pairwise disjunct, and this function is independent of Y because the sets W_j are Y-saturated. Hence $\mathcal{A} \models_{\mathsf{G}} \psi[V, \{f_{\exists x_{/Y}\, \psi}\} \cup F_\psi]$.

Proof ($\Leftarrow$).
We consider only the cases where $\exists$loise applies her strategy.

3. $\mathcal{A} \models_{\mathsf{G}} (\psi \vee \theta)\, [V, F_{\psi\vee\theta}]$.
   Let $V_1 = f^{-1}_{\psi\vee\theta}(L)$ and $V_2 = f^{-1}_{\psi\vee\theta}(R)$, so $V = V_1 \cup V_2$. Then for the substrategy from F_ϕ for ψ, viz. F_ψ, holds $\mathcal{A} \models_{\mathsf{G}} \psi[V_1, F_\psi]$. Analogously $\mathcal{A} \models_{\mathsf{G}} \theta[V_2, F_\theta]$. By induction hypothesis $\mathcal{A} \models_{\mathsf{V}} \psi[V_1]$ and $\mathcal{A} \models_{\mathsf{V}} \theta[V_2]$. Then V satisfies the conditions of clause 3 in definition 20, hence $\mathcal{A} \models_{\mathsf{V}} (\psi \vee \theta)[V]$.

4. $\mathcal{A} \models_{\mathsf{G}} (\psi \vee_{/Y} \theta)\, [V, F_{\psi\vee_{/Y}\theta}]$.
   Define V_1 and V_2 as in clause 3 above. Since $f_{\psi\vee_{/Y}\theta}$ is independent of Y in V, sets V_1 and V_2 are Y-saturated in V. So V_1 and V_2 satisfy the conditions of clause 4 in definition 20, hence $\mathcal{A} \models_{\mathsf{V}} (\psi \vee_{/Y} \theta)[V]$.

6. $\mathcal{A} \models_{\mathsf{G}} \exists x\psi[V, F_{\exists x\psi}]$
   Let B be the range of $f_{\exists x\psi}$ and define $V_b = f^{-1}_{\exists x\psi}(b)$ for each $b \in B$. Then $\mathcal{A} \models_{\mathsf{G}} \psi[\cup_b (V_b * \{x\colon b\}), F_\psi]$. By induction hypothesis we know that $\mathcal{A} \models_{\mathsf{V}} \psi[\cup_b (V_b * \{x\colon b\})]$. So $W = \cup(V_b * \{x\colon b\})$ satisfies clause 6 from definition 20. Hence $\mathcal{A} \models_{\mathsf{V}} \exists x\psi[V]$.

7. $\mathcal{A} \models_{\mathsf{G}} \exists x_{/Y}\, \psi[V, F_{\exists x_{/Y}\, \psi}]$.
   Let V_b be as in clause 6 above. Since $f_{\exists x_{/Y}\, \psi}$ is independent of Y in V, the sets V_b are Y-saturated in V. So $V = \cup_b V_b$ satisfies the conditions from clause 7 in definition 20. Hence $\mathcal{A} \models_{\mathsf{V}} \exists x_{/Y}\, \psi[V]$

15.7 Proofs

15.7.1 Renaming bound variables

[Caicedo and Krynicki, 1999] p. 26, present a result that bound variables under standard conditions can be renamed. They formulate a general version for formulas with free variables, and therefore a general notion of equivalence is needed. Their definition requires that the two expressions for all assignments agree on truth ($\models^+_{\mathsf{G}}$ what is our $\models_{\mathsf{G}}$) and falsehood ($\models^-_{\mathsf{G}}$). For completeness of information we also quote the falsehood part, but in our discussion only truth plays a role.

Quote 23. ([Caicedo and Krynicki, 1999] p. 24) *Formulas ϕ and ψ are* G-equivalent, *notated as $\phi \equiv_{\mathsf{G}} \psi$, if and only if for any set V of valuations on a fixed domain including $FV(\phi) \cup FV(\psi)$ in a structure $\mathcal{A}$, we have $\mathcal{A} \models^+_{\mathsf{G}} \phi\,[V] \iff \mathcal{A} \models^+_{\mathsf{G}} \psi\,[V]$, and $\mathcal{A} \models^-_{\mathsf{G}} \phi\,[V] \iff \mathcal{A} \models^-_{\mathsf{G}} \psi\,[V]$.*

The renaming theorem allows renaming by a fresh variable:

Quote 24. (Lemma 3.1a [Caicedo and Krynicki, 1999] p. 26) *Let Q be $\exists$ or $\forall$. Then: $Qx_{/Y}\ \phi(x) \equiv_{\mathsf{G}} Qz_{/Y}\ \phi(z)$, if z does not occur in $Qx_{/Y}\ \phi(x)$.*

Our counterexample is obtained from the two sentences which were used in the discussion in section 15.4, viz (15.9) and (15.10). We consider here the situation after ∀belard has made his choices for the universal quantifiers, and ∃loise has chosen the right disjunct.

Lemma 25. *Let $V = \{xyz\colon 110, 111, 000{,}001\}$. Then:*

$$\mathcal{A} \models_{\mathsf{V}} \exists s\, \exists u_{/x}\ [u = x \wedge s = z]\ [V], whereas \tag{15.22}$$

$$\mathcal{A} \not\models_{\mathsf{V}} \exists y\, \exists u_{/x}\ [u = x \wedge y = z]\ [V] \tag{15.23}$$

So renaming the bound variables may change the truth value of the formula.

Proof.
The winning strategy for (15.22) is $f_{\exists s} \equiv s := z$ and $f_{\exists u_{/x}} \equiv u := y$. The negative result will be proved using the interpretation with valuations.

Assume that

$$\mathcal{A} \models_{\mathsf{V}} \exists y\, \exists u_{/x}\ [u = x \wedge y = z]\ [\{xyz\colon 110, 111, 000, 001\}] \tag{15.24}$$

Then there must be a set $W \sim_y (V * \{y : A\})$ such that

$$\mathcal{A} \models_{\mathsf{V}} \exists u_{/x}\ [u = x \wedge y = z]\ [W] \tag{15.25}$$

The values of y and z will not change any further in the recursion to subformulas, so we have to restrict here our choice to valuations for which $y = z$ holds. So

W consists of the y-variants for which $y = z$ holds: $W = \{xyz\colon 100, 111, 000, 011\}$. Since (15.25) holds, there must be a collection W_i of in W x-saturated subsets such that $W = \cup_i W_i$. The x-saturated subsets of W are W, $W_1 = \{xyz\colon 100, 000\}$ and $W_2 = \{xyz\colon 111, 011\}$ (and the empty set). Now we have to find for each element of the collection a value for u such that the left conjunct $u = x$ becomes satisfied (the right conjunct is already satisfied by our choice for W). For none of the three candidates for the collection such a value for u can be found. So there is no collection W_i that satisfies the requirements for $\exists u_{/x}$. So our initial assumption (15.24) is incorrect, which proves (15.23).

15.7.2 Prenex normal form

Caicedo and Krynicki present the following rephrase of the quantifier extraction part of the prenex normal form theorem.

Quote 26. (Lemma 3.1.c,d [Caicedo and Krynicki, 1999] p. 26).
Let Q be $\exists$ or $\forall$. Let $\psi_{/x}$ denote the result of adding to all quantifiers in ψ the independence condition $_{/x}$. Then:

1. $[Qx_{/Y}\ \phi \vee \psi] \equiv_{\mathsf{G}} Qx_{/Y}\ [\phi \vee \psi_{/x}]$
2. $[Qx_{/Y}\ \phi \wedge \psi] \equiv_{\mathsf{G}} Qx_{/Y}\ [\phi \wedge \psi_{/x}]$

In section 15.4 we have presented two counterexamples. The first one showed that new signalling possibilities emerge by quantifier extraction. We consider that example after the first choice by $\forall$belard.

Lemma 27. *Let $V = \{z\colon 1, 0\}$. Then:*

$$\mathcal{A} \not\models_{\mathsf{V}} \forall x[x \neq z] \vee \exists u_{/z}[u = z \vee_{/z} u \neq z]\ [V] \tag{15.26}$$

$$\mathcal{A} \models_{\mathsf{V}} \forall x[x \neq z \vee \exists u_{/z,x}[u = z \vee_{/z} u \neq z]]\ [V] \tag{15.27}$$

So quantifier extraction as quoted in (26), may change the interpretation of a formula.

Proof.
The winning strategy for (15.27) is $f_{\vee} \equiv$ *if* $x \neq z$ *then* L *else* R, $f_{\exists u_{/z,x}} \equiv 0$ and $f_{\vee z} \equiv$ *if* $x = 0$ *then* L *else* R.

Next we prove the negative result using valuations. Assume that

$$\mathcal{A} \models_{\mathsf{V}} \forall x[x \neq z] \vee \exists u_{/z}[u = z \vee_{/z} u \neq z]\ [\{z\colon 1, 0\}] \tag{15.28}$$

It will be clear that $\forall x[x \neq z]$ will not be satisfied for any nonempty subset of $\{z\colon 1, 0\}$. Therefore

$$\mathcal{A} \models_{\mathsf{V}} \exists u_{/z}[u = z \vee_{/z} u \neq z][\{z\colon 1, 0\}] \tag{15.29}$$

must hold. Then we have consider the z-saturated subsets of $\{z\colon 1, 0\}$. That is only the set itself. So there must be a value a such that

$$\mathcal{A} \models_{\mathsf{V}} [u = z \vee_{/z} u \neq z][\{zu\colon 1a, 0a\}] \tag{15.30}$$

Again, there is only one way to divide the valuations into z-saturated subsets, the set itself and the empty set. But no matter what the value of a would be, neither $\mathcal{A} \models_{\mathsf{V}} u = z\ [\{zu\colon 1a, 0a\}]$ nor $\mathcal{A} \models_{\mathsf{V}} u \neq z\ [\{zu\colon 1a, 0a\}]$. Hence (15.28) cannot be true.

The second counterexample showed that signals from outside can be blocked by quantifier extraction. We consider that example after the initial choices by ∀belard en ∃loise.

Lemma 28. *Let* $V = \{yt\colon 00, 11\}$. *Then*

$$\mathcal{A} \models_{\mathsf{V}} \forall t\ [t \neq t] \vee \exists x_{/y}[x = y]\ [V]] \tag{15.31}$$

$$\mathcal{A} \not\models_{\mathsf{V}} \forall t\ [t \neq t \vee \exists x_{/y,t}[x = y]]\ [V] \tag{15.32}$$

So quantifier extraction as described in claim 26, may change the interpretation of a formula.

Proof.
The winning strategy for (15.31) is given by $f_\vee \equiv R$ and $f_{\exists x_{/y}} \equiv x := t$. We show (15.32) using the interpretation with valuations. So, suppose:

$$\mathcal{A} \models_{\mathsf{V}} \forall t\ [t \neq t \vee \exists x_{/y,t}[x = y]]\ [\{yt\colon 00, 11\}]. \tag{15.33}$$

Due to the meaning of $\forall t$ (15.34) follows, so (15.35) holds:

$$\mathcal{A} \models_{\mathsf{V}}\ [t \neq t \vee \exists x_{/y,t}[x = y]]\ [\{yt\colon 00, 01, 11, 10\}]. \tag{15.34}$$

$$\mathcal{A} \models_{\mathsf{V}}\ \exists x_{/y,t}[x = y]\ [\{yt\colon 00, 01, 11, 10\}] \tag{15.35}$$

The set of valuations in (15.35) has only itself and the empty set as y-saturated subset. So there must be a value a such that:

$$\mathcal{A} \models_{\mathsf{V}}\ x = y\ [\{xyt\colon a00, a01, a11, a10\}] \tag{15.36}$$

There is, however, no value which does so for all valuations in the set. Hence (15.32) is proven.

15.7.3 Slashed disjunction elimination

Quote 29. (Lemma 3.2, [Caicedo and Krynicki, 1999] p. 25)
$\phi \vee_{/Y} \psi \equiv_{\mathsf{G}} \exists u_{/Y}\, \exists s_{/Y,u}[[[u = s \wedge \phi] \vee [u \neq s \wedge \psi]] \vee \exists! u[u = u \wedge [\phi \vee \psi]]]$

Since the counterexample is in a domain with three elements, we omit the part after ∃! (that disjunct is than false).

Lemma 30. *Let* $\text{dom}(\mathcal{A}) = \{0, 1, 2\}$. *Then*

$$\mathcal{A} \not\models_{\mathsf{V}} \forall y \forall t[\exists x_{/t}[x = t] \vee_{/y} \exists x_{/t}[x = t]] \tag{15.37}$$

Proof.
Let $V = \{0, 1, 2\}^{\{y,t\}}$, and suppose that the formula mentioned in the lemma was true in $\mathcal{A}$. Then (15.38) must hold.

$$\mathcal{A} \models_{\mathsf{V}} [\exists x_{/t}[x = t] \vee_{/y} \exists x_{/t}[x = t][V] \tag{15.38}$$

Then there must be sets V_1 and V_2, both y-saturated in V, such that each satisfies one of the conjuncts. The three elementary candidates for these y-saturated sets are $V_i = \{i\}^t \times \{0, 1, 2\}^y$, where $i \in \{0, 1, 2\}$. The other candidates are the union of two or three of these. Assume that V_1 consists of one of the elementary sets, then V_2 must consist of the union of the other two. In that union two values for t occur, hence there is no value for x which for all valuations satisfies $x = t$. Analogously for the other combinations of y-saturated subsets.

Since we have already shown (section 15.4) that the rule for slashed disjunction elimination transforms (15.37) into a true formula, the proposed rule cannot be correct.

15.7.4 Conservative extension

Quote 31. ([Hintikka, 1996] p. 65) *Technically speaking IF first-order logic is a conservative extension of ordinary first-order logic.*

Lemma 32. *The IFG sentence (15.39) is not true.*

$$\forall x \forall y \exists u\ [u = x \wedge \forall x[x = y \vee [u = 1 \vee_{/u} u \neq 1]] \tag{15.39}$$

Proof.
Let $\mathcal{A}$ be a model with two elements $\{0, 1\}$. Assume

$$\mathcal{A} \models_{\mathsf{V}} \forall x \forall y \exists u\ [u = x \wedge \forall x[x = y \vee [u = 1 \vee_{/u} u \neq 1]]][\{\epsilon\}] \tag{15.40}$$

By definition 20 the following must hold:

$$\mathcal{A} \models_{\mathsf{V}} \exists u[u = x \wedge \forall x[x = y \vee [u = 1 \vee_{/u} u \neq 1]]]\ [\{xy\colon 11, 10, 01, 00\}] \tag{15.41}$$

Since u must be equal to x, we take for the u-variant only those valuations where $x = u$.

$$\mathcal{A} \models_{\mathsf{V}} [u = x \wedge \forall x[x = y \vee [u = 1 \vee_{/u} u \neq 1]]]\ [\{xyu\colon 111, 101, 010, 000\}] \tag{15.42}$$

The right conjunct has to be satisfied by the same set of valuations. The $\forall x$ adds all x-variants, so it must be the case that:

$$\mathcal{A} \models_{\mathsf{V}} [x = y \vee [u = 1 \vee_{/u} u \neq 1]]\ [\{xyu\colon 111, 011, 101, 001, 010, 110, 000, 100\}] \tag{15.43}$$

Let V_1 consist of all valuations for which $x = y$, then V_1 satisfies the left disjunct. Let V_2 consist of the other valuations, so $V_2 = \{xyu\colon 011, 101, 010, 100\}$. Then it should be the case that:

$$\mathcal{A} \models_{\mathsf{V}} [u = 1 \vee_{/u} u \neq 1]\ [V_2] \tag{15.44}$$

If we would have made another division, the V_2 would have been larger, but (15.44) still should hold. The u-saturated subsets of V_2 are $\{xyu\colon 011, 010\}$ and $\{xyu\colon 101, 100\}$. None of these subsets satisfies $u = 1$, and none satisfies $u \neq 1$. So (15.44) cannot be the case. Hence (15.40) is not true, so there is no winning strategy for (15.39).

Lemma 33. *IF logic is* not *a conservative extension of ordinary predicate logic.*

Proof.
Consider the IF formulation of sentence (15.39):

$$\forall x \forall y \exists u\ [u = x \wedge \forall x[x = y \vee [u = 1 \vee u \neq 1]]] \tag{15.45}$$

The previous lemma proves that there is no winning strategy for (15.39), hence not for (15.45), whereas classically (15.45) is valid.

15.8 Conclusions

Signalling is a tricky business. It disturbs several extrapolations from classical logic (change of bound variables, prenex normal form), and the interaction of signalling and implicit independence causes that Hintikka's IF is *not* a conservative extension of predicate logic.

Acknowledgements

We thank Xavier Caicedo, Harrie de Swart and Peter van Emde Boas for their comments on a previous version of this paper.

References

Caicedo, X., Dechesne, F., and Janssen, T. M. V. (to appear). Equivalence and quantifier rules in logics with imperfect information. Paper in preparation.

Caicedo, X. and Krynicki, M. (1999). Quantifiers for reasoning with imperfect information and Σ^1_1-logic. In Carnielli, W. A. and Ottaviano, I. M. L., editors,

Contemporary Mathematics, volume 235, 17–31. American Mathematical Society.

Hintikka, J. (1996). *The Principles of Mathematics Revisited*. Cambridge University Press.

Hintikka, J. and Sandu, G. (1997). Game-theoretical semantics. In van Benthem, J. and ter Meulen, A., editors, *Handbook of Logic & Language*, 361–410. Elsevier Science.

Hodges, W. (1997). Compositional semantics for a language of imperfect information. *Logic Journal of the IGPL*, 5(4), 539–563.

Janssen, T. M. V. (2002). Independent choices and the interpretation of IF-logic. *Journal of Logic, Language and Information*, 11, 367–387.

Väänänen, J. (2002). On the semantics of informational independence. *Logic Journal of the IGPL*, 10(3), 339–352.

Chapter 16

INDEPENDENCE-FRIENDLY LOGIC AND GAMES OF INCOMPLETE INFORMATION*

Ahti-Veikko Pietarinen
Department of Philosophy
University of Helsinki
pietarin@cc.helsinki.fi

16.1 Proem

Of late, studies on Independence-Friendly (IF) logics have burgeoned. The purpose of the present paper is to extend the core idea of 'IFing' different logics to a new domain. Whereas sentences of IF logics have thus far been associated with semantic games of *imperfect information* to capture the idea of informational independence between quantifiers and connectives, I will suggest another type of independence. The expressions of logic may, unlike in traditional IF logics in which the independence refers to hiding of information concerning the actions of players, be associated with games in which the players have restricted information concerning aspects of the formal structure of the game itself. Such a situation is referred to in game theory by speaking of games of *incomplete information*.

16.2 Independence-friendly logic and semantic games

Let us begin by reviewing the essentials of IF first-order logic. Its language L_{IF} is derived from sentences φ of first-order logic L by replacing quantifiers and connectives in φ with the following expressions ($j_n \in \{\mathsf{l}, \mathsf{r}\}$):

$$(\exists x/W), (\forall y/W), (\bigvee_{j_n} /W), (\bigwedge_{j_n} /W).$$

*Supported by the Academy of Finland (*Dialogical and Game Semantics*, project no. 1101687). My thanks to Denis Bonnay, Theo Janssen and Shahid Rahman for their comments.

J. van Benthem, G. Heinzmann, M. Rebuschi and H. Visser (eds.), The Age of Alternative Logics: Assessing Philosophy of Logic and Mathematics Today, 243–259.

The point is to remove the restrictions concerning the linear, asymmetric and transitive dependence relations between quantifiers and connectives. The sets W consist of variables in *Var*(φ) and indices ranging over finite integers that have occurred in the syntactically superordinate context within which the expressions slashed with W reside.[1]

For instance,

$$\forall x(\exists y/\{x\})\, Rxy \in L_{\mathbf{IF}},\ \exists x{\sim}\exists y{\sim}(\exists z/\{x\})\, Rxy \in L_{\mathbf{IF}}$$

and

$$\forall x \forall y(\bigvee_{j_1}/\{y\})(\bigvee_{j_2}/\{x\})\,(\psi)_{j_1 j_2} \in L_{\mathbf{IF}}.$$

If $W = \emptyset$, the slashes may be omitted. In case the semantic games (to be outlined below) are closed under reflexive information sets, it is further assumed that for all (Qx/W) $(Q \in \{\exists, \forall\})$, $x \notin W$, and for all (Q_{j_n}/W) $(Q \in \{\bigvee, \bigwedge\})$, $j_n \notin W$. This innocent-looking idea has many repercussions, some of them documented in (Hintikka, 1996).

$L_{\mathbf{IF}}$-sentences φ are interpreted through strictly competitive, non-variable-sum semantic games $G(\varphi, M)$ of imperfect information on a model M, between two teams of players, the team of Verifiers V and the team of Falsifiers F.[2] The games have imperfect information because the slashes indicate that the values chosen for the elements in W on the right-hand side of the slash are not visible to the player moving at the position signalled by the left-hand side of the slash.

The legitimate moves are determined by the following three game rules:

If $\varphi = (\bigwedge_{j_n}/W)\,(\psi)_{j_n}$ (resp. $\varphi = (\bigvee_{j_n}/W)\,(\psi)_{j_n}$), then V (resp. F) chooses $j_n \in \{\mathsf{l}, \mathsf{r}\}$, and depending on this choice, the game continues either as $G((\psi)_\mathsf{l}, M)$ or as $G((\psi)_\mathsf{r}, M)$.

If $\varphi = (\forall x/W)\,\psi$ (resp. $\varphi = (\exists x/W)\,\psi$), then V (resp. F) chooses a from the domain $|M|$ of M, and the game continues as $G(\psi[x/a], M)$.[3]

If $\varphi = {\sim}\psi$, then the current V becomes F and the current F becomes V, and the game continues as $G(\psi, M)$.

For ψ atomic $Rx_1 \ldots x_n$, if $M \models Rx_1 \ldots x_n$ in a given interpretation, then the current V wins, and if $M \not\models Rx_1 \ldots x_n$ in a given interpretation, then the current F wins.

---

[1]The infix connectives of L may be rewritten in the prefix notation in terms of restricted quantifiers with indices correlated with subformulas of φ.

[2]The reference to teams concerns the implementation of semantic games for cases in which there may be imperfect recall, in other words players forget their own choices or information (Pietarinen, 2001b).

[3]Notation $[x/a]$ means instantiating the value a for x.

Let $\mathcal{I}_j$ be a partition of non-terminal histories $h \in H \setminus Z$ of the extensive-form representation of $G(\varphi, M)$.[4] A strategy s_j for a player $j \in \{V, F\}$ is a function $s_j : I \to |M| \cup \{\mathsf{l}, \mathsf{r}\}$ that specifies a choice from the set of actions for any information set $I \in \mathcal{I}_j$ in which j is to make a decision. Actions are objects from the domain or from the set of two designated individuals. The terminal histories $h' \in Z$ are labelled with the atomic formulas and are associated with payoffs $u_j : Z \to \{1, -1\}$.

The extensive-form structure of semantic games is general enough to capture all independence and dependence patterns in IF logic, even the mutual dependencies.[5]

Truth and falsity are defined as the existence of a winning strategy for the appropriate player. For $|M|$ infinite and assuming the Axiom of Choice, φ is true in M iff there exists a winning strategy for the player who started the game as V, and φ is false in M iff there exists a winning strategy for the player who started the game as F. The solution concept of the winning strategy is a vector of Skolem functions.

I will assume a basic familiarity with IF logic and the associated semantic games. In the next section, I make a few observations concerning IF logic.[6]

16.3 Independence-Friendly logic, imperfect information and partiality

IF logic aims at disposing of any unanalysed scope conventions. Its expressions may be seen as forming networks or dependency graphs that channel semantic information in ways not regimented prior to their interpretation. The interpretation is in terms of games, which assign semantic attributes to the expressions while preserving the bindings and references between the values for the variables.

Aside from IF first-order logic, there are other IF logics. Slashing is a platform-independent method of relaxing the semantically linear dependency relations between logical expressions. Beyond partially-ordered quantifiers (Henkin quantifiers), this may involve non-transitive and cyclic dependencies. Since non-linearity is operationalised via games, IF logic is likely to arise whenever a coherent game-theoretic interpretation is arranged for expressions of a given logic.

[4]For more details, cf. Pietarinen, 2001c, 2004a, Sandu, and Pietarinen, 2001, 2003.

[5]Alpern, 1991 and Selten & Wooders, 2001 have studied cycles in extensive games. Hintikka, 2002 provides motivation for having symmetric dependence relations between quantifiers.

[6]Studies on IF logic include: Bonnay, 2004, Hintikka, 1996, 2002, Hintikka & Sandu, 1989, 2004, Janssen, 2002, Pietarinen, 2004a, Rebuschi, 2004 and Väänänen, 2002.

IF versions exist for sentential, modal and epistemic logics.[7] Moreover, IF logic appears useful for common and distributed knowledge (Pietarinen, 2002), polyadic and generalised quantifiers, bounded quantification, and many-sorted logics. For instance, bounded IF quantification is likely to arise from standard translations of certain IF modal logics to the background language that is the IF counterpart of a suitable bounded fragment of first-order logic.

Non-logical (descriptive) constant may also enjoy independence. But how to interpret such independence? What does it mean that a constant c is independent of a variable x? This is derivative of the question of what the game rules for non-logical constants are. Such rules would pertain to the interpretation of language. They state that, when an atomic formula ψ of an $L_{\mathbf{IF}}$-sentence φ is reached in $G(\varphi, M)$, a low-level atomic game $G^{\text{Atom}}(\psi, M)$ is evoked. This low-level rule is now simply that whenever a constant in encountered in ψ, a value is assigned to it.

As soon as we have such rules at our disposal, further things may be attained besides informational independence. For instance, low-level atomic games enable us to reverse the usual winning conventions:

- If V wins the play of $G^{\text{Atom}}(\psi, M)$, then ψ is associated with the payoff 1 in the parental game $G(\varphi, M)$.
- If F wins the play of $G^{\text{Atom}}(\psi, M)$, then ψ is associated with the payoff -1 in the parental game $G(\varphi, M)$.

Winning means that the assignment produces the right values as intended by the predicates and other non-logical constants of the assertion in question. Conversely, losing means the unintended values or failure to produce a value at all.

With such reverse conditions at hand, we can make game-theoretically meaningful distinctions between partially and totally interpreted languages. Partial interpretations arise if neither V nor F wins in $G^{\text{Atom}}(\varphi, M)$. This is an objective fact of the indefinite determinations of the universe of discourse. The parental games will in that case exhibit payoff structures of $u_V(h) = -1, u_F(h) = -1, h \in Z$.

Conversely, complete interpretations lack any such indefiniteness. A consequence of these cases is that the law of excluded middle fails either at the level of atomic formulas (partial interpretations) or at the level of complex formulas (complete or partial interpretations).

Moreover, 'non-standard' partial logics may now arise, both at the level of winning conventions and at the level of truth conditions:

[7] See Sandu & Pietarinen, 2001 and 2003 for sentential and Bradfield, 2000, Hintikka, 1996, Pietarinen, 2001a, 2002, 2003b, 2004c, and Tulenheimo, 2004 for modal formulations.

Weak-Verifier-Winning: If V wins the play of $G^{\text{Atom}}(\psi, M)$, then ψ is not false. If F wins the play of $G^{\text{Atom}}(\psi, M)$, then ψ is false.

Weak-Falsifier-Winning: If V wins the play of $G^{\text{Atom}}(\psi, M)$, then ψ is true. If F wins the play of $G^{\text{Atom}}(\psi, M)$, then ψ is not true.

Weak-Verifier-Truth: $L_{\mathbf{IF}}$-sentence φ is not false in M iff there exists a winning strategy for the player who started the game as V.

Weak-Falsifier-Truth: $L_{\mathbf{IF}}$-sentence φ is not true in M iff there exists a winning strategy for the player who started the game as F.

The players are perhaps best seen here as playing the roles of the Non-Falsifier and the Non-Verifier. Such non-standard clauses may be applied to sentences of IF as well as of non-IF logic.[8]

16.4 The logic of Payoff Independence

The types of independence in the literature on IF logics have concerned informational independence, operationalised by games of imperfect information. But also another type exists, not confined to *informational* independence. It need not refer solely to players' knowledge, information or ignorance concerning the choices made in the game. Accordingly, it need not refer solely to possible patterns of dependence and independence between quantifiers or connectives. For other uncertainties may exist in the game that affect players' strategic decisions. Most notably, players may lack information concerning the mathematical structure of the game. In fact, such games are part and parcel of game theory, known as incomplete information, or Bayesian, games.[9]

Lacking an off-the-peg term for such a logic, let us call it logic of *payoff independence* (PI logic). The reason for referring to payoffs is historical, going back to Harsanyi, 1967. What Harsanyi showed was that uncertainties concerning the structure of the game, including players' strategies, may be transformed into uncertainties concerning the values of the players' payoff functions.

What kinds of characteristics of incompleteness can we meaningfully have in logic? Purely game-theoretically, before Harsanyi's pioneering work it was thought that incompleteness concerns players' uncertainty about the rules of the game and thus they cannot act strictly according to the rules. I do not mean rule incompleteness. Players must follow the rules that define the legitimate moves, simply because at each position of the game, the corresponding expression

[8]In changing the game conventions here one is reminded here of the 'no-counterexample' interpretations proposed by Georg Kreisel in the early 1950s.

[9]Little hinges on the Bayesian paradigm in the present context, however, since the prior probabilities will be very simple. Moreover, in the case non-partitional information sets (such as in propositional IF logic, cf. Pietarinen, 2004c), Bayesian reasoning will in any case break down.

in the sentence completely determines what the legitimate moves of a player are. IF logics do not change this fact. Correlated with games of imperfect information, in IF sentences imperfectness affects players' strategies, not the degree of familiarity with the game rules.

Some examples of incomplete information are:

- Players of a semantic game may lack information about the *strategies* used in the game. This may concern adversaries' strategies as well as players' own strategies.
- Players may only know the *sorts* of previous choices made in the game but not the choices themselves.[10]
- Players may be uncertain about various parameters attached to players, including uncertainty about
  - the *number of agents* there may be in the opponent team;
  - the *size* of one's own team.
- Players may be uncertain about *payoffs*.

The uncertainties listed above provide examples of ignorance on the values of payoffs. They are thus instances of the Harsanyi transformation. The Harsanyi transformation asserts that games of incomplete information may be thought of as games of complete but imperfect information with random moves by Nature hidden from subsequent players. This is implemented so that Nature chooses *types* of players but only reveals to the players their own types. Even the assumption concerning awareness of one's own types may be relaxed.

I will ignore here the uncertainty about the cardinality of teams and about sorts. The former becomes relevant if the games of incomplete information — after the Harsanyi transformation has been performed — exhibit imperfect information and imperfect recall. Imperfect recall is generated by an unrestricted application of slashes, in which case uncertainties concerning the organisation and structure of these associated teams of agents may materialise.[11]

As to the uncertainty concerning strategies, I will limit the discussion to a simple incompleteness in which the (non-empty) type space is at most binary, containing two types corresponding to the two roles that the players may have. The reason for this limitation is that this type of uncertainty is reflected in a natural extension of IF logic to independent negations.

---

[10]This may happen in many-sorted logics. Take $\forall X(\exists y/X)\,RXx$, for instance, in which V, in choosing an individual for y is not informed about the set that was chosen for the universally quantified second-order variable X.

[11]About the team-theoretic outlook on IF logics. See (Pietarinen, 2001b, 2001c and 2004a).

This extension applies slash indicators to strong (dual) negations.[12] It is assumed that strong negations may occur on either side of the slash.[13] In addition to variables and connectives, a finite sequence of negations may exist on the right-hand side of the slash. These negations, as indeed all occurrences of strong negations, are indexed to distinguish between different tokens in a formula. The case is similar with hidden connective information (cf. Sandu & Pietarinen, 2001 and 2003). In contrast to hidden connectives, however, in negation independence we do not have restricted quantification at our disposal that would accomplish informational independence of binary connectives in terms of restricted quantifiers over indices.

More precisely, we take the language $L_{\mathbf{PI}}$ to consist of IF first-order logic plus instances of the following expressions, in which φ is any formula of IF first-order logic:

- $(\sim_j / W)\, \varphi, W \subseteq A = \{x_1 \ldots x_n, \sim_1 \cdots \sim_m\}, \sim_j \notin W, j \in \omega.$
- $(Qx/W)\, \varphi, Q \in \{\exists, \forall\}, W \subseteq A = \{x_1 \ldots x_n, \sim_1 \cdots \sim_m\}, x \notin W.$

Elements in A are those that are already visited in the clauses syntactically superordinate to the slashed expressions.

For example, $\sim_1 \sim_2 (\sim_3 / \{\sim_1\})\, \varphi \in L_{\mathbf{PI}}$, $\forall x \sim_1 \forall y (\exists z / \{x, \sim_1\})\, Rxyz \in L_{\mathbf{PI}}$ and $\exists x \sim_1 \bigwedge_{j_1} (\bigvee_{j_1} / \{j_1, \sim_1\})\, (\psi)_{j_1 j_2} \in L_{\mathbf{PI}}$.

The indexing schema may be taken to refer to all the distinct morphological manifestations of negative words (n-words) in natural language.

Negation independence may further be applied to modal contexts, by adding $\{\Box_1^1 \ldots \Box_n^m\}$ into the set of actions and extending $L_{\mathbf{PI}}$ by the applications of the clause

- $(\Box_j^i / W)\varphi, A \subseteq W, \Box_j^i \notin W.$

Likewise for the dual $\Diamond_j^i \varphi := \sim_n \Box_j^i \sim_m \varphi$.[14] To save space, modalities are omitted.

---

[12] Recall that in IF logics, contradictory negation ($\neg$) may only be introduced by using the meta-level rule $M \models \neg\varphi$ iff $M \not\models \varphi$. The denial of a proposition asserts truth precisely in those circumstances in which the proposition is false. It lacks the game-theoretic, processual notion of negation ($\sim$), which — unlike contradictory negation that reverses the polarities of the *partitions* denoted by the proposition — reverses the polarities of the *processes* correlated with the propositions.

[13] A quantifier or an epistemic operator being independent of negation was suggested already in (Hintikka & Sandu, 1989). The meaning that was assigned to such expressions was that the order of the quantifier or the modal operator and the negation is reversed. However, troubles begin for this interpretation as soon as there is more than one negation of which something is independent, or if the negation does not immediately precede the slashed expression and there are some intermediate negations or other constituents on which the slashed expression is dependent. In contrast, we want a general notion of *negation independence*.

[14] The subscripts refer to different agents, and so to different classes of accessibility relations, and the superscripts to different tokens of operators in a sentence. These modalities may be interpreted as pertaining to knowledge, belief, time, and so on. The associated semantic games leave some latitude as to what the exact meaning of the independence with respect to modalities is (cf. Pietarinen, 2001a, 2003b and 2004c).

The semantics for $L_{\mathbf{PI}}$ is through games of imperfect and incomplete information. However, by the applications of the Harsanyi transformation, the semantic games reduce to those of complete but imperfect information, with hidden chance moves. We let the indices k, l range over a two-element set $\mathfrak{T} = \{v, f\}$. At $h, h' \in Z$, the payoffs $u_V^k(h)$ and $u_F^l(h')$ for V^k and F^l will depend not only on strategies s_V^k and s_F^l but also on the types of players.

Also, let R be the role function $R\colon \Phi \times H \rightarrow \{\uparrow, \downarrow\}$ from a set of $L_{\mathbf{PI}}$-formulas Φ and a set of histories $H \setminus Z$ of the associated extensive game to the set of two values. The role function is merely a flip-flop register that changes its state every time a negation is encountered in a sentence.

The semantic games $G(\varphi, M)$ for $L_{\mathbf{PI}}$-sentences are like those of imperfect information but interspersed with a type space $\mathfrak{T}$ and three players instead of two. The third player, Nature, is, effectively, a probability generator. Let φ be an $L_{\mathbf{PI}}$-formula. The set of game rules are adjoined with the following three rules:

- If $\varphi = (\neg_j/W)\,\psi$ and $R(\varphi, h) = \uparrow$, then:
  $$\begin{cases} \textbf{If } \text{for any } i \text{ for which } \ \sim_i \notin W, \text{ then } R(\psi, h \frown \psi) = \downarrow. \\ \text{The game continues as } G(\psi, M); \\ \textbf{Else: } \text{Nature chooses from } \mathfrak{T} = \{v, f\}. \\ \text{The game continues as } G(\varphi', M), \varphi' = (\sim_j/W \setminus \{\sim_i\})\,\psi. \end{cases}$$

- If $\varphi = (Qx/W)\,\psi, Q \in \{\exists, \forall\}$, then:
  $$\begin{cases} \textbf{If } \text{for any } i \text{ for which } \ \sim_i \notin W, \text{ then} \\ \text{the game continues as } G((Qx/W)\,\psi, M); \\ \textbf{Else: } \text{Nature chooses from } \mathfrak{T} = \{v, f\}, \text{ and} \\ \text{the game continues as } G(\varphi', M), \varphi' = (Qx/W \setminus \{\sim_i\})\,\psi. \\ \text{Later choices for } Qx \text{ are independent of Nature's moves.} \end{cases}$$

- If $\varphi = (Qj_n/W)\,\psi, Q \in \{\bigvee, \bigwedge\}$, then:
  $$\begin{cases} \textbf{If } \text{for any } i \text{ for which } \ \sim_i \notin W, \text{ then} \\ \text{the game continues as } G((Qj_n/W)(\psi)_{j_n}, M); \\ \textbf{Else: } \text{Nature chooses from } \mathfrak{T} = \{v, f\}, \text{ and} \\ \text{the game continues as } G(\varphi', M), \varphi' = (Qj_n/W \setminus \{\sim_i\})\,(\psi)_{j_n}. \\ \text{Later choices for } Qj_n \text{ are independent of Nature's moves.} \end{cases}$$

The winning conventions are as before. The truth and the falsity of complex formulas φ of $L_{\mathbf{PI}}$ are now defined somewhat differently, however. By f and v-branches we mean those subgames G' of G the roots of which correlate with Nature's choices of the types f and v, respectively.

- The formula φ is true in M iff in the subgame G' of $G(\varphi, M)$ that has no f-branches, there exists a strategy profile S_1 of pure optimal strategies

$\{s_1^k, s_2^1 \dots s_2^m\}$ for the player who started the game $G(\varphi, M)$ as playing the role of V, and Nature chooses her type to be v.[15]

- The formula φ is false in M iff in the subgame G' of $G(\varphi, M)$ that has no v-branches, there exists a strategy profile S_2 of pure optimal strategies $\{s_2^1 \dots s_2^k, s_1^l\}$ for the player who started the game $G(\varphi, M)$ as playing the role of F, and Nature chooses his type to be f.

Nature's choices precede players' assessment of expected values of payoffs and the selection of optimal strategies. Because chance moves are concealed, and unless Nature decides to choose the right types for the players, there is not much a player can do to enforce a win, even if optimal strategies would exist. The aggregate of optimal strategies plus Nature's random choices comprise a winning strategy that agrees with the truth and falsity of $L_{\mathbf{PI}}$-formulas in M.

It follows that given such chance moves, the law of excluded middle fails even if there were no slashes in an $L_{\mathbf{PI}}$-formula φ. This holds irrespectively of whether the underlying language is completely or partially interpreted.

By choosing types, Nature's moves affect the strategy sets of players, their information partitions, and payoffs. What Nature thus chooses is a *state of the world.* The state may denote context, environment, time, mood, collateral information or the common ground according to which the particular plays of the game are to proceed. Such choices are not restricted to initial deals.

The law of double negation is valid for indexed negations in slashes to the right. It suffices to check whether the parity of negations in W is even or odd. Given the probabilistic nature of the system, the double negation $\sim_1 (\sim_2 / \{\sim_1\})$ φ reduces to φ with equal probability of reducing to $\sim_1 \sim_2 \varphi$. Likewise, the prima facie contradictory formulas such as $\sim_1 \varphi \wedge \sim_2 (\sim_3 / \{\sim_1\}) \varphi$ are associated with a conditional probability distribution determined by Nature.

The type, or the state of the world, carries information about the values of the payoff functions. Despite chance moves marking decisions between two options, the notion of a type is not the same thing as the notion of a role of the player. In general, we may think of roles consisting of finitely many types.

In extensive-form semantic games, four types of information may be discerned (Rasmusen, 1989).

1 *Certain information*: Nature does not move elsewhere than at the root of the game. Otherwise the game is one of uncertain information.

2 *Complete information*: Nature does not move first, or her initial move is observed by every player. Otherwise the game is one of incomplete information.

[15] Subgames are defined, standardly, as subtrees of the given finite extensive-form tree.

3 *Perfect information*: information sets of both players are singletons. Otherwise the game is one of imperfect information.

4 *Symmetric information*: both players get the same information in the game. Otherwise the game is one of asymmetric information.

Certainty thus means initial, unhidden chance moves by Nature.

Examples are provided by the following $L_{\mathbf{PI}}$-formulas:

$$\phi := \sim_1 (\sim_2 / \{\sim_1\}) \, \varphi \tag{16.1}$$

$$\phi := \sim_1(\sim_2/\{\sim_1\}) \bigvee_{j_n} (\varphi)_{j_n} \tag{16.2}$$

$$\phi := \sim_1 \forall x (\exists y / \{\sim_1\}) \, \varphi. \tag{16.3}$$

The game $G(\phi, M)$ correlated with (16.1) is of perfect, complete but uncertain information. $G(\phi, M)$ correlated with (16.2) is of imperfect, incomplete and uncertain information. $G(\phi, M)$ correlated with (16.3) is likewise of imperfect, incomplete and uncertain information. The third game shows the utility of the role function R: the player in $\{V, F\}$, when choosing for the existentially quantified variable y, may not possess the information that she was the Verifier who in fact chose the value for the universally quantified variable x; however, she might infer that she must have been located within the scope of an odd number of unslashed negations.

On the other hand, all extensive semantic games that are not undetermined and not correlated with formulas with mutual dependencies have asymmetric information.

For instance, the sentence (16.1) reduces to φ with a chance move by Nature. Likewise, (16.2) reduces to $\bigvee_{j_n} (\varphi)_{j_n}$ with a chance move. The sundry effect such reductions have is that they change the game of uncertain information into one of certain information. Other properties of the information structure are preserved.

16.5 Applications and amplifications

Let me suggest a few consequences of having incomplete information in logic.

First, given ignorance concerning payoffs, comparisons of actions under uncertainty vs. actions under risk become possible, since the rival player, although optimal maximiser, can conceal the costs that his or her moves may incur.

In conversational settings, for instance in those provided by game-theoretic formulations of Grice's maxims (Hintikka, 1986, Parikh, 2001), certain moves are costly so as to be charged against the payoffs of the player. But if these costs were concealed, strategic advantages to decision makers are inevitable.

Third, many heterogeneous and iconic systems of logic, including Peirce's existential graphs (Pietarinen, 2003a and 2004b), provide analogues to the partially-overlapping scopes of negations of $L_{\mathbf{PI}}$ in terms of non-partitional graphs, in which some of the cuts (the closed lines of separation representing negations) may overlap. In other words, given two cuts c_1 and c_2, the affirmative area lies within the double cut area $c_1 \cap c_2$, whereas the negative area is in their union $c_1 \cup c_2$.

Permitting overlaps is not as radical as it may seem. In semantic networks and conceptual graphs related to existential graphs, partitioned networks have provided some insight into knowledge representation (Hendrix, 1979). Similar idea can be extended to non-partitioned cases. Furthermore, following the recognition of the limitations of standard information structure of extensive games, a growing body of research is emerging on non-partitioned structures in imperfect-information games.[16]

In strict relation to the aforementioned point on heterogeneous systems, the formulas of $L_{\mathbf{PI}}$ are, indeed, more like semantic nets than syntactic constructions, with the dependency relations given by directed graphs.

Linguistically, such nets may depict collections of semantic constraints for possible interpretations of a sentence. They are thus connected with underdeterminacy of scope relations. The perspective of networks as semantic constraints is also applicable to theories of discourse interpretation.

There is plenty of linguistic motivation for negation independence. Among them is negation transportation (NEG-transportation) in belief sentences (Sandu, 1993). NEG-transportation entails expressions in which belief operator is independent of negation. For instance, given a supply of modal belief operators B_i, negation independence distinguishes not believing a proposition and believing a negation of it:

- John does not believe that ψ: $B_{\mathsf{John}} \sim_j \psi$.
- John believes that not ψ: $B_{\mathsf{John}}(\sim_j / B_{\mathsf{John}})\, \psi$.

For the negation has to be the same in all scenarios compatible with what John believes.

Moreover, PI logic provides the logical counterpart to the distinction between languages that have *double negations* (such as English) and languages that have *negative concord* (such as most Romance languages). Negative concord means that multiple occurrences of morphologically negative items compress into a single negative expression.

[16]See e.g. Bacharach, 1985, Geanakoplos, 1989 and Morris, 1994.

To come clean on the history of intellectual ideas, this phenomenon was described by Charles S. Peirce in 1905 as the distinction between logical languages and quantitative languages:

> There are some languages in which two negatives make an affirmative. Those are the logical languages. [...] In other languages, probably the majority, a double negative remains a negative. These are quantitative languages. We should expect the people who speak them to be more humane and more highly philosophical. The quantitative view of negation ... does not really involve any bad reasoning (Peirce, 1967, manuscript 283: 120–1, *The Basis of Pragmaticism*).

In a similar vein, Peirce noted in 1896: "Were ordinary speech of any authority as to the forms of logic, in the overwhelming majority of human tongues two negations intensify one another" (Peirce, 1931).

The subsequently established idea of drawing this division was in terms of categorising the n-words of such languages. Alternatively, the basic division has been drawn between single and multiple concord (Forget, *et al.*, 1998).

Even intralingually, single concord languages such as English are not devoid of negation independence. A case in point is ellipsis. In (16.4), the range of negation appears to be the left conjunct 'start crying' only (the conjunction is resultative):

I hope that the baby doesn't start crying and the neighbours wake up. (16.4)

However, the game-theoretic interpretation of (16.4) is such that, if the antecedent conjunct contains a negation, it is interpreted as ordinary static conjunction, in other word F makes a move. An unmarked reading of this sentence nevertheless is that the range of negation extends over the whole of conjunction, in other words also over the right conjunct *the neighbours wake up*. This interpretation is produced by another, tacit token of verbal negation of the same type in the latter conjunct, which is independent of the conjunction. Such a tacit token is needed, because the first occurrence of negation negates the assertion *baby starts crying* and as such has nothing to do with negating *neighbours wake up*. This independent negation is omitted in (16.4) with resultative *and*, the latter conjunct of which being an elliptical or truncated subclausal expression of the sentence (16.5):

$$\exists x \exists y((Bx \wedge_1 Ny) \rightarrow (\sim_2 Cx \wedge_3 (\sim_3/\wedge_3) Wy)) \tag{16.5}$$

(In representing (16.4) by (16.5), the initial indexical and the volitional context as well as the plural are omitted for simplicity, and the infix notation is used. The abbreviations are self-explanatory.)

Similarly, let us compare (16.6) and (16.7):

He didn't know that anything had happened and neither did she. (16.6)

*He knew that nothing had happened and neither did she. (16.7)

How do we explain the difference that amounts to the latter being ill-formed? My solution is that the same *neither*-tag which functions as a truncated reproduction of the antecedent (dynamic) conjunction in (16.7) yields an ungrammatical sentence if the downwards-transported negation is independent of the initial epistemic attitude. But if so, then — assuming that the same transportation affects the right conjunct — there will inevitably be a difference in meaning between the two *neither*-tags, namely with (16.6) and (16.7). Accordingly, we represent (16.6) and (16.7) by (16.8) and (16.9), respectively:

$$K_h \forall x \sim_1 Hx \wedge K_s \forall y \sim_2 Hy \tag{16.8}$$

$$K_h \forall x (\sim_1 / K_h) Hx \wedge K_s \forall y \sim_2 Hy. \tag{16.9}$$

We need no epistemic operators informationally independent of negation. What we need is a negation that goes together with the universal quantifier, *viz.* captures the negative expression *nothing*, and is also payoff independent of the epistemic operator. If this independent expression is copied to the antecedent *neither*-tag in (16.9), it remains dependent on the universal quantifier. In contrast, in (16.8) the two occurrences of negation are independent both of epistemic operators and of universally quantified variables.

In the semantic games for formulas of $L_{\mathbf{PI}}$ (restricted to negation independence and duplex type spaces), the players are uncertain about which of the two possible subgames are being brought out by Nature. The distribution is furthermore assumed to be fair, and there is an equal chance of playing either of these games. This is reflected in the truth-conditions so that, in addition to the existence of an optimal strategy and the expected values of payoffs, Nature has to 'side with' the player in order for a game to verify (i.e., tagging Truth to the sentence) or falsify (tagging Falsity to it) an $L_{\mathbf{PI}}$-sentence in M. The sole reason as to why I have restricted the focus to the simplified case of payoff independence is that it falls naturally from the outfit of IF logic via negation independence.

The assumption of common priors that the players have is also simple: the prior probabilities are the same for all elements of the type space, in other words just the probabilities of 0.5. We thus avoid the intricacies of Bayesian reasoning and Bayesian extensions of equilibrium, apart from what was used in the definition of the truth and falsity of formulas. Common priors being common knowledge has for a long been the standard, albeit by no means indispensable or the most realistic assumption in economics, since it looks away from genuine asymmetric higher-order knowledge and belief concerning type distributions.

Moreover, we may even think of there being differently weighed occurrences of dual negations, contributing to logical rules such as the law of double negation with varying degrees.

Also notable is the commonality of common priors as common knowledge with pragmatic theories of language and communication. Following H. Paul Grice, many have assumed the key element in the creation of the common ground being common knowledge among speakers and hearers. However, in my case tables can be turned on Grice and his followers, as communicative situations may, alternatively, be modelled by applying the Harsanyi doctrine, thus dispensing with any higher-order knowledge. Such communicative situations are far from recondite for the sheer reason that there are strategic advantages in hiding payoffs that are costly. Assuming common priors, Grice's conversational maxims would follow from the rationality assumptions involved in the Harsanyi doctrine. As far as I know, this possibility has not been pursued by natural-language pragmatists.

Given the probabilistic and context-dependent character of $L_{\mathbf{PI}}$ expressions, a new complication apparently accompanies the advocates of compositional semantics. Namely: how to extend the semantics that assigns — instead of sequences of assignments — sets of sequences of assignments to subformulas of a sentence of IF first-order logic, into a compositional semantics for $L_{\mathbf{PI}}$-formulas? This is a genuine complication, since also whether such sets or co-sets go with V or with F (and whether such sets for an $L_{\mathbf{PI}}$-sentence φ in M are co-sets for an $L_{\mathbf{PI}}$-sentence $\sim\varphi$ in M) are parts of imperfect information. This is no mug's game, either, since $L_{\mathbf{PI}}$ is not a fabrication of yet another artificial language, but a logical reflection of what goes on in very commonplace classes of games pursued in game theory.

My approach should not be confounded with that of Blinov, 1994 who also considered the possibility of introducing chance moves into game-theoretic semantics. My approach is very different. Blinov does not use type spaces, and his games are ones of complete information. He also restricts chance moves to initial positions of the game, which may be interpreted as Nature's deal of values for free variables. Blinov considers no incomplete information, as his goal is to find a game semantic correlate to that of supervaluations (van Fraassen, 1979).

16.6 Conclusions

Precisely the same ideas that have motivated the passage from traditional first-order logic to its IF extension motivate the move from traditional logics to their payoff-independent extensions, including the recognition of the ambiguous and restrictive notion of scope (Hintikka, 1997). Negations do have priority orders, too, which may be studied via payoff independence.

A key philosophical trait that needs to be re-tracked concerns our pre-theoretical notion of truth. The received apparatus of semantic games asserts that the existence of winning strategies agrees with the truth of assertions. But as noted, winning strategies may be decomposed into several parts, one of the components being the existence of probabilistic optimal strategies, which has to go hand in hand with Nature's random choices of states of the world. But how can truth or falsity be probabilistic?

One reply could be that the games considered here are no longer strictly semantic, and they codify some of the actual verificatory and falsificatory practices linked with strategies. I do not think this is right; there is similar probabilistic flavour already in customary imperfect-information games in terms of mixed strategies. Surely these strategies are not that different from pure strategies so as to introduce some entirely novel epistemic elements.

Furthermore, while certain applications of semantic games with chance moves may suggest dialogic, discourse-theoretic or conversational analogues to such games, the moderate aim of the present treatise is intended to lie firmly within the realm of semantics, even though the characteristics of such games may well take some unexpected turns.

References

Alpern, S., 1991. Cycles in extensive form perfect information games, *Journal of Mathematical Analysis and Applications* 159, 1–17.

Bacharach, M., 1985. Some extensions of a claim of Aumann in an axiomatic model of knowledge, *Journal of Economic Theory* 37, 167–190.

Blinov, A., 1994. Semantic games with chance moves, *Synthese* 99, 311–327.

Bonnay, D., 2004. Independence and games, to appear in *Philosophia Scientiae*.

Bradfield, J.C., 2000. Independence: logics and concurrency, *Lecture Notes in Computer Science* 1862, 247–261.

Forget, D. et al., (eds.), 1998. *Negation and Polarity: Syntax and Semantics*, Amsterdam: John Benjamins.

van Fraassen , J., 1969. Presuppositions, supervaluations, and free logic, in K. Lambert (ed.), *The Logical Way of Doing Things*, New Haven: Yale University Press, 67–91.

Geanakoplos, J., 1989. Game theory without partitions, and applications to speculation and consensus, *Cowles Foundation Discussion Paper* 914, Yale University.

Harsanyi, J., 1967. Games with incomplete information played by 'Bayesian' players. Part I: The basic model, *Management Science* 14, 159–182.

Hendrix, G. G., 1979. Encoding knowledge in partitioned networks, in Findler, N.V. (ed.), *Associative Networks: Representation and Use of Knowledge by Computers*, Orlando: Academic Press, 51–92.

Hintikka, J., 1986. Logic of conversation as a logic of dialogue, in R.E. Grandy and R. Warner (eds.), *Philosophical Grounds of Rationality*, Oxford: Clarendon Press, 259–276.

Hintikka, J., 1996. *The Principles of Mathematics Revisited*, New York: Cambridge University Press.

Hintikka, J., 1997. No scope for scope?, *Linguistics and Philosophy* 20, 515–544.

Hintikka, J., 2002. Hyperclassical logic (a.k.a. IF logic) and its implications for logical theory, *The Bulletin of Symbolic Logic* 8, 404–423.

Hintikka, J. & Sandu, G., 1989. Informational independence as a semantical phenomenon, in J.E. Fenstad, I.T. Frolov, and R. Hilpinen (eds.), *Logic, Methodology and Philosophy of Science*, Vol. 8, Amsterdam: North-Holland, 571–589.

Hintikka, J. and Sandu, G., 1997. Game-theoretical semantics, in J. van Benthem and A. ter Meulen (eds.), *Handbook of Logic and Language*, Amsterdam: Elsevier, 361–410.

Hodges, W., 1997. Compositional semantics for a language of imperfect information, *Logic Journal of the IGPL* 5, 539–563.

Janssen, T.M.V., 2002. On the interpretation of IF logic, *Journal of Logic, Language and Information* 11, 367–387.

Morris, S., 1994. Revising knowledge: a hierarchial approach, in Fagin, R. (ed.), *Proceedings of the 5th Conference on Theoretical Aspects of Reasoning about Knowledge*, Morgan Kaufmann, 160–174.

Parikh, P., 2001. *The Use of Language*, Stanford: CSLI.

Peirce, C.S., 1931–58. *Collected Papers of Charles Sanders Peirce*, 8 vols., ed. by C. Hartshorne, P. Weiss, and A.W. Burks. Cambridge, Mass.: Harvard University Press.

Peirce, C.S., 1967. Manuscripts in the Houghton Library of Harvard University, as identified by R. Robin, *Annotated Catalogue of the Papers of Charles S. Peirce* (Amherst: University of Massachusettes Press, 1967), and in The Peirce Papers: A supplementary catalogue, *Transactions of the C.S. Peirce Society* 7, 1971, 37–57.

Pietarinen, A.-V., 2001a. Intentional identity revisited, *Nordic Journal of Philosophical Logic* 6, 144–188.

Pietarinen, A.-V., 2001b. Varieties of IFing, in Pauly, M., and Sandu, G. (eds.), *Proceedings of the ESSLLI 2001 Workshop on Logic and Games*, University of Helsinki.

Pietarinen, A.-V., 2001c. Propositional logic of imperfect information: foundations and applications, *Notre Dame Journal of Formal Logic* 42, 193–210.

Pietarinen, A.-V., 2002. Knowledge constructions for artificial intelligence, in M.-S. Hacid, Z.W. Ras, D.A. Zighed and Y. Kodratoff (eds.), *Foundations of*

Intelligent Systems, Lecture Notes in Artificial Intelligence 2366, Springer, 303–311.

Pietarinen, A.-V., 2003a. Peirce's game-theoretic ideas in logic, *Semiotica* 144, 33–47.

Pietarinen, A.-V., 2003b. What do epistemic logic and cognitive science have to do with each other?, *Cognitive Systems Research* 4, 169–190.

Pietarinen, A.-V., 2004a. Semantic games in logic and epistemology, in D. Gabbay, J. P. Van Bendegem, S. Rahman and J. Symons (eds.), *Logic, Epistemology and the Unity of Science*, Dordrecht Kluwer.

Pietarinen, A.-V., 2004b. Peirce's magic lantern: moving pictures of thought, to appear in *Transactions of the Charles S. Peirce Society: A Quarterly Journal in American Philosophy*.

Pietarinen, A.-V., 2004c. Some games logic plays, in D. Vanderveken (ed.), *Logic, Thought and Action*, Dordrecht Kluwer.

Rasmusen, E., 1994. *Games and Information* (first edition 1989), Cambridge, Mass.: Blackwell.

Rebuschi, M., 2004. IF and epistemic action logic, the present volume.

Sandu, G., 1993. On the logic of informational independence and its applications, *Journal of Philosophical Logic* 22, 29–60.

Sandu, G. and Pietarinen, A., 2001. Partiality and games: propositional logic, *Logic Journal of the IGPL* 9, 107–127.

Sandu, G. and Pietarinen, A., 2003. Informationally independent connectives, in G. Mints and R. Muskens (eds.), *Games, Logic, and Constructive Sets*, Stanford: CSLI, 23–41.

Selten, R. and Wooders, M., 2001. Cyclic games: an introduction and examples, *Games and Economic Behavior* 34, 138–152.

Tulenheimo, T., 2004. *Independence-Friendly Modal Logic: Studies in its Expressive Power and Theoretical Relevance*, dissertation, University of Helsinki.

Väänänen, J., 2002. On the semantics of informational independence, *Logic Journal of the IGPL* 10, 339–352.

Chapter 17

IF AND EPISTEMIC ACTION LOGIC

Manuel Rebuschi*
University of Nancy 2/Archives Poincaré (CNRS)
manuel.rebuschi@univ-nancy2.fr

17.1 Introduction

In this paper I will explore several features of the interrelation between logic and games, as they are conceived of according to two radically different frames: Hintikka's IF first-order logic (IF-FOL) and van Benthem's Epistemic Action Logic (EAL). EAL appears to be a young and serious challenger for IF-FOL since it provides a very sharp account of imperfect-information games. One of the goals of this paper is to show that both approaches should be taken for complementary views likely to mutually enrich one another rather than as irreducibly rival conceptions.

After a short presentation of both logics (Sections 17.2 and 17.3), we will deal with the question whether IF-FOL is reducible to EAL in some sense, or not. From an IF-FOL sentence φ, a model (or 'game board') $\mathbf{M}$ and an assignment s, one can build the evaluation game of φ in the given model relative to s: $game(\varphi, \mathbf{M}, s)$. Standard EAL then enables to describe the corresponding game tree. Does EAL enable a description of every relevant property of the game? Unfortunately, it is in general not the case.

Fact 1. *Standard EAL cannot express that* there is a uniform winning strategy for the verifier in $game(\varphi, \mathbf{M}, s)$.

An example is provided where standard EAL can express that there is a winning strategy at the beginning of the game, but cannot define that there

*I am greatly indebted to Johan van Benthem's critical comments and constructive ideas, without which this paper could not have been written down. I also wish to thank Tero Tulenheimo for his very useful comments. All errors remain mine, though.

J. van Benthem, G. Heinzmann, M. Rebuschi and H. Visser (eds.), The Age of Alternative Logics: Assessing Philosophy of Logic and Mathematics Today, 261–281.

is no *uniform* winning strategy. In the game-theoretical frame, this is an important fact: the existence of a uniform winning strategy for the verifier in $game(\varphi, \mathbf{M}, s)$ indeed constitutes the truth-condition for φ.

Using Hintikka's idea that IF-FOL has a truth definition inside IF, but not inside standard FOL, I will propose to translate it to EAL and to consider an IF extension of EAL (Section 17.4). In this new logic one can assert that there is a *u*niform *w*inning *s*trategy for the verifier in *game*(φ, **M**, *s*). Let's denote by **uws**(*game*(φ, **M**, *s*)) this formula. What is expected is that **uws**(*game*(φ, **M**, *s*)) is true at the root of *game*(φ, **M**, *s*) iff φ is true at **M**, *s*:

$$game(\varphi, \mathbf{M}, s), \mathbf{root} \Vdash \mathbf{uws}(game(\varphi, \mathbf{M}, s)) \Leftrightarrow \mathbf{M}, s \vDash \varphi \qquad (17.1)$$

Of course, there is no standard way to evaluate IF-EAL formulas but there is a natural resort to games for IF languages in general. I will thus propose to use new evaluation games. Now there is an interesting fact about the evaluation game of **uws**(*game*(φ, **M**, *s*)):

Fact 2. *At the root of G =* $game(\varphi, \mathbf{M}, s)$*, the evaluation game of* **uws**(*G*) *is isomorphic to the original game G:*

$$game(\mathbf{uws}(G), G, \mathbf{root}) \cong G. \qquad (17.2)$$

Were both games only bisimilar, the conclusion to be drawn would have been that they (their roots) would share the same *standard* EAL formulas; as they are isomorphic, they also share IF-EAL formulas. As a consequence, **uws**(G) is true at the roots of *both* games, not only of the original one G; **uws**(G) is thus true at the root of its own evaluation game:

Fact 3. $G = game(\varphi, \mathbf{M}, s)$ *is enough – i.e. in order to see whether the verifier has a uniform winning strategy in* $game(\mathbf{uws}(G), G, \mathbf{root})$*, no more 'meta game' is needed.*

Hence IF-EAL can put an end to some fearsome infinite regression between IF and EAL. This is just the translation of Hintikka's idea that truth for an IF language can be defined within the very same language.

Besides this result, a few issues connected to IF-EAL will be discussed in Section 17.5. Thanks to IF-EAL we will obtain a kind of 'equivalence' between an IF-FOL sentence φ and a corresponding IF-FOEL (IF first-order *epistemic* logic) formula, **egof**(φ, **M**, *s*), stating that in the evaluation game *game*(φ, **M**, *s*) the verifier knows which strategy is a winning strategy for herself. If one takes games and players at face value, this equivalence appears to be a very natural one since it corresponds to the truth-definition for IF formulas, viz. the existence of a uniform winning strategy for the verifier, formulated in the frame of EAL.

17.2 IF First-Order Logic in a Nutshell

IF First-Order Logic (IF-FOL) was created by Hintikka and developed by Hintikka and Sandu in the 1990s as an extension of standard first-order logic (FOL). It is a quite natural extension when connected to Game-Theoretical Semantics (GTS). According to GTS, each FOL-formula φ is interpreted relatively to some model **M** through a specific game, *game*(φ, **M**), played between two abstract players, the initial verifier and the initial falsifier, s.t. the first player (resp. the second one) has a uniform winning strategy iff the formula is true (resp. false) in **M**. (A more fine-grained definition would add an assignment s and consider *game*(φ, **M**, s), but it is not essential here.) Such evaluation games are played according to the following rules:

- **(R.At)**. If A is a true atomic sentence (or identity), the verifier wins *game*(A, **M**) and the falsifier loses it. If A is a false atomic sentence (or identity), vice versa.
- **(R.$\vee$)**. In *game*($\varphi_1 \vee \varphi_2$, **M**) the verifier picks out an index $i \in \{1, 2\}$. The rest of the game is as in *game*(φ_i, **M**).
- **(R.$\wedge$)**. *game*($\varphi_1 \wedge \varphi_2$, **M**) is likewise, except that the choice is made by the falsifier.
- **(R.$\exists$)**. *game*($(\exists x)\ \varphi[x]$, **M**) begins with the choice by the verifier of a member of do(**M**) and of a name b; the rest of the game is as in *game*($\varphi[b]$, **M**).
- **(R.$\forall$)**. *game*($(\forall x)\ \varphi[x]$, **M**) is likewise, except that the falsifier makes the choice.
- **(R.$\sim$)**. *game*($\sim\varphi$, **M**) is like *game*(φ, **M**), except that the roles of the two players (as defined by these rules) are interchanged.

Games corresponding to standard FOL formulas are of course determined, so that the principle of excluded middle holds. Moreover, these are perfect-information games: both players know or remember what all the previous moves of the play are. The "natural extension" of FOL consists in considering *imperfect*-information games, i.e. games where the players lack some information about the actual play. Hintikka suggested considering the case where the initial verifier has to make (some of) her moves (i.e. according to (R.$\vee$) or (R.$\exists$)) while ignoring some prior moves of her opponent. Such informationally independent moves in the semantic interpretation are expressed by the slash-notation at the level of the language. For instance, while playing the game associated with $\forall x\ (\exists y/\forall x)\ \varphi[x, y]$, the initial verifier will have to choose a value for y independently from that (chosen by the falsifier) of x. Similarly, in the game correlated to $\forall x\,(\varphi_1\,(\vee/\forall x)\,\varphi_2)$, the verifier will choose a disjunct φ_i not knowing the value of x.

The introduction of the slash-notation at the syntactic level leads to a new logic, IF-FOL, which enables to tackle new patterns of mutual (in)dependence between quantifiers. A paradigmatic example is provided by the Henkin or branching quantifiers, such as:

$$\begin{smallmatrix}\forall x\exists y\\ \forall z\exists u\end{smallmatrix}\varphi[x,y,z,u] \tag{17.3}$$

Into IF-FOL this formula is rendered e.g. by $\forall x\exists y\forall z(\exists u/\forall x)\,\varphi[x,y,z,u]$, whereas it is not expressible in standard FOL. The new patterns can be made visible in the Skolem normal forms of first-order formulas, where existential quantifiers are replaced by function symbols whose variables are taken among the preceding universally quantified ones: an existential quantifier independent from some universal quantifiers will thus be replaced by a function without the corresponding variable. For example, whereas the Skolem normal form of $\forall x\exists y\forall z\exists u\,\varphi[x,y,z,u]$, (where φ is quantifier-free) will be $\forall x\,\forall z\,\varphi[x, \mathbf{f}(x), z, \mathbf{g}(x, z)]$, that of the IF formula $\forall x\,\exists y\,\forall z\,(\exists u/\forall x)\,\varphi[x, y, z, u]$ will be $\forall x\,\forall z\,\varphi[x, \mathbf{f}(x), z, \mathbf{h}(z)]$, i.e. the function replacing the independent quantifier is made independent from x.

Skolem functions and their analogues for disjunctions play a special role in IF-FOL, since they are natural candidates to encode the initial verifier's winning strategies. Now the GTS truth-conditions of IF or standard first-order formulas are straightforwardly expressible using Skolem normal forms by prefixing them with second-order existential quantifiers of the Skolem (or strategy) functions. For instance, the IF formula $\forall x\,\exists y\,\forall z\,(\exists u/\forall x)\,\varphi[x, y, z, u]$ is GTS-true in some model **M** iff there is a winning strategy for the initial verifier in the correlated game, which is expressed by the second-order and in fact Σ^1_1 sentence: $\exists\mathbf{f}\,\exists\mathbf{h}\,\forall x\,\forall z\,\varphi[x, \mathbf{f}(x), z, \mathbf{h}(z)]$.

Let's add a few comments. Besides what can be called the *model-denotation* of a FOL formula φ, (i.e. its standard model-theoretic semantic value: the set of models where φ is true), Hintikka's semantics thus provides a more fine-grained denotation, the *game-denotation* of φ which is the set of games associated with φ where the initial verifier has a uniform winning strategy. The latter notion is more fine-grained than the former one because two logically equivalent formulas will lead to two different classes of games. Furthermore, both kinds of denotations can be restrictively defined relative to a given model **M**: the model-denotation is thus a set of denotations, whereas the game-denotation is a restricted set of games.

To put it in a reversed perspective: given a model **M**, true FOL formulas *describe* it in the usual way but can hardly be said to 'describe' their evaluation games. However, the second-order formula *uws*(*game*(φ, **M**, s)), which states that there is a *u*niform *w*inning *s*trategy for the verifier in *game*(φ, **M**, *s*), can be said to describe this game. The distinction between the two meanings of 'denotation' is very clear for standard FOL formulas. But in IF-FOL, the situation appears to be different. Indeed, IF-FOL is an extension of standard FOL which exactly coincides with the Σ_1^1 fragment of second-order logic. As a consequence, game-theoretical truth-conditions *uws*(*game*(φ, **M**, *s*)) of first-order sentences φ can be translated into IF-FOL, and in particular *IF formulas describe their own (GTS) truth-conditions*. It means that if φ is IF, it is identical with *uws*(*game*(φ, **M**, *s*)).

An IF-FOL formula φ can thus be considered for itself – with two denotations – or as a (second-order) assertion about its game-denotation. It does not mean that model- and game-denotations of IF formulas coincide, for model-denotations are classes of usual models whereas game-denotations are classes of games. However, relatively to a model **M** an IF formula φ is simultaneously an assertion about **M** (via its model-denotation) and an assertion about the class of games $game(\varphi, \mathbf{M}, s)$. Of course, this combination is not the case for standard FOL formulas. (I will go back to this reflexive aspect of IF-FOL in Section 17.5 below.)

17.3 Epistemic Action Logic in a Nutshell

Van Benthem's EAL is a competing frame to deal with imperfect-information games. More precisely: Whereas standard (i.e. game-theoretical) semantic interpretation associates imperfect-information games to IF-formulas, EAL is a dynamic epistemic language specially designed to describe the properties of those games. In fact, the starting point is different and the whole perspective is reversed: IF-FOL uses (evaluation) games as good tools for semantic interpretation whilst EAL considers games for themselves and aims to provide interesting insights on their properties.

Syntax. Like Propositional Dynamic Logic, EAL is a dual language made of 'formulas' and 'actions' with mutual combination. The vocabulary of EAL will partly depend on the model on which the games described by the language are played. Hence relatively to some model **M** of domain do(**M**), formulas and actions can have the following syntactic forms:

- Atoms: $At = \{\text{win}_\mathbf{V}, \text{turn}_\mathbf{V}, \text{win}_\mathbf{F}, \text{turn}_\mathbf{F}\}$
(the verifier **V** is winning, it is the verifier's turn, and the same for the falsifier **F**)

- Basic actions: $B = \{x := a, x := b, \ldots, y := a, y := b, \ldots, \text{L}, \text{R}\}$
(object picking – where a, $b\ldots$ are non-logical individual constants designating elements of do(**M**) –, going left, going right)

- Actions: $A ::= B \mid A \cup A \mid A \,;\, A$
  (basic actions, choice, composition[1])

- Wffs: $F ::= At \mid \bot \mid \neg F \mid F \vee F \mid \langle A \rangle\, F \mid \mathrm{K_i}\, F$
  (atoms, contradiction, negation, disjunction, action modality, epistemic modality)

Let's add a few comments:

- $\pi_1 \cup \pi_2$ is the (complex and) non-deterministic action consisting of the execution of π_1 or of π_2.

- $\pi_1 \,;\, \pi_2$ is the action consisting of the execution of π_1, then of π_2.

- $\mathrm{K_i}\, \varphi$ should be read as "**i** knows that φ", where $\mathbf{i} \in \{\mathbf{V}, \mathbf{F}\}$.

- $\langle \pi \rangle\, \varphi$ should be read as "some execution of π from the current node leads to a node where φ is true."

- We can define its dual, $[\pi]\varphi := \neg\langle\pi\rangle\neg\varphi$, and read it as "every execution of π from the current node leads to a node where φ is true."

- There are as many object-picking basic actions in *B* as elements in the domain do(**M**); consequently, when the model is infinite, so is the set of (basic) modalities.

Models. Games in extensive form provide (regular) Kripke models for EAL:

$$\mathbf{G} = (\mathrm{W}, \{\mathrm{R}_\pi\}_{\pi \in A}, \{\sim_\mathbf{V}, \sim_\mathbf{F}\}, \mathrm{V})$$

where: W is a set of states (nodes); R_π is the binary accessibility relation encoding the transition for action of type π (with $\mathrm{R}_{\pi_1 \cup \pi_2} = \mathrm{R}_{\pi_1} \cup \mathrm{R}_{\pi_2}$ and $\mathrm{R}_{\pi_1 ; \pi_2} = \mathrm{R}_{\pi_1} \circ \mathrm{R}_{\pi_2}$); $\sim_\mathbf{V}$ and $\sim_\mathbf{F}$ are equivalence 'uncertainty' relations encoding the information sets for each player; V is a valuation function for the atoms.

Semantics. Truth of formulas and successful executions of actions must be defined simultaneously. The truth of an EAL-formula φ at a node *s* of a game **G** is denoted by: $\mathbf{G}, s \Vdash \varphi$. The fact that a successful execution of the action π in a game **G** corresponds to a transition from *s* to *t* is denoted by:

$$\mathbf{G}, s, t \Vdash \pi$$

[1]Kleene iteration 'π^*', and tests for formulas '$(\varphi)?$' are other kinds of actions. (Adding the 'demonic' duality-operation 'π^d' would yield a kind of Game Logic, see Pauly and Parikh 2003.) Here I consider a simplified version of EAL without these operations, which is sufficient for the purpose of this paper.

- Formulas:

$$
\begin{array}{lll}
\mathbf{G}, s \nVdash \bot & & \\
\mathbf{G}, s \Vdash p & \text{iff} & s \in \mathrm{V}(p) \text{ for } p \in At \\
\mathbf{G}, s \Vdash \neg\varphi & \text{iff} & \mathbf{G}, s \nVdash \varphi \\
\mathbf{G}, s \Vdash \varphi_1 \vee \varphi_2 & \text{iff} & \mathbf{G}, s \Vdash \varphi_1 \text{ or } \mathbf{G}, s \Vdash \varphi_2 \\
\mathbf{G}, s \Vdash \langle\pi\rangle\varphi & \text{iff} & \text{there exists a node } t \text{ s.t. } \mathbf{G}, s, t \Vdash \pi \\
 & & \text{and } \mathbf{G}, t \Vdash \varphi \\
\mathbf{G}, s \Vdash K_i\varphi & \text{iff} & \mathbf{G}, t \Vdash \varphi \text{ for all node } t \text{ s.t. } s \sim_i t, \\
 & & \text{where } i \in \{\mathbf{V}, \mathbf{F}\}.
\end{array}
$$

- Actions:

$$
\begin{array}{lll}
\mathbf{G}, s, t \Vdash \pi & \text{iff} & (s, t) \in \mathrm{R}_\pi \\
\mathbf{G}, s, t \Vdash \pi_1 \cup \pi_2 & \text{iff} & \mathbf{G}, s, t \Vdash \pi_1 \text{ or } \mathbf{G}, s, t \Vdash \pi_2 \\
\mathbf{G}, s, t \Vdash \pi_1; \pi_2 & \text{iff} & \text{there exists a node } u \text{ s.t. } \mathbf{G}, s, u \Vdash \pi_1 \\
 & & \text{and } \mathbf{G}, u, t \Vdash \pi_2
\end{array}
$$

The following equivalences are obtained in a straightforward manner:

$$
\begin{array}{lcl}
\langle\pi_1 \cup \pi_2\rangle\,\varphi & \Leftrightarrow & \langle\pi_1\rangle\,\varphi \vee \langle\pi_2\rangle\,\varphi \\
[\pi_1 \cup \pi_2]\,\varphi & \Leftrightarrow & [\pi_1]\,\varphi \wedge [\pi_2]\,\varphi
\end{array}
$$

Example 1. Let's consider the (GTS) evaluation game **G** of the standard first-order formula

$$\forall x \exists y\, (x \neq y) \tag{17.4}$$

on a two-element model, in extensive form (see the figure on page 268.)

In this example, one can easily check the following assertion:

$$\mathbf{G}, \mathbf{1} \Vdash [x := a \cup x := b]\, \mathrm{turn}_{\mathbf{V}} \tag{17.5}$$

It means that whatever move is initially made by the falsifier, it will be the verifier's turn. Similarly, at Node 2 (i.e. after the choice of a by the falsifier), the verifier is not ensured to win whatever value she chooses:

$$\mathbf{G}, \mathbf{2} \nVdash [y := a \cup y := b]\, \mathrm{win}_{\mathbf{V}} \tag{17.6}$$

but she *can* choose one value and win:

$$\mathbf{G}, \mathbf{2} \Vdash \langle y := a \cup y := b\rangle\, \mathrm{win}_{\mathbf{V}} \tag{17.7}$$

Moreover, one can express that there is a winning strategy for the verifier:

$$\mathbf{G}, \mathbf{1} \Vdash [x := a \cup x := b]\langle y := a \cup y := b\rangle\, \mathrm{win}_{\mathbf{V}} \tag{17.8}$$

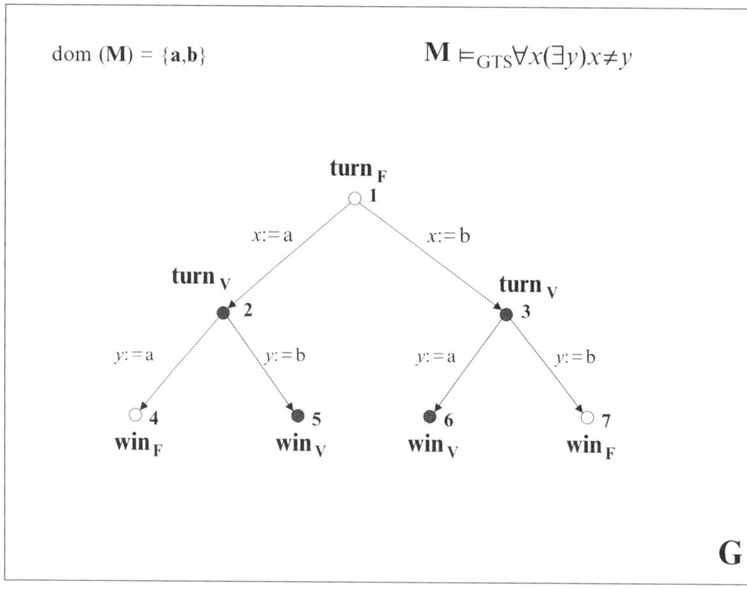

This can be generalized to more complex games, with more complex strategies: the existence of a winning strategy is thus expressed with more complex sequences of action diamonds and boxes.

Example 2. Now, if we introduce games of imperfect information, we can complete the illustration of EAL in a natural way. Consider the evaluation game **G'** of the IF-sentence:

$$\forall x(\exists y/x)(x \neq y) \tag{17.9}$$

The dotted line indicates the 'information set' for player **V**: it relates two states that are indistinguishable from the verifier's viewpoint. Information sets provide natural candidates for the accessibility relation of the epistemic operator $K_{\mathbf{V}}$.[2] As van Benthem explains, one can thus see that at Node 2 the verifier knows 'de dicto' that she has some winning strategy:

$$\mathbf{G'}, \mathbf{2} \Vdash K_{\mathbf{V}}(\langle y := a\rangle \text{win}_{\mathbf{V}} \vee \langle y := b\rangle \text{win}_{\mathbf{V}}) \tag{17.10}$$

because in every epistemic alternative to 2, namely 2 and 3, she has one:

$$\begin{array}{l} \mathbf{G'}, \mathbf{2} \Vdash \langle y := a\rangle \text{win}_{\mathbf{V}} \vee \langle y := b\rangle \text{win}_{\mathbf{V}} \\ \mathbf{G'}, \mathbf{3} \Vdash \langle y := a\rangle \text{win}_{\mathbf{V}} \vee \langle y := b\rangle \text{win}_{\mathbf{V}} \end{array} \tag{17.11}$$

[2]In fact, one has to consider the complete equivalence relation linking nodes undistinguishable for player V as the required accessibility (or 'alternativeness') relation, i.e. loops would have to be added at each node.

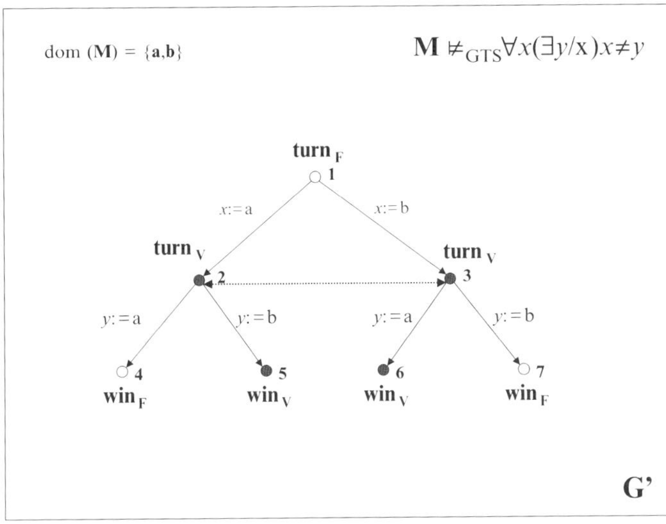

whereas she doesn't know 'de re' which strategy is the winning one:

$$\mathbf{G}', \mathbf{2} \Vdash \neg \mathrm{K}_{\mathbf{V}} \langle y := a \rangle \mathrm{win}_{\mathbf{V}} \wedge \neg \mathrm{K}_{\mathbf{V}} \langle y := b \rangle \mathrm{win}_{\mathbf{V}} \tag{17.12}$$

The contrast between the two kinds of knowledge for the verifier can hence be accounted for within EAL. Should we stop here and consider that one can get rid of IF languages thanks to this dynamic epistemic logic?

IF-FOL and EAL. As a competing frame, has EAL the same expressive power as IF-FOL? As such, the question is meaningless since EAL describes local properties of (evaluation-)games whereas IF-FOL (like standard FOL) describes usual models. To put it in other words: the model-denotation of EAL formulas correspond to the game-denotation of (IF-)FOL formulas. Hence lots of assertions about game trees can be made in EAL that have no counterpart in standard IF-FOL, such as the following:

- Basic actions are deterministic (which is valid in EAL):
  $\langle \pi \rangle \varphi \rightarrow [\pi] \varphi$, for all $\pi \in B$.

- Every turn of the falsifier is followed by one of the verifier:
  $\mathrm{turn}_{\mathbf{F}} \rightarrow [\pi]\, \mathrm{turn}_{\mathbf{V}}$, for all $\pi \in B$.
  (or: $\mathrm{turn}_{\mathbf{F}} \rightarrow [\cup_{d \in dom(\mathbf{M})} (x{:=}d) \cup (\mathrm{L} \cup \mathrm{R})]\, \mathrm{turn}_{\mathbf{V}}$)

  and so forth.

Another reason why EAL should not be directly compared to FOL or IF-FOL is the fact that EAL depends on a previously chosen model **M**: basic action modalities such as object picking obviously depend on the domain do(**M**).[3] EAL thus corresponds to an already interpreted language.

So it is interesting to compare the two languages not as wholes, but on specific formulas. I will concentrate on the kind of assertion that play a crucial role in GTS and IF logic, namely the assertion that there is a uniform winning strategy for the initial verifier in $game(\varphi, \mathbf{M}, s)$. Let's denote by **uws**$(game(\varphi, \mathbf{M}, s))$ the corresponding EAL formula, when it exists. For instance, if **M** is a two-element model with its domain do(**M**) = $\{a, b\}$, the existence of a uniform winning strategy for the verifier in $game((\exists x)\, x = x, \mathbf{M}, s)$ is defined by: **uws**$(game((\exists x)\, x = x, \mathbf{M}, s)) = \langle x{:=}a \rangle\, \mathrm{win}_{\mathbf{V}}$.

Unfortunately this easy case does not generalize:

Fact 1. *Standard EAL cannot express that there is a uniform winning strategy for the verifier in* $game(\varphi, \mathbf{M}, s)$.

It is not definable in EAL but in a modal fixed-point extension of EAL (see van Benthem 2000a, 2000b). A demonstration of FACT 1 could consist in showing that for some specific IF-FOL formula φ, the class of games $game(\varphi, \mathbf{M}, s)$ where there is a uniform winning strategy for the verifier is not definable in standard EAL. In the next section, I will only give an illustration of FACT 1 with the example of an IF formula such that the existence of a uniform winning strategy for the verifier in the correlated game is directly definable in an extension of EAL, but with no obvious counterpart in standard EAL.

17.4 IF modal logics and IF-EAL

In comparison with IF-FOL, EAL appears to give a new, more local and fine-grained approach of imperfect-information games. On the other hand, IF-FOL enables to express game-theoretical truth-conditions of FOL formulas, and this cannot be grasped within EAL (see FACT 1). I will now propose a kind of compromise: an extension of EAL which preserves the sharp insight of EAL while increasing its expressive power.

'Slashing' some modal language, i.e. considering its IF version, is one interesting way to extend it. Tulenheimo 2004 is the first systematic work on this issue and it contains several important results. What I will consider here is the 'uniformity interpretation' of the slash-notation for modal languages.[4] This interpretation is grounded in a game-theoretical semantics for modal logics in the same manner as IF-FOL is based on GTS for standard FOL.

---

[3]The usual quantifiers can then be construed as abbreviations for action modalities ($\forall x =_{def} [\cup_{d \in dom(\mathbf{M})} x := d]$; $\exists x =_{def} \langle \cup_{d \in dom(\mathbf{M})} x := d \rangle$).

[4]Tulenheimo 2004 proposes two other interpretations, namely the Backwards-Looking Operations interpretation (BLO) and the Algebraic one (ALG).

Tulenheimo's IF modal logic of k modality types (IFML[k]) is an extension of basic modal logic ML[k] where modal operators are allowed to be independent from specified other modal operators. In other words, formulas such as the following are allowed:

$$[A]_1\ [A]_2\ (<C>/[A]_2)\ \varphi, \text{ where } \varphi \text{ is a standard ML}[k]\text{-formula}$$

whereas others such as the following, where modalities are independent from connectives, are not:

$$(<A>_1/\wedge)\ \varphi_1 \wedge (<B>_2/\wedge)\ \varphi_2$$

The latter is a formula of another logic developed by the same author, *Extended* IF modal logic of k modality types (EIFML[k]).

For the purpose of this paper, I will choose the 'extended' mode of slashing modal logic instead of the restricted mode, since it appears to be easier to handle according to our intuitions about the epistemic operators. However, Tulenheimo demonstrated that IFML[k] is translatable into standard FOL, whereas EIFML[k] is not – it is second-order. One interesting issue would be to check whether the existence of a uniform winning strategy for the verifier is definable in the restricted IF extension of EAL (let's denote it by: IF*-EAL), and more generally, what about imperfect-information games cannot be said with IF*-EAL and requires the extended version, IF-EAL.

GTS for (IF-)EAL. EAL is a propositional (multi-)modal language: as such, it can have a game-theoretical interpretation. For that purpose, one needs to choose a model before playing. This model is in fact a game. A GTS-interpretation of an EAL-formula leads thus to the construction of a meta-game, a game 'about' the original game. Let's recall that a model for EAL is a tuple:

$$\mathbf{M} = (W, \{R_\pi\}_{\pi\in A}, \{\sim_\mathbf{V}, \sim_\mathbf{F}\}, V)$$

where W is a set of states, $\{R_\pi\}_{\pi\in A}$ the set of accessibility relations corresponding to actions, $\sim_i$ is the accessibility (equivalence) relation for the epistemic operator K_i, and V a valuation function for atomic formulas. (Models for IF-EAL will be the same.)

Now, we can give natural GTS rules for the (basic) action modalities:

- **(G.$\langle\pi\rangle$).** If the game is of the form $game(\langle\pi\rangle\varphi, \mathbf{M}, s)$, then the verifier picks out, if possible, a state t resulting from the execution of π (i.e. $R_\pi st$); the rest of the game is as in $game(\varphi, \mathbf{M}, t)$; if she cannot choose such a state, she loses and the falsifier wins.

- **(G.[π]).** $game([\pi]\varphi, \mathbf{M}, s)$ is likewise, except that the falsifier makes the choice.

- **(G.$\mathrm{K_V}$).** If the game is of the form $game(\mathrm{K_V}\varphi, \mathbf{M}, s)$, then the falsifier picks out an epistemic alternative of the current state for the verifier (i.e. a state t s.t. $s \sim_{\mathbf{V}} t$); the rest of the game is as in $game(\varphi, \mathbf{M}, t)$. (*Remark*: As it is assumed that the alternativeness relation is reflexive, the falsifier can always pick out an alternate to the current state.)

We can also add rules for complex action modalities:

- **G($\langle\cup\rangle$).** $game(\langle\pi_1 \cup \pi_2\rangle\varphi, \mathbf{M}, s)$ starts with the choice of an index $i \in \{1, 2\}$ by the verifier, and the rest of the game is as in $game(\langle\pi_i\rangle\varphi, \mathbf{M}, s)$;
- **G([$\cup$]).** $game([\pi_1 \cup \pi_2]\varphi, \mathbf{M}, s)$ is likewise, except that the falsifier makes the choice;
- **G($\langle;\rangle$).** $game(\langle\pi_1; \pi_2\rangle\varphi, \mathbf{M}, s)$ is like $game(\langle\pi_1\rangle\langle\pi_2\rangle\varphi, \mathbf{M}, s)$.

So we get, for any IF-EAL formula φ:

$\mathbf{M}, s \Vdash_{\mathrm{GTS}} \varphi \quad \Leftrightarrow \quad$ there is a winning strategy for the verifier in $game(\varphi, \mathbf{M}, s)$.

IF-Epistemic Action Logic. Game-theoretically interpreted, EAL can be extended to cases of imperfect-information and lead to Independence-Friendly Epistemic Action Logic (IF-EAL). The motivation for such an extension rests on the ability of IF epistemic logic to account for the distinction between knowledge de dicto and knowledge de re (*knowing-that* vs. *knowing-what, who, which* ... in Hintikka's terminology). As we have already seen, in standard EAL one can express the knowledge de dicto of the existence of a winning strategy by means of a propositional disjunction

$$\mathbf{G}', \mathbf{2} \Vdash \mathrm{K_V}(\langle y := a\rangle \mathrm{win_V} \vee \langle y := b\rangle \mathrm{win_V}) \tag{17.10}$$

or by means of some complex action modality involving the union symbol:

$$\mathbf{G}', \mathbf{2} \Vdash \mathrm{K_V}(\langle y := a \cup y := b\rangle \mathrm{win_V}) \tag{17.13}$$

We can check the extensive game for (17.10) (see next figure):

Then the ignorance de re of the same winning strategy can be accounted for in IF-EAL with the slash notation, applied either to the disjunction:

$$\mathbf{G}', \mathbf{2} \nVdash \mathrm{K_V}(\langle y := a\rangle \mathrm{win_V}(\vee/\mathrm{K_V})\langle y := b\rangle \mathrm{win_V}) \tag{17.14}$$

or to the union symbol:

$$\mathbf{G}', \mathbf{2} \nVdash \mathrm{K_V}(\langle y := a(\cup/\mathrm{K_V})y := b\rangle \mathrm{win_V}) \tag{17.15}$$

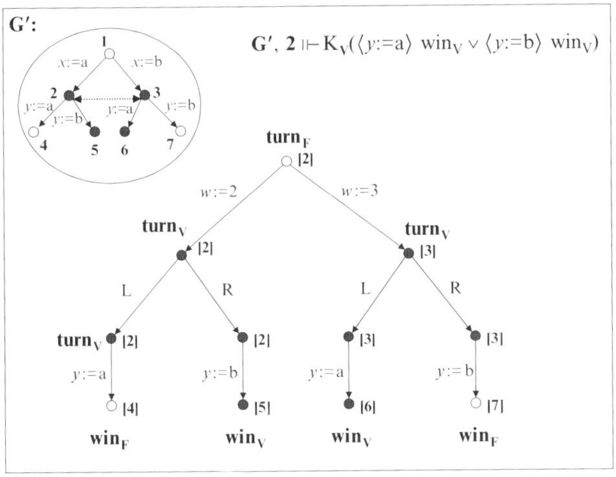

(As was announced before, these formulas clearly belong to the *extended* IF version of EAL.) Both formulas (17.14) and (17.15) mean that the choice of picking *a* or picking *b* is independent of the knowledge of the verifier. (Formula (17.15) indicates a new kind of complex actions, whose status is not clear at first glance!). In order to evaluate the IF-EAL sentences using GTS, we need not introduce new rules: the rules for standard EAL can do the job.

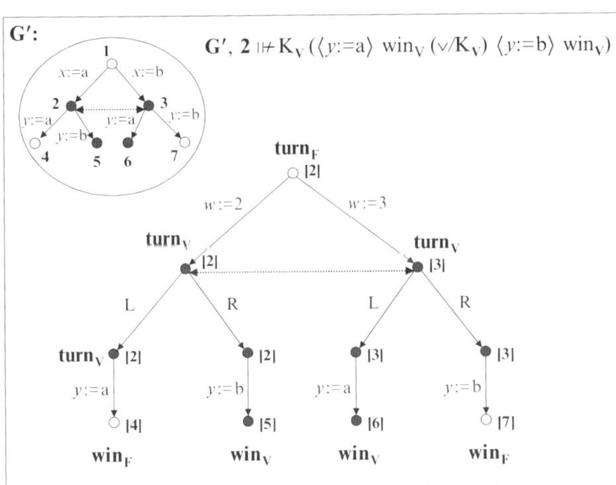

Generally speaking, in order to obtain IF-EAL we have to allow several patterns of independence between operators – and in fact, lots of them – together with the correlated formula forms:

- $(\langle \pi_2 \rangle / [\pi_1])$, $(\langle \pi \rangle / \mathrm{K_V})$…
- $(\vee / [\pi])$, $(\vee / \mathrm{K_V})$, $(\vee / [\pi], \mathrm{K_V})$

and lots of new (complex action) modalities:

- $\langle \pi_1 \,(\cup / [\pi])\, \pi_2 \rangle$, $\langle \pi_1 \,(\cup / \mathrm{K_V})\, \pi_2 \rangle$, $\langle \pi_1 \,(\cup / [\pi], \mathrm{K_V})\, \pi_2 \rangle$…
- $[\pi_1 (\cup / [\pi])\, \pi_2]$, $[\pi_1 (\cup / \langle \pi \rangle)\, \pi_2]$, $[\pi_1 (\cup / \mathrm{K_V})\, \pi_2]$…

IF-EAL thus seems to provide a good account of different kinds of verifier's knowledge through the game process. Now, as our main concern is evaluation games we have to look at what comes about at the root, that is, at Node 1 of the initial game.

At the root of game **G'**, there is no uniform winning strategy for the verifier: this can be expressed with each of the following equivalent formulas:

$$\begin{aligned} &\mathbf{G}', \mathbf{1} \nVdash_{\mathrm{GTS}} [x := a \cup x := b](\langle y := a \cup y := b \rangle / [x := a \cup x := b])\mathrm{win_V} \\ &\mathbf{G}', \mathbf{1} \nVdash_{\mathrm{GTS}} [x := a \cup x := b](\langle y := a(\cup / [x := a \cup x := b])y := b \rangle \mathrm{win_V} \end{aligned} \tag{17.16}$$

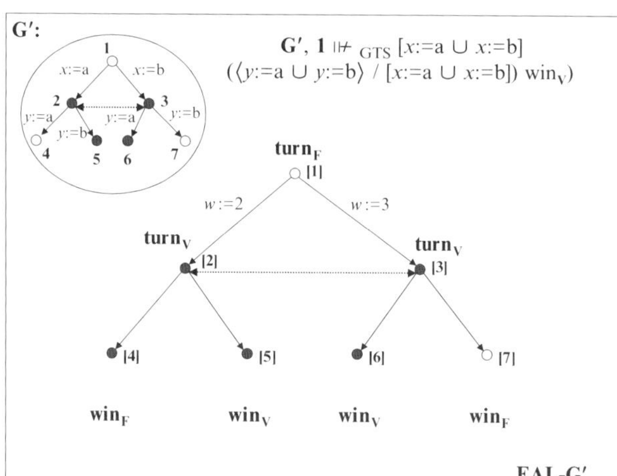

But in fact, the following formula which states that there is a winning strategy for the verifier in the corresponding perfect information game **G**, still holds in **G'**:

$$\mathbf{G}', \mathbf{1} \Vdash_{\mathrm{GTS}} [x := a \cup x := b]\langle y := a \cup y := b \rangle \mathrm{win_V} \tag{17.17}$$

Indeed, there is still a winning strategy for the verifier in the imperfect information game (17.17), but it is not a uniform one (17.16). Or to put it in other words: the verifier still has a winning strategy, but it is no more available to her. The contrast between the EAL formula in (17.17) ("there is a winning strategy") and the IF-EAL formula in (17.16) ("there is no uniform winning strategy") constitutes an interesting illustration of FACT 1: no obvious standard EAL formula appears that would do the job of the IF-EAL formula about the uniform strategy.

17.5 Game comparison

Isomorphism. One can compare the evaluation game $\mathbf{G}'$ of our original IF-sentence (17.9):

$$\forall x(\exists y/x)(x \neq y) \tag{17.9}$$

in the model $\mathbf{M}$, with the evaluation game $\mathbf{EAL\text{-}G}'$ of the assertion of the existence of some uniform winning strategy for the verifier in game $\mathbf{G}$:

$$\mathbf{M} \nvDash_{\text{GTS}} \forall x(\exists y/x)(x \neq y) \tag{17.18}$$

$$\mathbf{G}', \mathbf{1} \nVdash_{\text{GTS}} [x := a \cup x := b](\langle y := a(\cup/[x := a \cup x := b])y := b\rangle \text{win}_{\mathbf{V}} \tag{17.16}$$

It's worth noting the following: There is an obvious *bisimulation* between G′ and EAL-G′ relating the roots. Consequently: *The roots of the games G′ and EAL-G′ verify the same EAL formulas* (van Benthem 2000b, 162).

But this result is still limited: the roots of the 'object-game' and of the 'meta-game' verify the same *standard* EAL formulas, and of course this does not mean that they share every *IF*-EAL formula. As IF multi-modal logic is strictly more expressive than the corresponding standard fragment (see Tulenheimo 2004), the bisimulation relating the roots of the two games is not enough to ensure that the roots verify the same IF-EAL formulas, especially those stating that there's a winning strategy for the verifier in the evaluation of an IF first-order formula. Therefore, in order to extend the equivalence of some first-order sentence with its epistemic GTS-oriented form to *IF* first-order sentences, we need more than bisimulation.

Fortunately we have here a higher, and in fact the highest degree of similarity between the two games, namely *isomorphism*, and this can easily be generalized to other (IF or standard) first-order sentences φ:

Fact 2. *At the root of G =* $game(\varphi, \mathbf{M}, s)$*, the evaluation game of* **uws**(*G*) *is isomorphic to the original game G:*

$$game(\mathbf{uws}(G), G, \mathbf{root}) \cong G. \tag{17.2}$$

Indeed, let's consider such a formula φ in prenex normal form, and its transformation **uws**(*game*(φ, **M**, *s*)) into a IF-EAL formula, stating that there is a uniform winning strategy for the verifier in the game associated with φ:

(i) Replace the (independent) connectives by quantifiers, e.g.: $\forall x\,(\psi_1\,(\vee/\forall x)\,\psi_2)$ will be transformed into: $\forall x\,(\exists i/\forall x)\,\psi(i)$, where $\psi(i) = \psi_i$.

(ii) φ is now of the form $Q_0 x_0 (Q_1 x_1/W_1)(Q_2 x_2/W_2)\ldots(Q_n x_n/W_n)\,\psi$, where Q_i is a quantifier, W_i the set of quantifiers Q_i is independent from ($W_i \subseteq \{Q_0, \ldots, Q_{i-1}\}$), and ψ the matrix. Each quantifier $Q_i x_i$ can be translated in the following way:

- if it is a universal quantifier ($\forall_i x_i$), then replace it by the "box": $[\bigcup_{i\, d \in dom(\mathbf{M})}(x_i := d)]$,
- if it is an existential one ($\exists_i x_i$), then replace it by the "diamond": $\langle\bigcup_{i\, d \in dom(\mathbf{M})}(x_i := d)\rangle$.

Such a translation is to be effected also for quantifiers in the sets W_i.

(iii) Replace the matrix ψ by $\text{win}_\mathbf{V}$.

For instance, from the first-order sentence: $\forall_0 x_0 (\exists_1 x_1/\forall_0 x_0)\,\psi(x_0, x_1)$ on a model with two elements (dom(**M**) = $\{a, b\}$), we will reach the IF-EAL formula:

$$[(x_0 := a) \cup_0 (x_0 := b)](\langle(x_1 := a) \cup_1 (x_1 := b)\rangle/[(x_0 := a) \cup_0 (x_0 := b)])\,\text{win}_\mathbf{V} \quad (17.19)$$

According to this transformation, the evaluation game of a first-order sentence φ and that of the corresponding formula **uws**(*game*(φ, **M**, *s*)) are obviously isomorphic: this could be proved by a straightforward induction on the complexity of (the prefix of) φ.

Stop the regression. As the evaluation game of φ, $game(\varphi, \mathbf{M}, s)$, and its meta-game (i.e. the evaluation game of $\mathbf{uws}(game(\varphi, \mathbf{M}, s))$) are isomorphic, their roots verify the same *IF*-EAL formulas. It leads to the following fact:

Fact 3. *G = $game(\varphi, \mathbf{M}, s)$ is enough – i.e. in order to see whether the verifier has a uniform winning strategy in $game(\mathbf{uws}(G), G, \mathbf{root})$, no more 'meta game' is needed.*

This explains how to stop the headlong rush apparently threatening the whole enterprise: While extending EAL (which was designed to escape from IF) into an IF version, we are not led to build a new language to speak about the new games. IF-EAL is enough: $\mathbf{uws}(game(\varphi, \mathbf{M}, s))$ does not only state that there is a winning strategy for the verifier in the evaluation game of φ, $\mathbf{uws}(game(\varphi, \mathbf{M}, s))$ also states that there is a winning strategy for the verifier *in its own evaluation game.*

17.6 Discussion

Some advantages of IF-EAL. Our IF extension of EAL is not superfluous, as EAL is expected to account for the winning strategies of the evaluation games (among other things). Thanks to informational independence, the enriched version of EAL can account for imperfect information evaluation games in a straightforward way.

1 IF-EAL enables to formulate the contrast between knowledge de dicto and ignorance de re in a way which is more natural than standard EAL: this can be seen with the EAL formula (17.12), that is literally rendered by: "(At Node 2) the verifier doesn't know whether choosing a is a winning strategy, and she doesn't know whether choosing b is a winning strategy." By contrast, formula (17.15) is directly read as: "(At Node 2) the verifier doesn't know which choice is a winning strategy". And the gap between (17.10) and (17.12) – expressing the difference between knowledge de dicto and knowledge de re – should similarly be compared to the distinction between (17.10) and (17.14) (or between (17.13) and (17.15)).

2 As was already mentioned, what formulas (17.16)-(17.17) reveal is that the non-existence of uniform winning strategy for the verifier in the whole game is not expressible in a direct way in standard EAL. And we will see below that the *knowledge* of the verifier is of no help in such cases.

3 Some IF-EAL formulas cannot be translated into standard EAL formulas. An example is provided by the following schema:

$$[\,]_1 \mathrm{K}_{\mathbf{V}} [\,]_2 (\langle\rangle / \mathrm{K}_{\mathbf{V}})\, \varphi \tag{17.20}$$

where the diamond is independent from the epistemic operator, but still dependent from the boxes. One could meet such a schema in the evaluation game of e.g.:

$$\forall x \forall y \exists z (x + y = z) \tag{17.21}$$

stating that whatever value is chosen for x by the falsifier, the verifier will know *de re* what is her winning strategy – and this is certainly true!

Epistemic statements. Let's go back to what happens with the epistemic operator at the root of the games. Relatively to the alternativeness relation ($\sim_{\mathbf{V}}$), evaluation games of IF first-order sentences start with a reflexive singleton – because informational independence (namely, independent quantifiers) cannot occur just at the beginning of a sentence, but only 'inside' it.

As a result, we have the following implication and equivalence for any formula of IF-EAL:

$$\mathbf{G}, \mathbf{1} \Vdash_{\mathrm{GTS}} \varphi \quad \Rightarrow \quad \mathbf{G}, \mathbf{1} \Vdash_{\mathrm{GTS}} \mathrm{K}_{\mathbf{V}}\varphi \tag{17.22}$$

$$\mathbf{G}, \mathbf{1} \Vdash_{\mathrm{GTS}} \mathrm{K}_{\mathbf{V}}\varphi \quad \Leftrightarrow \quad \mathbf{G}, \mathbf{1} \Vdash_{\mathrm{GTS}} \mathrm{K}_{\mathbf{V}}(\varphi/\mathrm{K}_{\mathbf{V}}) \tag{17.23}$$

where (φ/$\mathrm{K}_{\mathbf{V}}$) is the IF-EAL sentence resulting from φ by the replacement of each action diamond ($\langle\pi\rangle$/W) by the $\mathrm{K}_{\mathbf{V}}$-liberated corresponding one, ($\langle\pi\rangle$/W, $\mathrm{K}_{\mathbf{V}}$), and the same for each disjunction ($\vee$/W). Moreover as the alternativeness relation is reflexive, the knowledge property ($\mathrm{K}_{\mathbf{V}}\varphi \to \varphi$) holds in our frame: the implication (17.22) actually leads to an equivalence:

$$\mathbf{G}, \mathbf{1} \Vdash_{\mathrm{GTS}} \varphi \quad \Leftrightarrow \quad \mathbf{G}, \mathbf{1} \Vdash_{\mathrm{GTS}} \mathrm{K}_{\mathbf{V}}\varphi \tag{17.24}$$

and, combined with (17.23), we obtain an interesting equivalence between any IF-EAL formula and its 'epistemic' version:

$$\mathbf{G}, \mathbf{1} \Vdash_{\mathrm{GTS}} \varphi \quad \Leftrightarrow \quad \mathbf{G}, \mathbf{1} \Vdash_{\mathrm{GTS}} \mathrm{K}_{\mathbf{V}}(\varphi/\mathrm{K}_{\mathbf{V}}) \tag{17.25}$$

Hence at the root, *the epistemic operator cannot provide any new and interesting description* of the game. However we can raise an interesting question with this result: In what sense can a first-order sentence φ be said equivalent to what I shall call its *epistemic game-oriented form*, i.e. to the IF epistemic formula asserting the knowledge *de re* of a winning strategy by the verifier in the evaluation-game of the original sentence? Let's denote by **egof**(φ) the *epistemic game-oriented form* of φ: **egof**(φ) belongs to IF-FOEL (IF first-order epistemic logic); it is like $\mathrm{K}_{\mathbf{V}}(\varphi/\mathrm{K}_{\mathbf{V}})$, where ($\varphi$/$\mathrm{K}_{\mathbf{V}}$) is the IF sentence resulting from φ by the replacement of each existential quantifier $(\exists_i x_i/\mathrm{W}_i)$ by the $\mathrm{K}_{\mathbf{V}}$-liberated corresponding one, $(\exists_i x_i/\mathrm{W}_i, \mathrm{K}_{\mathbf{V}})$. For instance:

$$\mathbf{egof}(\forall x \exists y(x \neq y)) = \mathrm{K}_{\mathbf{V}}\forall x(\exists y/\mathrm{K}_{\mathbf{V}})(x \neq y) \tag{17.26}$$

We can now compare the respective 'translations' of φ and **egof**(φ) into IF-EAL, *i.e.*, **uws**(*game*(φ, **M**, *s*)) and **uws**(*game*(**egof**(φ), **M**, *s*)) respectively, stating that there is a uniform winning strategy for the verifier in the game associated with φ, **M**, *s*, and the same with **egof**(φ). That **uws**(*game*(**egof**(φ), **M**, *s*)) = $\mathrm{K}_{\mathbf{V}}$(**uws**(*game*(φ, **M**, *s*))/$\mathrm{K}_{\mathbf{V}}$) in IF-EAL is readily verified. Consequently, and according to (17.25), for every IF first-order formula φ (**G** being isomorphic to its evaluation game):

$$\begin{array}{c}\mathbf{G}, \mathbf{1} \Vdash_{\mathrm{GTS}} \mathbf{uws}(game(\varphi, \mathbf{M}, s)) \\ \Leftrightarrow \\ \mathbf{G}, \mathbf{1} \Vdash_{\mathrm{GTS}} \mathbf{uws}(game(\mathbf{egof}(\varphi), \mathbf{M}, s))\end{array} \tag{17.27}$$

This means that there is a uniform winning strategy for the verifier in the game associated with a specific formula φ if and only if there is one in the game associated with the 'epistemic game-oriented form' of φ. Now, the right side of the equivalence, $\mathbf{uws}(game(\mathbf{egof}(\varphi), \mathbf{M}, s))$ can be read in the following two ways: (i) it can mean that the verifier in $game(\varphi, \mathbf{M}, s)$ knows (de re) which is the winning strategy for herself (this is the reason why it is equivalent to $\mathbf{uws}(game(\varphi, \mathbf{M}, s))$); (ii) it can also be understood as meaning that there is a winning strategy for the verifier of $\mathbf{egof}(\varphi)$ in the evaluation game $game(\varphi, \mathbf{M}, s)$ of φ in $\mathbf{M}$ (which leads to the intended equivalence). So (17.27) exactly states that φ is GTS-true iff $\mathbf{egof}(\varphi)$ is GTS-true: this is the expected equivalence.

To sum up: any IF first-order formula φ is equivalent to a correlated formula whose meaning is "The verifier (of the evaluation game of φ) *knows* de re *which* strategy is a winning strategy for herself". This reflexive feature of IF logic, usually claimed in an informal way, can be established within the EAL frame which – against Hintikka – takes evaluation games and their players' knowledge and powers at face value. However, this result is established thanks to the application of two Hintikkian ideas to EAL: IF extension, and the epistemic concept of *knowing-wh.*

Is it Genuine knowledge? The equivalence (17.23) given above implies that there will be no more distinction between the verifier's knowledge 'de re' of her (uniform) winning strategy, and her corresponding knowledge 'de dicto'. This would threaten the whole construction of our epistemic logic, if it were to hold in general, but here, of course, it is not the case (as can be seen e.g. at Node 2 in the preceding examples). In fact, this equivalence can be read in a more 'positive' way: (17.23) means that the verifier's knowledge of the existence of some winning strategy implies her knowledge of that strategy. If the frame employed here is a suitable one, it means that the verifier in evaluation games is a '*perfect knower*' in some sense. This meets the requirement that players of such games be ideal players. What is more: The equivalence (17.27) between a sentence and its epistemic game-oriented form strongly reinforces the idea that the truth of a sentence is a property of the 'game-board' rather than of the game course.

However, we have reached an interesting phenomenon with IF-EAL. The sentences designed to describe the evaluation game of (IF or standard) first-order sentences describe their own evaluation games. This reflexive feature is in fact independent from any 'epistemic' property of the players: what we needed to arrive at it is only dynamic logic, with no epistemic operator.

17.7 Conclusion

Dealing with imperfect-information games, we are usually faced with two competing frames: IF-FOL and EAL. After having observed that the two logics provide complementary views on games, I proposed to consider an extension of EAL: IF-EAL, based on some game-theoretical semantics for dynamic logic. Thanks to this new IF multi-modal, dynamic and epistemic language, we can express for any IF-FOL formula φ the existence of a uniform winning strategy for the verifier of some corresponding evaluation game with a formula **uws**(*game*(φ, **M**, *s*)) which does not belong to standard EAL. In general, different epistemic assertions about the players appear to be more intuitive in the extended version than in the original one.

Moreover, we showed that **uws**(*game*(φ, **M**, *s*)) *constitutes its own truth-conditions*, since it coincides with the assertion of the existence of some winning strategy in its own GTS evaluation game. This is an EAL-correlate of a well-known 'reflexive fact' in IF-FOL, namely that the truth-conditions of a formula can be formulated in the same language, using the very same formula. Another correlate of the same equivalence was established in IF-FOEL, φ being equivalent to **egof**(φ), its epistemic game-oriented form. Finally, asserting a formula and asserting that the initial verifier knows which is the winning strategy in its evaluation game, are the same assertion.

The important fact about these equivalences which all reflect Hintikka's idea that IF languages can define their own truth predicate, is that it can stop the indefinite regression IF/EAL/IF/EAL... Van Benthem indeed created EAL to escape from IF logic. Taking evaluation games seriously, EAL gives a local and precise perspective on features of games that were neglected from the global viewpoint of FOL, IF or not. What is more: EAL reduces informational independence to dynamic and epistemic features of players of evaluation games. Then 'slashing' EAL seems going back to the prior situation.

However this procedure is not worthless. IF-EAL formulas asserting the existence of a uniform winning strategy in a game G have the nice property that their own evaluation game is G. They are simultaneously about G, and evaluated by G. The hierarchy of games and meta-games thus stops with IF-EAL formulas.

References

Benthem, Johan van: 1996, *Exploring Logical Dynamics*, CSLI Publications, Stanford.

Benthem, Johan van: 1999, 'When are Two Games the Same?', Technical report ILLC. Appeared as 'Extensive Games as Process Models', *Journal of Logic, Language and Information* **11** (2002), 289-313.

Benthem, Johan van: 2001a, 'Games in Dynamic-Epistemic Logic', *Bulletin of Economic Research* **53**:4, 219-248 (invited lecture LOFT-4 2000, Torino).

Benthem, Johan van: 2001b, 'Logic in Games', Lecture Notes, ILLC Amsterdam.

Benthem, Johan van: 2002, 'The Epistemic Logic of IF Games', Tech Report PP-2002-24, ILLC Amsterdam. To appear in Lewis Hahn, ed., Hintikka Volume, *The Library of Living Philosophers*.

Blackburn, P., M. de Rijke and Y. Venema: 2002, *Modal Logic*, Cambridge University Press.

Hintikka, Jaakko: 1996a, *The Principles of Mathematics Revisited*, Cambridge University Press.

Hintikka, Jaakko: 1996b, 'Knowledge Acknowledged: Knowledge of Propositions vs. Knowledge of Objects', *Philosophy and Phenomenological Research* **61**, 251-273.

Hintikka, Jaakko: 2001, 'Post-Tarskian Truth', *Synthese* **126**, 17-36.

Hintikka, Jaakko: 2001a, 'Intuitionistic Logic as Epistemic Logic', *Synthese* **127**, 7-19.

Hodges, Wilfrid: 1997, 'Compositional Semantics for a Language of Imperfect Information', *Journal of the IPGL* **5**, 539-563.

Pauly, Marc and Rohit Parikh: 2003, 'Game Logic – An Overview', *Studia Logica* **75**, 165-182.

Pietarinen, Ahti: 1999, 'Informational Independence in Epistemic Logic', Preprint.

Pietarinen, Ahti and Gabriel Sandu: 2000, 'Games in Philosophical Logic', *Nordic Journal of Philosophical Logic* **4**, 143-173.

Tulenheimo, Tero: 2003, 'On IF Modal Logic and its Expressive Power', in P. Balbiani *et al.* (eds.): *Advances in Modal Logic*, Vol. 4, King's College Publications, London: 475-498.

Tulenheimo, Tero: 2004, *Independence-Friendly Modal Logic. Studies in its Expressive Power and Theoretical Relevance*, Philosophical Studies from the University of Helsinki, 4, 2004.

V

DIALOGUE AND PRAGMATICS

Chapter 18

NATURALIZING DIALOGIC PRAGMATICS

Gerhard Heinzmann
University of Nancy 2/Archives Poincaré (CNRS)
Gerhard.Heinzmann@univ-nancy2.fr

18.1 Naturalism and holism in the contemporary discussion between nominalists and realists†

Since the seminal joint paper of Quine and Goodman about nominalism, publications against object-platonism grew constantly, especially in the United States. Very different from European anti-platonism, — if mentioned, only Michael Dummett seems to be known — the guiding desideratum of anti-platonism is to avoid revisionism in mathematics. Shunning at the same time constructivism and instrumentalism, one of the current debates in the Foundation of Mathematics is the discussion between nominalists (no quantification over "abstract" objects) and new realists (structuralists) on the basis of a jointly supported naturalistic thesis. Hartry Field's book *Science without numbers* is perhaps the most known title, but there are others: Charles Parsons, Charles Chihara, Geoffrey Hellman, Penelope Maddy, Michael Resnik and Stewart Shapiro, for example, who analysed semantic, ontic and epistemic questions (if mathematical truth is independent of the mathematician's mind or not, if objects of a given kind exist or if it is only possible that these objects or objects with their structures exist and how we know their existence, their logical possibility or the truth of propositions).

According to the naturalists, the foundationalists share a common false presupposition, that is the normative approach of epistemic questions. In fact,

† I would like to thank Manuel Rebuschi for his critics of a draft of this paper.

J. van Benthem, G. Heinzmann, M. Rebuschi and H. Visser (eds.), The Age of Alternative Logics: Assessing Philosophy of Logic and Mathematics Today, 285–297.

formalists, constructivists and object-Platonists failed to found convincingly mathematical truth on account of purely abstract semantical or empirical criteria totally divorced from the science as practised. The rational reconstruction program had to be given up with the indeterminacy thesis: there is no clear synonymy-criterion concerning the relation between the concept to reconstruct and the means used in reconstruction. It seems also consequent to suspect, with Quine, not only the absolute distinction of analytic and synthetic, but also the validity of the alternative "descriptive" or "normative". Indeed, if on the one hand, mathematics loses its pure analytical status, the concept of mathematical progress will deserve attention, because knowledge would be vulnerable to defeat by future experience. No statement could be preserved from revision. The cumulative view of mathematical progress could no longer be supported (cf. Kitcher 1988, 519). On the other hand, what consequence should one draw from Quine's second suspicion? By excluding a normative "philosophy-first" position, one has a choice between a "philosophy-in-between" standpoint ("Philosophy and mathematics are intimately interrelated, with neither one dominating the other" (cf. Shapiro 2000, 15), or the identification of epistemology with cognitive psychology.

And in fact, Naturalism means different things:

(1a) Every theory of knowledge is an empirical theory of cognition (cf. Bieri 1994, 55). To reduce normative (*de iure*) questions to descriptive (*de facto*) ones or to eliminate the former, leads to a causal explication of knowledge, not very convincing for mathematics although it may exclude, trivially, epistemological platonism which postulates a non-causal relation between subjects and abstract entities.

(1b) The epistemic justification of opinions and statements has its source in the practice of science. This signifies, as Maddy formulated it, that current mathematical "practice need not be taken as gospel, but as a starting point [...] subject to ordinary scientific critique [...] which differs from the ordinary scientists' method in perspective not in the evidential standard" (Maddy 1998, 278). To this, moreover, one should add that mathematical practice depends on norms, influenced themselves from its development which gives the standards of ontological, epistemological and semantic questions. A similar position was already supported in the thirties by Jean Cavaillès, but largely ignored by the scientific community.[1]

The further discussion of the double naturalistic thesis depends on the question if one accepts or not a overall holistic position and its sister, the indispensability thesis:

(2a) Holistic thesis: We cannot separate mathematics (and philosophy) from the web of belief of scientific knowledge.

[1]Cf. Cavaillès 1962, where he speaks in this context of a dialectical concatenation of concepts.

A particular mathematical system is trivially under-determinated relative to experience. The procedure of abstraction whose results are the first mathematical theories, might well **start** with empirical observations and manipulations of so called concrete entities, but to return to them requires coordinative definitions so that mathematics, on the one hand, can only be confirmed together with physical hypotheses. But on the other hand, commode mathematics are indispensable for science. For this reason one associated holism with an indispensability principle so that mathematics took on a modal character:

(2b) Indispensability thesis: Mathematics is an indispensable part of scientific practice. A mathematical theory is accepted on account of its relation to a description of the empirical world.

Now, if one is willing to accept some form of thesis 1 and 2, scientific practice commits science to mathematical objects and truth and it seems clear "that the *standard* account of mathematical truth forces us to believe in mathematical entities" (cf. Field 1989, 53, *underlined by myself*), that is to be a realist. Once accepted that the distinction of abstract and concrete is one of kind and not of degree, the new realists tried then to avoid the well-known difficulties with abstract mathematical objects: how should one establish the reference to the particular abstracta if one takes into account "that some universals cannot be instantiated if others are, just as the existence of the arbitrary subsets of every set precludes the universal set" (cf. Resnik 1997, 77)? Since Helmholtz and Poincaré we are quite familiar with the thesis that relations are the only intersubjectively existing objects. With respect to mathematics, this thesis is in the contemporary discussion reinforced and expressed by structuralism: mathematical objects exist only as positions in structures. The existence of individual mathematical objects as numbers or sets are then only relative to the existence of a structure postulated by the structural-realist to exist in its own right.

As Resnik underlines (cf. Resnik 1997, 52), nominalists have then three options to refute the point of view of realists: They could

(a) show that existence and/or truth can be understood in nominalistic terms;

(b) show that using mathematical formalism in science need not commit one to mathematical objects and truth;

(c) show that the mathematical formalism is not necessary for doing science.

And in fact, Hellman showed that existence and truth could be understood in a nominalistic modal structural interpretation and he developed "translation patterns of mathematical theories into suitable modal theories — capable of standing independently of set theory—and then [justified] these as equivalent for mathematical purposes" (Hellman 1989, 8). By interpreting mathematics in anti-realistic terms, Chihara and Kitcher showed how we can reaffirm the indispensability thesis and mathematized science without committing ourselves to mathematical objects (cf. Resnik 1997, 59). In parrying the indispensability thesis, Field combined the strategy of eliminating some branches of mathematics from science with showing how to use others (e.g. metamathematics) without presupposing mathematical objects. In other terms, Field showed that "mathematics needn't be true but only conservative" and that "nominalistic resources are adequate to the statements of good scientific theories" (cf. Field 1989, 128).

18.2 Some difficulties

Without taking sides, we can verify that realists *and* nominalists are together confronted with one or the other of the following difficulties, well known in the literature. Now, there may be for all challenges a single answer. Indeed, in the next paragraph, I will argue that all of them could be surrounded in the spirit of a pragmatic approach.

P1 Indispensability thesis

"If our grounds for accepting mathematics are limited to those we have for accepting our best scientific theories, then we might be justified in accepting very little mathematics" (cf. Resnik 1998, 233): mathematical revisionism will be set up in its rights.

P2 Holism

What about the relation between evidence and theory?

(a) There is the following dilemma: suppose our cognitive abilities were the subject of empirical cognitive science. If evidence then means the sensory event consisting in a sensory stimuli, how should such a happening be the reason of an eventual conceptual accommodation? But if experience has itself a conceptual character, then in order to be meaningful it should be part of semantic holism (cf. Esfeld 2001, 205).

(b) There is a tension in holism between on the one hand confirmation concerning a culturally shared knowledge, and on the other hand, confirmation concerning statements held true by individuals (Esfeld 2001, 204).

(c) "A really good reason to consider a theory falsified will be given only if a better theory is available which explains why the adherents of the old theory were unsuccessful. In other words, an isolated theory should not be given up" (cf. Müllhölzer 1995, 204), that is we should pursue holism at a higher level.

P3 Realistic structuralism

Thus what is characterized by the postulate of first-order axiom system is generally not a determinate structure, but a species of structures. How should one identify in a unique way a position in a such a species apart from accepting from the beginning the inconvenience (semantical incompleteness) of a second order language?

P4 Nominalistic structuralism

1 With respect to modal structuralism Resnik underlines that "it seems [that] the difference between saying that something might be (or that it is

consistently describable) and saying that it exists, [...] is reasonably clear when it comes to possible concreta such as unicorns", [once the distinction of abstract and concrete is accepted as a distinction of kind]. But when it comes to sets or structures like the iterative hierarchy, which cannot be concretely realized, the difference threatens to be merely verbal" (cf. Resnik 1997, 77).

2 With respect to predicative foundations from a structuralist standpoint, the natural-number-type-structure is predicative only if the notion of finite set of individuals is assumed (cf. Parsons 1992, chap. I).

18.3 Dialogic pragmatics

Under pragmatism I understand an approach in the tradition of Charles Sanders Peirce, Ferdinand Gonseth, Paul Bernays and Kuno Lorenz: I adopt above all Lorenz's insight that "pragmatics has became the modern heir of ontology with semiotics being its counterpart as the heir of epistemology" (cf. Lorenz 1994, 103).

Such a position implies, firstly, that the distinction of concrete and abstract objects is an unjustified myth; the distinction is only justified to a relative degree. Every domain of objects must be introduced as a domain of abstract objects if one will preserve a common access to meaning and the possibility of identification. The myth of the abstract-concrete distinction and the myth of the innocent view of a crude fact form a pair.

Secondly, we call "intuitive level" a level where language is used in a non-reflexive manner. Such a common sense level is the starting-point of a schematic abstraction procedure. It is its ontological engagement. In fact, Strawson gives a strong argument against the opinion that there is a common sense *theory* which is in disagreement with a scientific theory (cf. Strawson 1988). It goes in the same direction as Russell's distinction between knowledge of acquaintance and knowledge by description without being identical to it: according to Russell, we do not genuinely know common sense objects like tables and books, because they are "inferred" in a certain sense from shape and colors with which we are acquainted. Strawson supported the thesis that the common sense view cannot be *scientific* knowledge and, more importantly, that the contrast between common sense and other knowledge cannot simply be that between scientific and non-scientific *knowledge*. The common sense sentences that Wittgenstein considers to be beyond justification are also and for that reason something which cannot be said to be *known*: "I should like to say", he writes in *On Certainty*, § 151, "Moore does not *know* what he asserts he knows, but it stands fast for him, as also for me [...]". Commonsense sentences as $2 + 2 = 4$ is a sort of evidence in a life-situation, in a 'Lebenswelt' (Husserl), in a language-game in the largest sense (Wittgenstein), in a 'situation intuitive préalable' (Gonseth) and I will call it the object-side for theory-building. But in fact, by means of the Wittgenstein-Strawson argument we have seen that the very question does not concern the distinction between the common sense theory (the theory about geometrical phenomenal pictures and number-theoretic operations) and the scientific theory but concerns the distinction between common sense and scientific theory. In other words, there is not an elementary or evident first level *theory* on figures and numbers, independent from all scientific activity, and which is only afterwards substantiated and developed by mathematical explanation. Rather, *mathematical* explanation already concerns the articulation and the expansion of the common sense schema. More exactly, mathematical activity concerns not only activities regarding knowledge by description (this is the realist's standpoint) but also activities regarding "knowledge" by acquaintance, called more adequately by K. Lorenz *object-competence*, that is, in mathematics we are trying to get acquainted with further objects. Whereas in a first common sense perspective the sentence $2 + 2 = 4$ could be understood in Quine's sense as a holophrastic association in a given situation (what seems impossible on the same level for, say, $73 + 24 = 97$), the same sentence is in the mathematical retrospective understood word by word. Nevertheless, in the same way as the work of mathematical abstraction progresses and as the theoretical framework develops, new intuitive levels with new domains as their ontological engagement emerge. The backward reference to an intuitive language use also has nothing to do with last foundations; it only signifies that to arise a validity claim presupposes a practice relative to which the question of validity does not arise.

Thirdly, one accepts a scientific theory because its resulting consequences are useful in a given context and well entrenched in Goodman's sense. What about the primacy of proof in mathematical knowledge? The context of justification is pragmatically connected with the logical construction of the genesis because "to understand" mathematics means: to learn their development. So, if S has a true belief for p, justified by a formal proof, it does not follow that he understands p. Justified belief may be equivalent to an abstract proof structure (cf. e.g. Helman 1992) but not with mathematical knowledge. Genuine mathematical knowledge is in general not the result proved by means of a series of analytical deductions.

Or, formulated in other terms, understanding a proof cannot be reduced to being able to checking a linguistic type, whose tokens may be printed in books (cf. Kitcher 1984, 36), but requires the awareness of the acquisition of an action schema. Indeed, when a mathematical proof has been shown to conform to the explicitly formulated rules or principles of logical inference we usually consider it valid. This precisely would constitute the problem the logicians are trying to solve. But there may be something awry with the problem formulated by the logicians : because what it means to *follow* correctly the defined rules as, for example, for the step-by-step process of substitution which makes up the atomic elements of proof, is only determined "within the established practices of working with" the substitution expression (Stenlund 1996, 469). We once again find ourselves in the tradition of the philosophy of the later Wittgenstein, where language has lost its role of being something available on the metalevel with respect to the level of formalism. The awareness of a mastery of a schema (=the execution of an action in a schematic perspective) is called intuitive and cannot itself be formalised without committing a *petitio principii.* Stenlund remarks rightly that "formalisation presupposes and applies the mathematical calculus of finite sequences [...] which we do not acquire until we learn elementary mathematics, [...] because the notion of a finite sequence is not the same one as the everyday notion of a list of concrete individual signs" (Stenlund 1996, 476). Suppose we have learned by intuition to follow the rules of a formalism. This formalism may be considered justified either because generating interesting problems within formalism or because it is considered to clarify or to simplify informal reasoning. In the latter case, justification consists in connecting a presupposed practical familiarity with informal reasoning with its formal characterisation. We have to ascertain that the formal characterisation is the translation, in the sense of precision, of the former. The trouble is that, according to Wittgenstein and Quine, the problem is quite insoluble insofar as there are simply two systems of rules without criterion to compare them. This insight constitutes, so to speak, the "conception"-day of general proof theory: the somewhat misleading metaphysical program of comparing ideal language with ordinary language in hopes of explaining occult properties of the former is replaced by the study of different formal languages with respect to their deductive connections. Then, naturally, there exist proof-theoretic criteria of comparison, for example, identity of normal form for identity of proofs (Martin-Löf/Prawitz). Now, you can even give criteria for the passage by formalisation from a body of mathematics M to a formal theory T. According to Feferman, for example, every concept, argument and result of M has to be represented by a concept, proof or theorem of T. But the Non-Standard models show that such a formalisation of a body of mathematics M goes far beyond what is actually needed to represent M (difficulty **P3**) (cf. Feferman 1992, 15). This confirms Poincaré's feeling that one has to be skeptical about logical consequences as a sufficient guide to exhaust the domain of truth in mathematics at least as complex as elementary arithmetic.

Poincaré holds that the varieties of formal logical theories don't express the proof-theoretical structure essential for understanding mathematics (Cf. Poincaré 1908, 149 (159)). He insists on the non-invariance of mathematical reasoning with regards to its contents; he promotes, so to speak, a "local" conception of mathematical reasoning according to which "a 'gap' is no longer a logical gap but, rather, a gap in *mathematical understanding.* [...]

The elimination of gaps thus no longer calls for the *exclusion* of topic-specific information in an inference" but for the inclusion of what Detlefsen calls a *epistemic condenser* "to fill what would otherwise be a mathematical gap between the premises and the conclusion" (Detlefsen 1992, 366, 360). Indeed, what does it mean exactly to seek for an element of condensation? I have given in (cf. Heinzmann 1999) an example in topology where deduction enclosed the non logical switch between different semiotic levels considered in part on the same level of notation.

According to this line, one should take into account the relation between the descriptive and the normative mode. In order to avoid the difficulty resulting, since the time of Aristotle, from the requirement that the "criteria of justified belief must be formulated on the basis of descriptive or naturalistic terms alone" (cf. Kim 1997, 34), one should not *describe* but *present* the interrelation of normative and descriptive modes. Mathematical symbols can be read with respect to different contents. There are iconic diagrams in the Peircian sense. The semiotic ambiguity involved cannot be checked on the level of notation but requires the acquisition of a practice. Such procedures do not concern, to be sure, mathematical reasoning in its totality. Nevertheless they may be predominant in some fields and constitute then a lack of logical rigour.

Fourth, we accept with Ferdinand Gonseth (and Penelope Maddy) naturalism in the second sense. In fact, the development in philosophy and mathematics prove the insufficiency of both, naturalism in the first sense and the variants of nominalistic positions. The first attaches too important a role to the psychological relations of mental states, the second excludes them in favour of logical relations between propositions. In confining mathematical activity, on the one hand, to a causal constitution of mathematical objects in a holistic framework and, on the other hand, to a description of a formalism of modalities, both approaches neglect the relation between the determination of objects and the guarantee of statements involving these objects. The feature of the here defended **dialogic pragmatism** can be seen in the simultaneity of object construction and object description, inserted in a process of socialization. Its naturalistic characteristic consists in the fact that theoretical means always depend on practical appropriateness.

From all this results a process of understanding and explication which Gonseth summarized very well in four principles:

• The *principle of duality* emphasizes a dialogical interplay between reason and experience: the "horizon of experience" and the "horizon of theory" should be always developed simultaneously. Theoretical terms are partially defined in terms of observational vocabulary, and the vague understanding of observational terms is only possible in view of sortal (theoretical) terms.

• The *principle of revisability* emphasizes that all scientific statements should be open for revision, including statements of logic.

• The *principle of technicity* serves to counterbalance that of revisability: it imposes "a limit on what counts as a legitimate reason for starting a revision", that means the technical progress involved should be adequate. (cf. Esfeld 2001, 8.)

• The *principle of solidarity* (or integrality) is the expression of a methodical holism concerning every level of reflection.

With these principles Gonseth pursued two epistemological goals: The comprehension of the **development** and the **validity** of mathematical propositions. The interrelation of these aims is warranted by a dialectical process depending on sociological parameters: it results in an action-process leading to syntheses between empirical, theoretical and pragmatical aspects of different levels. The criterion of mathematical progress may thereby consist in the successful modification of "the existing practice of mathematics so as to maximize the chances of attaining the two goals" (cf. Kitcher 1988, 531).

Gonseth neither promotes the identification between conventional acceptation and belief no separates them entirely (cf. Da Costa, 617). The pertinency of the normativity of the principles has no idealistic touch, because the principles are, as actions and action-signs, at the same time tool and subject in the dialogue. The only transcendental element is the dialogue itself.

Without a doubt, the principle of duality is the expression of an anti-dualistic position in theory of knowledge. It implicates namely the thesis that spontaneity and receptivity are quite inseparable. Comprehension is a capacity and not the result of a theory about factual data.

Furthermore, the juxtaposition of the principles of duality and revisability leads to a pragmatic transformation of the epistemological terms in question: on the one hand, the empirical (in its revisited signification) is no longer opposite to the rational in general, but to systematisation; and on the other hand, the rational (in its revisited signification) is not opposite to the empirical, but to the capacity to conceive contentual impressions, that means to conceive a domain of subjects (cf. Bernays 1937, 289). Hence the possibility to speak of mental experience (cf. Bernays 1952, 131) with regard to mathematics. The Kantian opposition between concept and intuition is replaced with form and content, which are conceived to stand, as in the neo-Kantian tradition, in a functional relation (cf. Cassirer 1990, 343). The invariability founding the analogy between experience and mental experience consists in the fact that both are conceived as contents standing in a functional opposition to forms.

These remarks should be sufficient in order to see how the principle of duality solves difficulty 2b and how the principle of solidarity solves difficulty 2c. Furthermore, difficulty 2a resembles the circle that psychologism introduced in the theory of knowledge. Indeed, I think it is possible to avoid it by choosing a pragmatic position which abrogates the rigid hierarchy of justification: the solution of point 2a consists in the fact, that it would be an illusion if one wanted to return to a non-conceptual basis. There is no pure information so that all objects and the relations between them are always schematic constructions or, in other words "des horizons de réalités", open to experience. Structures don't yet exist in their own right: they are always constructed "post rem" in a language in which the objects and relations under consideration have names; on this level the axiomatic aspect consists in sharpening this language. However, to consider structures "in their own right" means to abstract from their genesis so that the original objects and relations do not occur independently, but only as links "in an overall structure — they occur merely in their grammatical role, as it were — and the axiomatic system makes assertions about this overall structure" (cf. Bernays 1970, 182). On the contrary, we have idealized and not postulated structures. This pragmatic solution of point **P3** has much more in common with the tradition of continental structuralism (Cavaillès, Gonseth, Bernays) than with is Anglo-Saxon sister.

What remains are difficulties **P1** and **P4**.

Concerning the former, let us remember that nominalists and new realists often support the thesis affirming that it is hopeless to look for a relation between mathematics and experience because this relation has been broken. I would like to reply that what was in fact broken was only the relation between mathematics and a first tangible reality, for example, the Euclidean one. To introduce an absolute cut between mathematics and experience would signify that one disregards the problem of foundation by neglecting the history and the evolution of mathematical thinking.

There is, doubtlessly, not only evolution of mathematical content but even, at the same time, evolution of the means of knowledge which made it possible to invent new syntheses with regard to which the cut is inexistent: the intuitive reference is not necessarily an initial reference. In this manner, one should for example interpret the extension of the concept of space or the geometrisation of contents. We should apply mathematics to mathematics. So we could perhaps solve problem **P1**: fictive mathematics should be applied to well confirmed mathematics. Nevertheless, the existence of a completely autonomous domain of mathematics would constitute a counter-argument. But, personally, I'm not acquainted with such a domain.

Concerning difficulty 4a, the principle of duality emphasized that the difference of concrete and abstract is one of degree. In fact, the knowledge of concreta always has, considered from outside, a modal character, because the concrete object is only identifiable as an abstract schema whose actualisations or references are only incompletely specified. One can also agree plainly with Felix Müllhölzer's dictum that "as soon as one uses the word 'theory' in the manner of the practising scientist, the relation between theory and experience no longer appears as an affair which can be described by a short formula like 'logical implication of observation categoricals' " (cf. Müllhölzer 1995, 204).

Now, one can find an analogues situation in pure mathematics. It consists in the fact that mathematical symbols can be read with respect to different mathematical contents without the possibility of checking this ambiguity on the level of the given notation. Here we have a typical case of partial information. Poincaré suggests that this difficult situation including semantic ambiguity should be overcome by the introduction of aesthetic feeling in mathematics. The mastering of simultaneous reasoning about different contents, provoked by the lack of perfect information in one field, requires the acquisition of a practice. So, it seems to me that it would well coincide with Gonseth's idea to determine a criterion of admissibility of sentences by substituting truth-functional logic by one of the logics of vagueness and to introduce on a certain level a degree of clarity. Indeed, this approach, just as modal ones, can only made explicit by model-theoretic means, but this would be for our philosophical interests circular: all model theory depends on truth definitions. As long as these definitions can only be given on second order level or in set theory, then model theory depends on second order logic or set theory (Hintikka 1996, VIII).

Now, if Hintikka's IF-Logic is really a first order logic which performs that it promises, i.e. that does "model theory" in first order logic, it will be a formal way out. Then the most interesting research program concerns the study of modal logic in IF-Logic. On the other hand, the situation with mathematical theories is not so different from the situation concerning scientific theories in general: even mathematical "theories in the scientists' sense reveal [. . .] to be very flexible entities, [starting often with different contradictory hypotheses so that there are] not sharply defined sentence-buildings, and the answer to the question of whether a certain [. . .] outcome contradicts a theory depends on the skill with which [mathematicians] make use of the theory" (cf. Müllhölzer 1995, 204). From this point of view, the application of paraconsistent logic in mathematics itself should have some success in the future. Now, we have seen it during this symposium, all the mentioned non-classical approaches find their natural frame in dialogic or game theoretic logics which have philosophical pragmatism as their background.

Finally, to favour predicative analysis (4b) because it supplies enough mathematics for applications in physics, seemed first to have been a technical trick because Feferman's constructions presuppose the impredicative domain of natural numbers (Kreisel). But in the new Feferman/Hellman system (cf. Feferman/ Hellman 2000) the *initial* assumption of the finite set is not assumed to have the usual properties of a finite set but can be afterwards defined within the structure. In this point their system agrees very well with a pragmatic evolution.

References

Bernays, Paul: 1937, "Grundsätzliche Betrachtungen zur Erkenntnistheorie", *Abhandlungen der Friesschen Schule*, neue Folge 6 (Heft 3/4), 279–290.

Bernays, Paul: 1952, "Dritte Gespräche von Zürich", *Dialectica* 6, 130–136.

Bernays, Paul: 1970, "Die schematische Korrespondenz und die idealisierten Strukturen", *Dialectica* 24, 53–66, reproduced in Bernays, *Abhandlungen zur Philosophie der Mathematik*, Darmstadt: Wissen schaftl. Buchgesellschaf, 176–188.

Bieri, Peter: 1994, *Analytische Philosophie der Erkenntnis*, Frankfurt: Athenäum.

Burgess, John P./Rosen, Gideon: 1997, *A Subject with no Object*, Oxford: Clarendon Press.

Cassirer, Ernst: 1990, *Substanzbegriff und Funktionsbegriff*, Darmstadt: Wissenschaftliche Buchgesellschaft.

Cavaillès, Jean: 1962, *Philosophie mathématique*, Paris, Hermann.

Da Costa, Newton C. A./Otavio Bueno/Steven French: 1998, "The Logic of Pragmatic Truth", *Journal of Philosophical Logic* 27, 603–620.

Detlefsen, Michael: 1992, "Poincaré against the Logicians", *Synthese* 90, 349–378.

Esfeld, Michael: 2001, "Gonseth and Quine", *Dialectica* 55, 199–219.

Feferman, Solomon: 1992, "What rests on what? The Proof-Theoretic Analysis of Mathematics", *15th Int. Wittgenstein Symposium*, Kirchberg/ Wechsel, Manuscript.

Feferman, Solomon/ Hellman Geoffrey: 2000, "Challenges to predicative Foundations of Arithmetic", in: Sher,G./ Tieszen, R., *Between Logic and Intuition. Essays in Honor of Charles Parsons*, Cambridge: University Press, 317–338.

Field, Hartry: 1989, *Realism, Mathematics & Modality*, Oxford: Basil Blackwell.

Greffe, J.L./Heinzmann, G./ Lorenz, K.: 1996, *Henri Poincaré. Wissenschaft und Philosophie*, Berlin, Paris: Akademie Verlag, Blanchard.

Heinzmann, Gerhard: 1999, "Poincaré on Understanding Mathematics", *Philosophia Scientiae* 3 (2), 43–60.

Hellman, Geoffrey: 1989, *Mathematics without Numbers*, Oxford: Clarendon Press.

Helman, Glen: 1992, "Proof and Epistemic Structure", in: Detlefsen (ed.), *Proof, Logic and Formalization*, London, New York: Routledge, 24–56.

Hintikka, Jaakko: 1996, *The Principle of Mathematics Revisited*, Cambridge: University Press.

Kim, Jaegwon: 1997, "What is 'Naturalized Epistemology'?", in: Hilary Kornblith, *Naturalizing Epistemology, Cambridge* (Mass), London: MIT Press.

Kitcher, Philip: 1984, *The Nature of Mathematical Knowledge*, New York, Oxford: Oxford University Press.

Kitcher, Philip: 1988, "Mathematical Progress", *Revue Internationale de Philosophie* 167, 518–540.

Lorenz, Kuno: 1994, "Pragmatics and Semiotic: The Peircean Version of Ontology and Epistemology", in: G. Debrock/H. Hulswit (eds.) *Living Doubt*, Dordrecht: Kluwer, 103–108.

Lorenz, Kuno: 2002, "Le dialogue comme sujet et méthode de la philosophie", *Manuscript*.

Maddy, Penelope: 1998, "Naturalizing Mathematical Methodology", in: Schirn, Matthias (ed.) *The Philosophy of Mathematics Today*, Oxford: Clarendon Press, 175–193.

Müllhölzer, Felix: 1995, "Science without Reference?", in: U. Majer/ H-J. Schmidt, *Reflections on Spacetime. Foundations, Philosophy, History*, Dordrecht, Boston, London: Kluwer, 203–222.

Parsons, Charles: 1992, "The Inpredicativity of Induction", in: Detlefsen, Michael (ed.), *Proof, Logic and Formalization*, London, New York: Routledge, 139–161.

Poincaré, Henri: 1908, *Science et méthode*, Paris: Flammarion 1908; translation: *Science and Method*, London/ New York: Thomas Neldon & Sons, s. d.

Resnik, Michael, D.: 1997, *Mathematics as a Science of Patterns*, Oxford: Clarendon Press.

Resnik, Michael, D.: 1998, "Holistic Mathematics", in : Schirn, Matthias (ed.) *The Philosophy of Mathematics Today*, Oxford: Clarendon Press, 227–246.

Shapiro, Stewart: 2000, *Thinking about Mathematics. The Philosophy of Mathematics*, Oxford: University Press.

Stenlund, Sören: 1996, "Poincaré and the Limits of Formal Logic", in: Greffe *et al.* 1996, 467-479.

Strawson, Peter: 1988, "Perception and its Objets", in: J. Dancy (ed.), *Perceptual Knowledge*, Oxford: University Press, 92-112.

Wittgenstein, Ludwig: 1969, *On Certainty*, Oxford: Blackwell.

Chapter 19

LOGIC AS A TOOL OF SCIENCE VERSUS LOGIC AS A SCIENTIFIC SUBJECT

Kuno Lorenz*
University of Saarland
klorenz@rz.uni-saarland.de

There is nothing spectacular in turning activities into objects of investigation such that the procedures of investigation show up as certain second order objects, and climbing up in this way, the semantic ascent, may well be continued, if thought to be necessary for some purposes. The working scientist is doing this regularly. But what about the converse of dissolving an object into activities such that perspectives of the object will take the place of the object? In this case we are concerned with the substitution of an object by its properties, i.e., a set of second order objects. Again a familiar procedure, so it seems. But, usually, it is not realized, or taken for granted, that properties of an object – relational ones included – relate in a systematic fashion to the internal structure of the object, i.e., its being a whole out of parts. This implies that something which can only be said – by using predicates – matches in a rarely scrutinized way something which can only be shown – by using rules of action. Semiotics and pragmatics – hence, being an object and being a tool – are intimately bound together. For example, the property of being even, in the case of natural numbers, is equivalent to the number two being a (multiplicative) part of the respective natural number, though such an equivalence is stated without trying to relate the (multiplicative) part-whole-relation between numbers which is an *external* relation where numbers remain >indivisible< units, that is, >individuals<, to

*Part of this contribution overlaps with material contained in the author's paper 'Pragmatic and Semiotic Prerequisites for Predication' in: D. Vanderveken (ed.), *Logic, Thought and Action*, Dordrecht: Springer 2005, 343-357.

J. van Benthem, G. Heinzmann, M. Rebuschi and H. Visser (eds.), The Age of Alternative Logics: Assessing Philosophy of Logic and Mathematics Today, 299–310.

the *internal* structure of natural numbers as composites out of units. The well-established (minimal) logical structure of elementary propositions – a general term or, rather, a propositional function, is applied to (particular) objects represented by singular terms – is considered to be a kind of rock bottom on which to build theories without further attempts, neither by linguists nor by logicians, to give a foundation to such a bifurcation of terms. After Frege had used the mathematical terminology of function and argument for giving a functional account of (elementary) propositions that eventually allowed to read the sequence >singular term<, copula, general term' equally well as the expression of a set-theoretic membership relation by turning the general term into a singular term denoting a logically second order particular, it seemed to be superfluous to question the Fregean account. It had been – and still is – neglected that there is no chance in this way to get rid of the peculiar *internal* relation between particulars and what is said about them as established in a proposition. Furthermore, whoever attempts to eliminate the copula that articulates the internal relation in question, will lose the opportunity to explicate how the copula is mirrored by an internal part-whole-relation, as well. Wittgenstein's challenge to Frege's treatment of propositions as names (cf. T 3.143) that is a fatal blow also to a referential theory of truth, has never been widely accepted, certainly not among the majority of mathematicians. Whoever was convinced of the necessity of treating propositions in a pragmatic context and not in a semiotic one, did this by embedding propositions in speech acts, e.g., assertions, and not by considering a pragmatic approach to propositions themselves.

I would like to turn your attention, now, to the interplay of pragmatic and semiotic features in setting up propositions, both elementary and logically compound ones. You will notice that instead of 'activity of investigation' and 'object of investigation' I prefer to use the terms 'pragmatic' and 'semiotic' with respect to activities in general. I do this, because they are more appropriate to the general claim connected with the approach I want to sketch. In this approach I am concerned with conceptual clarifications beneath logic proper rather than with new scientific results. My starting point is derived from Peircean and Wittgensteinean ideas: A dialogue-situation conceptualized as a two-person-game being a generalized Wittgensteinean language-game without explicit linguistic activity in the beginning, will serve to model the acquisition procedure of an action-competence. Dialogical constructions will lead from modelling simple activity to modelling the growth of more complex activities up to elementary verbal utterances and eventually to logically compound propositions.

At first some general remarks: In accordance with C. S. Peirce, I consider pragmatics to have become the modern heir of ontology with semiotics being its counterpart as the modern heir of epistemology. Yet, in this context both disciplines should not be understood as two newly established empirical sciences, but as ways of investigation where empirical procedures are combined with reflexive procedures. Using such a broader perspective both actions and sign-actions are not only treated as *objects* of research and representation, as, e.g., in Ch. Morris' and U. Eco's approach, but also as a *means* or tool of research and representation. You not only observe and describe these entities according to certain standards, but you also produce them in a perspicuous fashion in order to arrive at some kind of approximating reconstruction of what you take to be available, already.

Hence, the constructions serve cognitive purposes in the sense of delineating the very areas of (particular) objects one proceeds afterwards to investigate in the more usual way. Language-games as well as the generalized ones of acquiring simple action competences exhibit a semiotic function if understood as icons in the sense of Peirce. An area of internally structured objects is found by inventing a prototype.

Thus, even the distinction of action and sign-action which still is prevalent in Wittgensteinian language-games where simple action competence is presupposed, has to be relativized in view of a purely functional account of both what it means to be an object and what it means to be a sign (of an object).

The two Aristotelian categories, ποιεῖν and πάσχειν, doing and suffering, will enjoy a lively comeback – they did this in Dewey, already – as the two sides we are concerned with when doing something: you do it yourself (active) and you recognize others (including yourself!) doing the same (passive [with respect to the content of recognition]). These two sides reoccur in the model of an elementary dialogue-situation with two agents being engaged in the process of acquiring an action-competence. At each given instant just one of the agents is active – a >real< agent – and the other agent – the >potential< agent or >patient< – is passive. The agent in active role is performing an action, i.e., he is able to produce different tokens of the same type, while the agent in passive role is recognizing an action, i.e., he sees different tokens as belonging to the same type. One has learned an action, if one is able to play both roles: While acting you know what you are doing, or, conversely, if you don't know what you are doing, you don't act. Another way of saying this would be: Each action appears in two perspectives, in the I-perspective by performing the action (=producing an action token) – it should be called the *pragmatic* side of an action, or its >natural< side – and in the You-perspective when recognizing the action (= witnessing an action type) which should be called its *semiotic* or >symbolic< side. We have come across the first step to execute the program of >naturalizing language< and other symbol systems, and, at the same time, of >symbolizing world<, in order to bridge the alleged gap between the two.

Peirce has sketched a way of deriving signs out of objects in more or less the same manner as I just did and was looking for something which is a sign of itself, that is, which combines object status and sign status, or better: which functions both ways. The basic point of his pragmatic foundation of semiotics or, rather, of the interdependence of pragmatics and semiotics, was to give an account of the process of separation between sign and its object within the framework of his Pragmatic Maxim.[1]

Now, within the model of acquisition of an action-competence by an elementary dialogue-situation some further distinctions are obligatory. They are based on the observation that producing an action-token and witnessing an action-type, i.e., I-perspective and You-perspective of an action, are inseparably bound together and cannot be treated in isolation from each other. The model of acquisition of action-competence is a model of actions as a means and not yet of actions as objects which, in order to be accessible, will in turn be dependent on other actions as a means of dealing with objects. Dialogical construction as a means of study asks for self-application such that the interdependence of the status of being-a-means and the status of being-an-object, hence of >epistemology< and >praxeology< on the one hand, and of >ontology< on the other hand, is laid bare. Actions as a means are characterized by their two sides as they arise from the two perspectives, from *singular* performance in I-perspective and *universal* recognition in You-perspective. Yet, when performing is understood to be a case of producing (an action-token) and, analogously, recognizing to be a case of witnessing (an action-type), the action in question is treated as an object, in fact, sometimes even as two objects, the token as an external or >corporeal< particular and the type as an internal or >mental< particular. But, even if action particulars, i.e., individual acts, are treated uniformly without being split into external and internal entities, particularity is to be kept strictly distinct from singularity and universality. Usually, in the terminology of type and token, where types are treated logically as generated >by abstraction< out of tokens, and where tokens originate >by concretion< from types, both types and tokens are (individual) objects, yet of different logical order, which are related in standard notation as sets to their elements. At the lowest level, if there is one, the final universe of discourse is located, i.e., a world of elementary individual objects, the *particulars*, to which everything else will have to be reduced. Such an account, by neglecting the distinction between particularity on the one hand and singularity as well as universality on the other hand, violates the inseparability of (producing a) token and (witnessing a) type in the context of actions as a means, or, rather, it exhibits an equivocation in the use of 'type' and 'token'. It is necessary to relinquish both the equivalence of 'performing an action' with 'producing an action token' and the equivalence of 'recognizing an action' with 'witnessing an action type'.

Instead, performance is performance of something singular and recognition is recognition of something universal, whereas producing (a token) together

---

[1] Cf. B. M. Scherer, *Prolegomena zu einer einheitlichen Zeichentheorie*: Ch. S. Peirces Einbettung der Semiotik in die Pragmatik, Tübingen: Stauffenburg Verlag 1984.

with its twin activity of witnessing (a type) occur with respect to something particular. Now, if tokens and types are not construed as particulars that are produced or witnessed, respectively, they should be identified, in tune with action as a means, with universal features and singular ingredients of particulars that are exhibited by actions which deal with them. Particulars together with the situations (of acting) of which they occupy the foreground are *appropriated* by performing an action which deals with them, and they are *objectified* by recognizing such an action. It should be noted that neither universal features nor singular ingredients have object status by themselves; they remain means with respect to (particular) objects. Universals cannot be appropriated and singulars cannot be objectified.

Hence, in performances of an action that is dealing with a particular you (pragmatically) *present* one of the (singular) token ingredients of this particular, whereas in recognitions of an action that is dealing with a particular you (semiotically) *represent* one of its (universal) type features. Switching from the language of means – pragmatic means are singular, semiotic means are universal – to the language of objects (= particulars) you may say that it is individual acts that provide both services, of presentation with respect to its performance perspective and of representation with respect to its recognition perspective. In appropriation as well as in objectivation of particulars of arbitrary category, like individual acts, individual things or events, groups of individuals or other non-individual particulars, etc., the actions of dealing with particulars are used as a means, of presentation (of singular tokens – the way a particular is present) in the case of appropriation, and of representation (of universal types – the way a particular is identified) in the case of objectivation.

Particulars may be said to act as appearances of >substances<, i.e., some part of the whole out of like singular tokens is a *part* of the particular, and as carriers of >properties<, i.e., the particular is an *instance* of a universal type.[2] Therefore, in order to avoid misunderstandings, instead of 'perform' we will, henceforth, say 'actualize', and we say 'schematize' instead of 'recognize'.

Within the model of an elementary dialogue-situation where two agents are engaged in the process of acquiring an action-competence, the activities of actualizing and schematizing should not be understood as performances of two separate actions; it is one action the competence of which is acquired by learning to play both the active and the passive role. Active actualization makes the action appear in I-perspective, passive schematization lets it appear in You-perspective. Any action as a means is characterized by its pragmatic and its

[2] A particular wooden chair, for example, acts as a carrier of all the properties conceptualized by 'wooden', and as an appearance of the substance >wood<, inasmuch as a part of >the whole wood< may be considered to be a part of the particular wooden chair; cf. the entry 'Teil und Ganzes' in: *Enzyklopädie Philosophie und Wissenschaftstheorie* IV, J. Mittelstraß (ed.), Stuttgart-Weimar: Metzler 1996, 225-228.

semiotic side, and it doesn't make sense as yet to speak of the action as an >independent< object(-type) split into particulars, i.e., some set of individual acts. In order to achieve the switch from action as a means to action as object, it is essential to iterate the process of acquiring an action-competence by turning the two sides of an action into proper actions by themselves, i.e., into actions of dealing with the original action under its two perspectives such that the (secondary) action-competences additionally required will have to be modelled in turn by means of (now non-elementary) dialogue-situations. Such a further step may be looked at as an application of the *principle of self-similarity*.

What has to be done is to schematize and to actualize the elementary dialogue-situation, i.e., to create a He/She-perspective towards the I / You-situation such that, on the one hand, He/She becomes a (secondary) You-perspective with respect to I/You as I, and, on the other hand, He/She becomes a (secondary) I-perspective with respect to I/You as You. In the first case you gain an >exterior view< of the original action by acquiring a second level action (with respect to the original action) which functions as one of the indefinitely many *aspects* of the original action: The You-perspective is turned into the schema of a second level action out of an indefinite series of second level actions. In the second case you gain an >interior view< of the original action by acquiring a second order action (with respect to the original action) which functions as one of the indefinitely many *phases* of the original action: The I-perspective is turned into an actualization of a second order action out of an indefinite series of second order actions. The semiotic side of an action is split into a multiplicity of aspects or (secondary) You-perspectives, and the pragmatic side of an action likewise into a multiplicity of phases or (secondary) I-perspectives.

By (dialogical) construction, it is in its active role that an aspect-action is I-You-invariant and, in this sense, >objective<, whereas a phase-action is I-You-invariant in passive role, only. Hence, by applying the principle of self-similarity once again to aspects and to phases, the pragmatic side of an aspect-action is split into a multiplicity of objective *articulations* or *sign-actions*, while the semiotic side of a phase-action is split into a multiplicity of objective *mediations* or *partial actions*. Any one of the sign-actions is a means to designate the original action, and any one of the partial actions is a means to partake of the original action, where designating and partaking function with the proviso that the original action itself is turned from a means into an object. In fact, an action as object – things, events, and other categories of entities are included among actions by identifying an entity[-type] with the action[-type] of dealing with the entity – is constituted, on the one hand >formally<, by *identification* of the schemata of the aspect-actions, i.e., of their >subjective< semiotic side, and, on the other hand >materially<, by *summation* of the actualizations of the phase-actions, i.e., of their >subjective< pragmatic side. On the one side, through identification, an action as object is a semiotic (abstract) invariant of which one partakes by means of a partial action, and on the other side, through summation, it is a pragmatic (concrete) whole which one designates by means of a sign-action. With respect to the additional dialogue-situations modelling the acquisition of second-order-action-competences as well as second-level-action-competences the original action as object occurs within a situation which, in fact, is responsible for individuating the original action as object.

The move of objectivation from action as a means to action as object is accompanied by a split of the action into (action-)particulars such that the respective invariants may be treated as *kernels* (=form) of the schemata of aspects (=universalia), and the respective wholes correspondingly as *closures* (=matter) of the actualizations of phases (=singularia).

Dialogical construction of particulars being dependent on the identification of schemata of aspects and on the summation of actualizations of phases, implies the establishment of mutual independence between objectival foreground and situational background. In order to achieve this, a specially chosen articulation has to act as a substitute for arbitrary aspects with respect to some partial action – such a function of *substitution* may be articulated by *rules of translation* among aspects – and will be called *symbolic articulation*. Constant foreground and variable background will thus become independent of each other. Analogously, any mediation will have to acquire the function of having the phase to which it belongs extended by arbitrary other phases with respect to some sign-action – such a function of *extension* may be articulated by *rules of construction* for phases – and will be called *comprehensive mediation*. In this case, constant background and variable foreground are made independent of each other. The two constructions together guarantee that particulars contrast with their surroundings.[3] By symbolic articulation that is a symbolic sign-action, you arrive at a semiotically determined particular in actualized situations, i.e., the particular is *symbolically represented*, whereas by comprehensive mediation that is a comprehensive partial action, you arrive at pragmatically determined particulars in a schematized situation, i.e., the particulars are *symptomatically present*.

The semiotic side of partial actions (>what you do<) and the pragmatic side of sign-actions (>how you speak<), together they make up the *ways of life* (of the agents). Correspondingly, the pragmatic side of partial actions (>how you act<) and the semiotic side of sign-actions (>what you say<), together they make up the *world views* (of the agents).

Articulation, on the semiotic side and not as a mere activity, is signified canonically by the result of a sign-action, an *articulator*, that has to be taken

[3] For an explicit dialogical construction of both identification and summation, cf. my 'Rede zwischen Aktion und Kognition', in: A. Burri (ed.), *Sprache und Denken. Language and Thought*, Berlin-New York: de Gruyter 1997, 139-156, p 145ff.

as a (verbal) type, in a speech situation. And if it is treated as functionally equivalent with any other way of articulation, including non-verbal ones, it acts as a *symbolic articulator*. Again semiotically, i.e., as a *sign*(-action), it shows its two sides, a pragmatic one and a semiotic one. The pragmatic one is to be called *communication*, or the side with respect to persons, and the semiotic one is to be called *signification*, or the side with respect to (particular) objects. By iteration, communication splits into (content of) *predication* on the semiotic side, and *mood* (of predication) on the pragmatic side, whereas signification splits into (intent of) *ostension* on the pragmatic side, and *mode of being given* on the semiotic side. Any predication can take place only by using a mood, and any ostension is effected only by using a mode of being given. We have strictly to distinguish: content and mood of predication, intent and mode of ostension. The moods of predication are, of course, speech acts, and only with respect to a mood a predication contains a claim, e.g., a truth claim.

Without second order articulation of mood and mode, we have arrived at one-word sentences '*P*' (pragmatically in a mood and semiotically using a mode of being given) by uttering the articulator 'P'. They combine predication within communication and ostension within signification by just one utterance (in a speech-situation).

With the next step we introduce the separation of significative and communicative function, two functions that coincide with showing and saying in the terminology of Wittgenstein's *Tractatus*. Separation may be executed in two ways by using operators for neutralizing one of the two functions: (1) with respect to predication, i.e., the semiotic side of communicative function; separation leads to: $\delta \mathrm{P} \varepsilon \mathrm{P}$ (this P [=something done] is P[-schematized]), or, alternatively, to $\sigma \mathrm{P} \pi \mathrm{P}$ (the universal P [= something imagined] is P-actualized), (2) with respect to ostension, i.e., the pragmatic side of significative function; here, separation leads to: $\delta \mathrm{P} \zeta \mathrm{P}$ (this P belonging to P), or $\kappa \mathrm{P} \xi \mathrm{P}$ (the whole P [=something intuited] being P-exemplified).

The operators: demonstrator 'δ' and attributor 'ε' (=copula), respectively, neutralize the communicative function and the significative one; hence, 'δ' keeps the significative function and 'ε' keeps the communicative one, with the result that 'δP' plays a singular role and 'εP' a universal one. In the terminology of logic or semiotics, 'δP', which is used to >ostend< P, is an *index* of an actualization of the action articulated by 'P', whereas 'εP', which is used to >predicate< P, is a predicator serving as a *symbol* of the schema of action P.

Predication εP and *ostension* δP with its respective associates: form of a proposition '$__\varepsilon$P' and form of an indication 'δP$__$', are the modern equivalents of the traditional >forms of thinking< and >forms of intuition<. It wood have been possible to proceed dually by using two operators, univeralisator 'σ' and presentator 'π', with switched roles as already mentioned – there is no time to discuss this, too. In the second case of separation with respect to ostension which invokes the pragmatic distinction 'active-passive' and not the semiotic distinction 'singular-universal', either demonstrator 'δ' and partitor 'ζ', or, dually, totalisator 'κ' and exemplificator 'ξ', serve the same purpose: 'δ' and 'κ' keep the significative function in active and passive role, respectively; vice versa with the other two.

What is not yet available up to now and what would not even make sense, are >propositions< like $\delta P\varepsilon Q$ and >indications< like $\delta Q\zeta P$. The reason why these expressions don't make sense, is simply the following: 'δP' is not the kind of expression to occupy the empty place in a propositional form '$\_\_\varepsilon Q$' with $Q\neq P$, and 'ζP' is not the kind of expression to occupy the empty space in an indicational form '$\delta Q\_\_$' with $Q\neq P$. Instead, we introduce *individuators* 'ιP' in order to refer to particulars, i.e., the situation-dependent units of the action articulated by 'P'; >things< as well as objects of other categories, any one (type) of them being identified with the action(-type) of arbitrary dealings with an object(-type), hence, any of the so-called >natural kinds<, are, of course, included among the P.

Particulars, be they individual things or events, individual acts or processes, are composed out of kernels of schemata of aspects: $\sigma(\iota P)$ (=invariants), together with closures of actualizations of phases: $\kappa(\iota P)$ (=wholes). Hence, particulars may be considered to be half thought and half action. Using individuators we, now, may write down *eigen-popositions* $\iota P\varepsilon P$ as well as *eigen-indications* $\delta P\iota P$ (short for: $\delta P\zeta\iota P$), and it is possible to render these versions of saying and showing in the following traditional way:

1. In the case of saying ($\iota P\varepsilon P$): the universal σP is *predicated of* a P-particular by means of 'εP' (or: within the proposition $\iota P\varepsilon P$, the individuator is a sign of an indication, and, hence, functions as a nominator of a P-particular, i.e., within the proposition $\iota P\varepsilon P$, *nomination by 'ιP'* is shown) , and
2. In the case of showing ($\delta P\iota P$): *ostending* the whole κP *at* a P-particular by means of 'δP' (or: within the indication $\delta P\iota P$, the individuator is a sign of a proposition, and, hence, functions to say that participation at a P-particular holds, i.e., within the indication $\delta P\iota P$, *participation at ιP* is said).

Hence, *reference* to particulars ιP includes both nomination of $\kappa(\iota P)$, i.e., of the *matter* of ιP, and participation at $\sigma(\iota P)$, i.e., at the *form* of ιP. As a remark, it may be added that nominating is the articulation of designating by symbolic articulation, and, analogously, participating is the articulation of partaking by comprehensive mediation.

The composition of P, e.g., wood, and Q, e.g., chair, is a result of separating speech-situation and situation-talked-about. It can be realized by analyzing and reconstructing what happens when, e.g., in a Q-situation you are uttering 'P'. In the foreground of the situation-talked-about which is articulated by 'P', there are two particulars to be welded. It may come about in either of two possible ways:

1 An aspect (with its schema being) out of σ (ιP) coincides with a phase (actualizations of which being) out of $\kappa(\iota Q)$, e.g., sitting on a wooden chair as a phase-action with respect to chair is simultaneously an aspect-action >sitting on the wood of the chair< with respect to wood;

2 A phase out of $\kappa(\iota P)$ coincides with an aspect out of $\sigma(\iota Q)$.

In the first case you may articulate the coincidence *predicatively* by εP_Q (= is a wood of [a] chair), in the second case *ostensively* by δ(QP) (=this wood with the form of [a] chair). Instead of $\delta P_Q \varepsilon P_Q$ we may write $\iota Q \varepsilon P$ (= ιQ *is* P, or: this [particular] chair is wooden), and likewise, instead of $\delta(QP)\zeta$ (QP), it is possible to write $\delta P \iota Q$ (short for: $\delta P \zeta \iota Q$) (= δP *at* ιQ, or: this dealing with wood belonging to this [particular] chair). Hence, 'εP' acts as a *symbol* for the result of schematizing ιQ, whereas 'δP' acts as an *index* for the result of actualizing ιQ.

The introduction of compound articulation Q*P such that $\varepsilon(Q{*}P) = \varepsilon P_Q$ and $\delta(Q{*}P) = \delta(PQ)$ – these two ways of specialization are *relativization* of 'P' by 'Q', yielding 'P_Q' [i.e., P of Q] and *modification* of 'Q' by 'P', yielding 'PQ' – is achieved by again using dialogical construction, and it is successful in case such specializations (wood of [a] chair and wooden chair, respectively) >make sense<; the details of this procedure I have to skip here, unfortunately.[4]

An *indication* $\delta P \iota Q$ shows that the *substance* κP is ostended at ιQ by means of 'δP'; a *proposition* $\iota Q \varepsilon P$ says that the *property* σP is predicated of ιQ by means of 'εP'. In short: ιQ consists both of phases such that the closure of their actualisations is $\kappa(\iota Q)$, and of aspects such that the kernel of their schemata is σ (ιQ), i.e., of form and matter in traditional terminology.

As a further historical remark, it may be added that the two sides of a particular ιQ, the concrete whole $\kappa(\iota Q)$ and the abstract invariant $\sigma(\iota Q)$, correspond neatly to >body< or >phenomenon< and >soul< or >fundament< of a monad as it is conceived in the *Monadologie* of Leibniz.[5] It may also be useful to observe that the identification of $\delta P_Q \varepsilon P_Q$ with $\iota Q \varepsilon P$, i.e., the introduction of (one-place) elementary propositions, is closely related to Reichenbach's transition from a thing-language to an event-language articulated with the help of an asterisk-operator which moves the predicative ingredients of a subject term of an (one-place) elementary proposition into its predicate term, e.g., from 'this man is smoking' you arrive at 'smoking of [this particular] man', or: $(\iota Q \varepsilon P)^* = P_Q$.

We have reached the usual account of (one-place) predication where >general terms< 'P' or, rather, propositional functions 'εP' in the sense of

[4]Cf. K. Lorenz, Sinnbestimmung und Geltungssicherung. Ein Beitrag zur Sprachlogik, in: G.-L. Lueken (ed.), *Formen der Argumentation*, Leipzig: Leipziger Universitätsverlag 2000, 87-106.

[5]Cf. for further corroboration various essays in: *Leibniz and Adam*, M. Dascal/E. Yakira (eds.), Tel Aviv: University Publishing Projects 1993.

Frege, serve to attribute properties to particulars of an independently given domain of Q-objects, in the simplest case referred to by deictic descriptions 'ιQ' that are special cases of >singular terms< [another use of 'singular'!]. The meaning of such an elementary proposition $\iota Q \varepsilon P$ will of course be defined with respect to the significative function of P_Q under its mode of being given, provided the compound articulation works, whereas the validity of $\iota Q \varepsilon P$ is tantamount to the existence of a particular $\iota(P_Q)$ such that the kind of existence is defined by the mood of the elementary proposition. In the assertive mood we speak of real existence and equate validity with truth. Now, the strategy to introduce compound propositions, especially logical composition, follows the procedure for introducing compound articulators, because neither move, the one of reducing propositions to primary constituents of the set-theoretic type $\alpha \in \beta$, or, alternatively, as in constructivism, to derivability propositions $\vdash_K \alpha$ with respect to some calculus K, is going to work outside special areas. What has to be done is to guarantee that compound articulations of whatever kind will in fact be articulations again, i.e., will have both a significative and a communicative function.

The special case of logical composition of propositions has been successfully handled by dialogical logic, and as the details are well known I may restrict myself to a few final remarks: The significative function of a logically compound proposition A being equal with the significative function of the compound articulator A* ($\delta A^* \varepsilon A^* \leftrightharpoons A$), is given by the rules of an open finitary two-person zero-sum game, i.e., a dialogical game with A in initial position, whereas the communicative function, again of A*, with respect to the assertive mood of A is a (material) truth claim which may be fulfilled by presenting a winning strategy for A. You will be aware that in case of, e.g., logically compound arithmetical propositions – its basis in constructive arithmetic is the >arithmetical< calculus for deriving sequences of strokes or another primitive figure – a theory of winning strategies will need a theory of (constructive) ordinals to handle it. At this point we reach present day research, especially in proof theory, that is beyond my concern today.

References

Scherer, B. M., 1984. *Prolegomena zu einer einheitlichen Zeichentheorie*: Ch. S. Peirces Einbettung der Semiotik in die Pragmatik, Tübingen: Stauffenburg Verlag 1984.

Lorenz, K., 1996. "Teil und Ganzes" in: J. Mittelstraß (ed.), *Enzyklopädie Philosophie und Wissenschaftstheorie* IV, Stuttgart-Weimar: Metzler 1996, 225–228.

Lorenz, K., 1997. "Rede zwischen Aktion und Kognition", in: A. Burri (ed.), *Sprache und Denken. Language and Thought*, Berlin-New York: de Gruyter 1997, 139–156, p 145ff.

Lorenz, K., 2000. "Sinnbestimmung und Geltungssicherung. Ein Beitrag zur Sprachlogik", in: G.-L. Lueken (ed.), *Formen der Argumentation*, Leipzig: Leipziger Universitätsverlag 2000, 87–106.
Dascal, M., and E. Yakira (eds.), 1993. *Leibniz and Adam*, Tel Aviv: University Publishing Projects 1993.

Chapter 20

NON-NORMAL DIALOGICS FOR A WONDERFUL WORLD AND MORE

Shahid Rahman
University of Lille 3
shahid.rahman@univ-lille3.fr

At the end of the 19th century Hugh MacColl (1837-1909), the father of pluralism in formal logic, attempted in the north of France (Boulogne sur mer) to formulate a modal logic which would challenge the semantics of material implication of the post-Boolean wave. It seems that in some of his various attempts MacColl suggested some systems where the rule of necessitation fails.[1] Moreover, the idea that no logical necessity has universal scope – or that no logic could be applied to any argumentative context – seems to be akin and perhaps even central to his pluralistic philosophy of logic.[2] Some years later Clarence Irwin Lewis furnished the axiomatics for several of these logics and since then the critics on the material implication have shown an increasing interest in these modal logics called "non-normal". When Saul Kripke studied their semantics of "impossible worlds" as a way to distinguish between "necessity" and "validity" these logics reached a status of some respectability.[3] As is well known, around the 70s non-normal logics were associated with the problem of omniscience in the epistemic interpretation of modal logic, specially in the work of Jaakko Hintikka and Veikko Rantala.[4] Actually impossible worlds received a intensive study and development too in the context of relevant and paraconsistent logics – specially within the "Saint-Andrews-Australasian connection" in the work

[1]Unfortunately he does not seem to have succeeded. Read 1998, differs from Storrs MacCall's (1963 and 1967) argues that the reconstruction of MacColl's modal logic yields T and not one of the non-normal logics.
[2]Cf. Grattan-Guinness 1998, Rahman 1997, 1997, 1998, 2000, Read 1998 and Wolenski 1998.
[3]Cf. Kripke 1965.
[4]Cf. Hintikka 1975 and Rantala 1975. See too Cresswell 1972 and Girle 1973.

J. van Benthem, G. Heinzmann, M. Rebuschi and H. Visser (eds.), The Age of Alternative Logics: Assessing Philosophy of Logic and Mathematics Today, 311–334.

of such people as Graham Priest, Stephen Read, Greg Restall and Richard Routley-Sylvan. Nowadays, though the association with omniscience seems to have faded out, the study of non-normal logics has received a new impulse motivated through the study of counterlogicals. The aim of the paper is to offer a dialogical interpretation of non-normal modal dialogics which will suggest some explorations beyond the concept of non-normality. This interpretation will be connected to the discussion of two issues, namely:

1 Counterlogicals as a minimalist defense of logical pluralism (pluralism for a monist) following the path prefigured by MacColl and

2 The difficulties involved in the application of the so-called *Hintikka strategy* and *hybrid languages* while constructing tableau systems for non-normal modal logics.

20.1 Pluralism for a Monist and the case of the counterlogical

Convincitur ergo etiam insipiens esse vel in intellectu . . .
Anselm of Canterbury,
Proslogion, capitulum II, Ps 13, 1, 52, 1
(Thus, even he who knows no better will be convinced that at least it is in the intellect. . .)

20.1.1 Would the real logic please stand up?

Conceiving situations in which not every mathematical or logical truth holds is a usual argumentation practice within formal sciences. However, to formulate the precise conditions which could render an adequate theory of logical arguments with counterpossibles in formal sciences is a challenging issue. Hartry Field has felt the need to tackle this challenge in the context of mathematics. Field writes:

> It is doubtless true that nothing sensible can be said about how things would be different if there were no number 17; that is largely because the antecedent of this counterfactual gives us no hints as to what alternative mathematics is to be regarded as true in the counterfactual situation in question. If one changes the example to "nothing sensible can be said about how things would be different if the axiom of choice were false", it seems wrong : if the axiom of choice were false, the cardinals wouldn't be linearly ordered, the Banach-Tarski theorem would fail and so forth. (Field 1989, 237)

These lines actually express the central motivation for a theory of counterpossibles in formal sciences. Namely, the construction of an alternative system where e.g. the inter-dependence of some axioms of a given formal system could be studied. If we were able to conceive not only a counterpossible situation where some axioms fail to be true but also even an alternative system without the axioms in question, then a lot of information could be won concerning the original "real" system. By the study of the logical properties of the alternative system we could e.g. learn which theorems of our "real system" are dependent on axioms missing in the alternative one.[5] Moreover, I would like to add that a brief survey of the history of mathematics would testify that this usage of counterpossibles seems to be a common practice in formal sciences.

The case of the study of counterpossibles in logic called *counterlogicals* is an exact analogue of the case of mathematics and motivates the study of alternative systems in the very same way. We learned a lot of intuitionistic logics, even the *insipiens* classical logical monist learned about his system while discussing with the antirealist. This seems to be a generally accepted fact, but why should we stop there? From free logics we learned about the ontological commitment of quantifiers, from paraconsistent logic ways of distinguishing between triviality and inconsistency;[6] from connexive logics the possibility of expressing in the object language that a given atomic proposition is contingently true; from relevance logics that it is not always wise to distinguish between metalogical and logical "if, then"; from IF and epistemic dynamic logic we learned about arguments where various types of flow of information are at stake, for linear how to reason with limited resources, and so forth.

Are these alternative logics "real" or even the "true" logic? Well actually to motivate its study the mere mental construction of them is enough, the mere being *in intellectu*, provided such a construction is fruitful. I would even be prepared to defend that as a start it is enough if they teach us something about the logic we take to be the "real" one. The construction of alternative logics, which in the latter case is conceived as resulting from changes in the original "real" logic, can be thought of as following a substructural strategy: changes of logic are structural changes concerning logical consequence.

In the next chapter I will offer a dialogical interpretation of non-normal logics which should offer the first steps towards such a minimalist defense of logical pluralism. In this interpretation the pair *standard–non-standard* will be added to the pair "normal"–"non-normal". Furthermore, the adjectives *standard* and *non-standard* will qualify the noun *logic* rather than *world*, e.g. I will write "the standard logic L_k in the argumentative context **m**". *Normal* will qualify those contexts, which do not allow the choice of a logic other than the standard one. Non-normal contexts do allow the choice of a new logic underlying the modalities of the chosen context. Before we go into the details let us distinguish between the following different kinds of counterlogical arguments:

1 Assume an intuitionist logician who puts forward the following conditional: *If tertium non-datur were valid in my logic, then the two sides of de Morgan Laws would hold (in my logic) too.*

[5] See too Read 1994, 90–91 and Priest 1998, 482.

[6] Already Aristoteles used counterlogical arguments while studying the principle of non-contradiction, which he saw as the principal axiom of logic.

2 We take here once more our intuitionist: *If tertium non-datur were valid in the non-standard logic L_k, then the two sides of de Morgan Laws would hold in L_k too.*

In the first case the alternative logic here classical logic might be thought of as a conservative extension of the standard one here intuitionistic logic i.e. any valid formula of the standard logic will be valid too in the non-standard logic. In the second case this seems to be less plausible: L_k could be a logic which is a combination of classical logic with some other properties very different from the intuitionistic ones. The situation is similar in the following cases where it is assumed that the standard logic is a classical one and the alternative logic can be a restriction:

3 *If tertium non-datur were not valid in my logic, then one side of de Morgan Laws would fail (in my logic).*

4 *If tertium non-datur were not valid in the non standard logic L_j, then one side of de Morgan Laws would fail (in L_j).*

Because of this fact it seems reasonable to implement the change of logics by means of a substructural strategy (akin to the concept of dialogics) – i.e. a strategy where the change of logics involves a change of the structural properties.[7] Now in these examples the precise delimitation of a logic is assumed as a local condition. However the conditional involved in the counterlogical seems to follow another logic which would work as a kind of a metalogic that tracks the changes of the local assumption of a given logic while building arguments with such conditionals. The point here is that in this type of study classical logic has no privileged status. Classical logic might be "the metalogic" in many cases but certainly not here.

20.1.2 Non-normal dialogics

Motivation. Let us call *non-standard* such argumentation contexts (or "worlds") where a different logic holds relative to the logic defined as *standard*. Thus, in this interpretation of non-normal modal logic the fact that the law of necessitation does not hold is understood as implementing the idea that no logically valid argument could be proven in such systems to be unconditionally necessary (or true in any context and logic). Logicians have invented several logics capable of handling logically arguments that are aware of such a situation.

[7] This strategy, as developed in Rahman/Keiff 2003, could be implemented either implicitly or explicitly. The implicit formulation presupposes that the structural rules are expressed at a different level than the level of the rules for the logical constants which are part of the object language. The explicit formulation renders a propositionalisation of the structural rules using either the language of the linear logicians or hybrid languages in the way of Blackburn 2001.

The main idea of their strategy is simple: logical validity is about standard logics and not about the imagined construction of non-standard ones; we only have to restrict our arguments to the notion of validity involved in the standard logic. Actually there is a less conservative strategy: namely, one in which a formula is said to be valid if it is true in all contexts whether they are ruled by a standard or a non-standard logic. The result is notoriously pluralistic: no logical argument could be proven in such systems to be unconditionally necessary.

Anyway if we have a set of contexts, how are we to recognize those underlying a standard logic? The answer is clear in modal dialogics if we assume that the players can not only choose contexts but also the (non-modal)logic which is assumed to underlie the chosen context. In this interpretation the Proponent fixes the standards, i.e. determines which is the (non-modal) standard logic underlying the modalities of a given context. However under given circumstances the Opponent might choose a context where he assumes that a (non-modal) logic different from the standard one is at work. Now, there are some natural restrictions on the Opponent choices. Assume that in a given context **O** has *explicitly conceded* that **P** fixes the standards. In other words, the Opponent concedes that the corresponding formulae are assumed to hold under those structural conditions which define the standard logic chosen by the Proponent: we call these contexts *normal*. Thus, **O** has conceded that the context is normal or rather, that the conditions in the context are normal. In this case **O** cannot choose the logic: it is **P** who decides which logic should be used to evaluate the formulae in question, and as already mentioned, **P** will always choose the logic he has fixed as the standard one. That is what the concession means: **P** has the choice. c which is assumed to underlie the chosen context. In this interpretation the Proponent fixes the standards, i.e. determines which is the (non-modal) standard logic underlying the modalities of a given context. However under given circumstances the Opponent might choose a context where he assumes that a (non-modal) logic different from the standard one is at work. Now, there are some natural restrictions on the Opponent choices. Assume that in a given context **O** has *explicitly conceded* that **P** fixes the standards. In other words, the Opponent concedes that the corresponding formulae are assumed to hold under those structural conditions which define the standard logic chosen by the Proponent: we call these contexts *normal*. Thus, **O** has conceded that the context is normal or rather, that the conditions in the context are normal. In this case **O** cannot choose the logic: it is **P** who decides which logic should be used to evaluate the formulae in question, and as already mentioned, **P** will always choose the logic he has fixed as the standard one. That is what the concession means: **P** has the choice. Notice once more that "standard" logic does not really simply stand for "normal": normality, in the usual understanding of non-normal modal logic, is reconstructed here as a condition which when a context **m** is being chosen restricts the choice of the logic underlying the modalities of **m**.

Dialogics for S0.5, S.2 and S3. The major issue here is to determine dynamically – i.e., during the process of a dialogue – in which of the contexts may the Opponent not have to conceded that it is a non-normal one and allowing him thus to choose a non-modal propositional logic different from the standard one. This must be a part of the dialogue's structural rules (unless we are not dealing with dialogues where the dialogical contexts with their respective underlying propositional logics are supposed to have been given and classified from the start). I will first discuss the informal implicit version of the corresponding structural rules and in the following chapter we will show how to build tableaux which implement these rules while formulating the notion of validity for the non-normal dialogics. Let us formulate a general rule implementing the required dynamics but some definitions first:

Definitions:

• *Normality as condition:*
We will say that a given context **m** is normal iff it does not allow to choose a (propositional) logic underlying the modalities of **m** other than the standard one. Dually a context is non-normal iff it does allow the choice of a new logic.

• *Standard logic:*
P fixes the standards, i.e. **P** fixes the (propositional) logic which should be considered as the standard logic underlying modalities and relative to which alternatives might be chosen.

• *Closing dialogues:*
No dialogue can be closed with the moves (**P**)a and (**O**)a if these moves correspond to games with different logics.

• *Particle rules for non-normal dialogics:*
The players may choose not only contexts they may also choose the propositional logic underlying the modalities in the chosen contexts:

$\Box, \Diamond$	*Attack*	*Defence*
$\Box A\ m$ ($\Box A$ is stated in context m underlying a logic L_k)	$?_{\Box/n_{L_j}}\ m$ (in the context m the challenger attacks by choosing an accessible context n and logic L_j)	$A_{L_j}\ n$
$\Diamond A\ m$ ($\Diamond A$ is stated in context m underlying a logic L_k)	$?_{\Diamond}\ m$	$A_{L_j}\ n$ (the defender chooses the accessible context n and the logic L_j)

Or in the more formal notation of state of game (see appendix):

• $\Box$-*particle rule*: From $\Box A$ follows $< R, \sigma, A, \lambda_{A/L_j,n} >$, responding to the attack $?_{\Box/n_{L_j}}$ stated by the challenger at m (underlying the logic L_k) and where $\lambda_{A/L_j,n}$ is the assignation of context n (with logic L_j) to the formula A, and n and L_j are chosen by the challenger.

• $\Diamond$-*particle rule*: From $\Diamond A$ follows $< R, \sigma, A, \lambda_{A/L_j,n} >$, responding to the attack $?_\Diamond$ stated by the challenger at m (underlying the logic L_k) and where $\lambda_{A/L_j,n}$ is the assignation of context n (with logic L_j) to the formula A, and n and L_j are chosen by the defender.

The accessibility relation is defined by appropriate structural rules fixing the global semantics (see appendix). To produce non-normal modal dialogic we proceed by adding the following (structural) rule:

(SR-ST10.05) (S0.5-rule):

- **O** may choose a non-standard logic underlying the modalities while choosing a (new) context n with an attack on a Proponent's formula of the form $\Box A$ or with a defense of a formula of the form $\Diamond A$ stated in m if and only if m is non-normal.
- **P** chooses when the context is normal and he will always choose the standard logic but he may not change the logic of a given context (generated by the Opponent). Furthermore, **P** may not choose a context where the logic is non-standard.
- The logic underlying the modalities of the initial context is assumed to be the standard logic.

Three further assumptions will complete this rule:

S0.5 assumptions:

(i) The dialogue's initial context has been assumed to be normal.

(ii) The standard logic chosen by **P** is classical logic L_c.

(iii) No other context than the initial one will be considered as being normal.

The dialogic resulting from these rules – combined with the rules for T – is a dialogical reconstruction of logic is known in the literature as **S0.5**. In this logic validity is defined relative to the standard logic and has the constraint that any newly introduced context could be used by **O** to change the standards. Certainly $\Box(a \vee \neg a)$ will be valid. Indeed, the newly generated context, which has been introduced by the challenger while attacking the thesis, has been generated from the normal starting context and thus will underlie the classical structural rule **SR-ST2C** (see appendix). The formula $\Box\Box(a \vee \neg a)$ on the contrary will not be valid. **P** will lose if **O** chooses in the second context, e.g., the intuitionistic structural rule **SR-ST2I**:

Contexts	O				P		Contexts
					$\Box\Box(a \vee \neg a)$	0	$\mathbf{1}_{\{Lc\}}$
$\mathbf{1}_{\{Lc\}}$	1	$<?_{\Box/1.1}>$	0		$\Box(a \vee \neg a)$	2	$\mathbf{1.1}_{\{Lc\}}$
$\mathbf{1.1}_{\{Lc\}}$	3	$<?_{\Box/1.1.1_{Li}}>$	2		$a \vee \neg a$	4	$\mathbf{1.1.1}_{\{Li\}}$
$\mathbf{1.1.1}_{\{Li\}}$	5	$<?_{\vee}>$	4		$\neg a$	6	$\mathbf{1.1.1}_{\{Li\}}$
$\mathbf{1.1.1}_{\{Li\}}$		a	6		—		$\mathbf{1.1.1}_{\{Li\}}$

The Proponent loses playing with intuitionistic rules. **O** wins by playing in 3 the structural rule, which changes the standard logic into an intuitionistic logic.

Let us produce a dialogical reconstruction of another logic, known as **S2**, where we assume not only that the logic of the first context is normal and in general **SR-ST10.05**, but also:
(SR-ST10.2) (S2-rule):

- If **O** has stated in a context m a formula of the form $\Box A$ (or if **P** has stated in m a formula of the form $\Diamond A$), then the context m *can* be assumed to be normal. Let us call (**O**)$\Box A$ and (**P**)$\Diamond A$ *normality formulae*.
- **P** will not change the logic of a given context (and he may not choose a context where the logic is non-standard) but he might induce **O** to withdraw a choice of a non-standard logic by forcing him to concede that the context at stake is a normal one.
- A normal context can only be generated from a(nother) normal context.

The first two points establish that a formula like $\Box B$ could be stated by **P** under the condition that another formula, say $\Box A$, holds. In this case **O** will be forced to concede that the context is normal and this normality will justify the proof of B within the standard logic. The third point of the rule should prevent this process of justification from becoming trivial: formulae such as (**P**)$\Box\Diamond A\ m$, and (**O**)$\Diamond\Box A\ m$ should not yield normality if m is not normal: the normality of m should be "outside" the scope of (**P**)$\Box \ldots m$ and (**O**)$\Diamond \ldots m$.

This is, for our purpose, a more appealing logic than **S0.5** because it makes of the status of the contexts at stake a question to be answered within the dynamics of the dialogue. One can even obtain certain iterations such as $\Box(\Box(a \rightarrow b) \rightarrow (\Box a \rightarrow \Box b))$ which is not valid in **S0.5**, but is in **S2**: the first context underlies the standard classical logic by the second **S0.5** assumption, the second context too because **O** will concede $\Box a$ there. Now, because the second context has been L_c-conceded by **O**, he cannot choose a logic different from the classical one, and **P** will thus win. Adding transitivity to **S2** renders **S3**.

Dialogics for E0.5, E2 and E3. The point of the logics presented in the preceding section was not to ignore the non-standard logics, but only to take into consideration the standard one while deciding about the validity of a given argument. We will motivate here a less conservative concept, namely, one in which a formula is said to be valid if it is true in all contexts whether they are ruled by a standard or a non-standard logic. These logics are known as **E**. In no **E** system will $\Box A$ be valid for any formula A.

Suppose one modifies **S0.5** in such a way that no context is assumed to be normal and thus every modality will induce a change of logic. This logic, called **E0.5**, is unfortunately not of great interest: a formula will be valid in **E0.5** iff it is valid in non-modal logics (think of $\Box(a \rightarrow b) \rightarrow (\Box a \rightarrow \Box b)$, which in this logic cannot be proven to be valid). Modality seems not be of interest there, and this logic can be thought of as a kind of a modal lower limit.

Now the elimination of the assumption that the first context is normal in **S2** that is, take **SR-ST10.05** and **SR-ST10.2** but drop the first and third **S0.5** assumptions yields an interesting dialogic for our purposes. $\Box(a \rightarrow b) \rightarrow (\Box a \rightarrow \Box b)$ is valid there, signalizing a more minimal structural condition for the validity of this formula than **K** (for it does not even assume, as **K** does, that validity concerns only contexts with the same kind of logic). Similarly one could produce **D** versions, etc. Indeed **E2** seems to be the appropriate language where the logical pluralist might explore the way to formulate statements of logical validity which do not assume a universal scope.

In fact, up to this point, this interpretation only offers a way to explore the scope of the validity of some arguments when confronted with counterlogical situations, where no middle term is to be conceived between what is to be considered standard and what not. Moreover, that a central aim of this dialogic is to explore fruitful counterlogicals seems not to have been implemented yet. In the next chapter I would like to suggest some further possible distinctions in order to perform this implementation.

Beyond non-normality. Let us take once more the following example, where the standard logic is classical logic:

> *If tertium non-datur were not valid in my logic, then one sense of double negation would fail (in my logic).*

One possible formalisation consists of translating not-valid by "non-necessary". Now the problem with this example is that, if **P** does not change the logic he can win the (negative) conditional in, say, **S2** in a trivial way. Indeed, **O** will attack the conditional conceding the protasis, **P** will answer with the apodosis and after the mutual attacks on the negation **P** will win defending *tertium non-datur* in classical logic. But then the argument seems not to be terribly interesting. This follows from the fact that in the interpretation displayed above **P** may not change the standard logic once it has been fixed. In general this is sensible because validity should be defined relative to one standard and we cannot leave it just open to just any change. Moreover, though there is some irrelevance there this irrelevance concerns only the formula conceded at the object language: in our case double negation. But what is relevant and is used is the concession that the standard logic is the one where the classical structural rule applies. Finally why should **P** change the logic if he can easily win in the one he defined as standard?

However, in order to implement the dialogic of counterlogicals, one could leave some degree of freedom while changing the logical standard without too much complexity and inducing a more overall relevant approach: a given standard logic may change into a restriction of this logic. In other words, the standard logic may be changed to a weaker logic where any of its valid formulae are also valid in the stronger one **P** first defined as standard. True, the problem remains that it does not seem plausible that **P** will do it on principle: on principle he wishes to win, and if the proof is trivial all the better for him. There are two possibilities:

One is to build a dialogue under conditions determining from the start which contexts are played under the standard logic and which are the ones where the restriction of the standard logic hold (fix a model).

The other is to leave **O** to choose a conservative restriction of the logic **P** first defined as standard.

(SR-ST10.2*):

- If **O** has stated in a context m a formula of the form $\Box A$ (or if **P** has stated in m a formula of the form $\Diamond A$), then the context m *can* be assumed to be normal. In these cases **O** might choose once a restriction of the standard logic and **P** must follow in his choices the restrictions on the standard logic produced by **O**.
- A normal context can only be generated from a(nother) normal context.

In our example **O** will choose intuitionistic logic and there **P** will need the concession of double negation if he wants to prove *tertium non-datur*. One way to see this point is that **O** actually tests if in the substructural rules defining the standard logic there are not some redundancies. Perhaps a sublogic might be enough.

For the example of this chapter this seems enough but one could even allow such restrictions in the case of the initial context in **S0.5**. Moreover one could even drop the second **S0.5** assumption and let **P** choose an arbitrary standard logic. Take for example the case

> *If transitivity were not holding in my logic, then* $\Box\, a \rightarrow \Box\Box\, a$ *would fail too (in my logic).*

Suppose the standard logic is **S4**. We should use a notation to differentiate the modality which defines the standard logic and which is normal from the modalities which are used within the corresponding non-normal logic. Let us use "Δ" (resp. "∇") for necessity (resp. possibility) in the standard logic.

Furthermore let us use Blackburn's hybrid language to "propositionalise" the properties of the accessibility relation. We could thus write

> $\neg\Box(\nabla\nabla\nu_i \rightarrow \nabla\nu_i)$ *(transitivity) (in my **S4** logic)* $\rightarrow \neg\Box(\Delta a \rightarrow \Delta\Delta a)$ *(in my **S4** logic).*

If **SR-ST10.2*** applies then the Opponent will choose, say, the logic **K** and the Proponent will win. In these types of dialogue the Opponent functions more constructively than in the sole role of a destructive challenger. In fact, the Opponent is engaged in finding the minimal conditions to render the counterlogical conditional. Actually there has already been some work done concerning the dialogic adequate for seeking the minimal structural conditions for modal logic. These dialogues are called *structure seeking dialogues* (SSD) and have been formulated in Rahman/Keiff 2003. In these dialogues, the "constructive" role of the Opponent is put into work explicitly.[8]

Here is another kind of example:

> *If the principle of non-contradiction were not valid in my logic, then one sense of double negation would fail (in my logic).*

One other way to formalize this would be to put the negation inside the scope of the necessity operator:

> *If it were necessary that the principle of non-contradiction does not hold, then it would be necessary that one sense of double negation will fail.*

If we assume here too that **SR-ST10.2*** applies then the Opponent will choose some sort of paraconsistent logic (such as Sette's P1). Certainly, the Opponent will lose, anyway but other choices would lead to a trivial winning strategy of the Proponent.

If, instead of using **SR-ST10.2*** we leave the choice of the standard logic open, **P** might choose any logic as standard and then it would seem that almost anything goes. It is perhaps not the duty of the logician to prevent this but the application of **SR-ST10.2*** and the corresponding SSD can help there, leaving the Opponent to search for the "right" the structural conditions under which the formula should be tested.

The point may be put in a different way. In the dialogues of the preceding chapters the role of the Opponent is to test if the thesis assumes surreptitiously

[8] In the context of the SSD with the thesis, say, A, the Proponent claims that he assumes that a determined element δ_i (of a given set Δ of structural rules) is the minimal structural condition for the validity of A. Informally, the idea is that structural statements can be attacked by the challenger in two distinct ways. *First*, by conceding the condition δ_i, claimed by the player X to be minimal, and asking X to prove the thesis. *Second*, by (counter)claiming that the thesis could be won with a (subset of) condition(s) of lesser rank in Δ. In that case, the game proceeds in a subdialogue, started by the challenger who now will claim that the formula in question can be won under the hypothesis δ_j, where δ_j is different from δ_i and has a lesser rank as δ_i. Since the challenger (Y) starts the subdialogue *he now has to play formally*. See details in Rahman/Keiff 2003.

that its validity holds beyond the limits of the standard logic. In this role the Opponent may choose any arbitrary logic without any constraints. Let us now assume, that the Opponent, still in the role already mentioned, comes to the conclusion that the thesis of the Proponent holds as it is. The Opponent can then play a slightly different role and explore the possibilities of another strategy: he might try to check if the standard logic chosen is not too strong concerning the thesis at stake. The latter is the aim of the structure seeking dialogues.

The preceding considerations hardly settle the matter of the ways the change of logics can be studied dialogically. There are many other possible variations one could for example think that the SSD would be activated when some problematic assumption of the standard logic arises which might not actually concern the thesis. This will do for the present though.

20.2 Tableaux

The aim is to discuss the failure of the so-called *Hintikka strategy* concerning the implementation of the accessibility relation while constructing tableau systems for non-normal modal logics. This problematic seems to apply too to the "propositionalisation" techniques of frame conditions such as practised in hybrid languages.

Let us first present the tableaux which result from our dialogic.

20.2.1 Dialogical tableaux for non-normal modal logics

As discussed in the appendix, the strategy for dialogical games introduced above furnishes the elements for building a tableau-notion of validity where every branch of the tableau is a dialogue. Following the seminal idea at base of dialogic, this notion is attained via the game-theoretic notion of *winning strategy*. X is said to have a winning strategy if there is a function, which, for any possible Y-move, gives the correct X-move to ensure the winning of the game.

Indeed, it is a well known fact that the usual semantic tableaux in the tree-shaped structure due to Raymond Smullyan are directly connected with the tableaux for strategies generated by dialogue games, played to test validity in the sense defined by these logics. E.g.

O-cases	**P**-cases
$\Sigma, (\mathbf{O})A \rightarrow B$	$\Sigma, (\mathbf{P})A \rightarrow B$
$\Sigma, (\mathbf{P})A, \ldots \mid \Sigma, <(\mathbf{P})A> (\mathbf{O})B$	$\Sigma, (\mathbf{O})A,$ $\Sigma, (\mathbf{P})B$

The vertical bar "|" indicates alternative choices for **O**; **P** must have defense strategies for the two possibilities (dialogues).
Σ is a set of dialogically signed expressions.

Matching pairs of "$<$" and "$>$" enclose formulae which could not be attacked. The elimination of expressions like $<(\mathbf{P}A)>$ and the substitution of **F(alse)** for **P** and **T(rue)** for **O** yield signed standard tableau for the conditional.

However, strictly speaking, as discussed in Rahman/Keiff 2003, the resulting tableaux are not quite the same. A special feature of dialogue games is the notorious formal rule **SR-ST4** which is responsible for many of the difficulties of the proof of the equivalence between the dialogical notion and the truth-functional notion of validity. The role of the formal rule, in this context, is to induce dialogue games which will generate a tree displaying the (possibly) winning strategy of **P**, the branches of which do not contain redundancies. Thus the formal rule actually works as a filter for redundancies, producing a tableau system with some flavour of natural deduction. This role can be generalised for all types of tableau generated by the various dialogics. Once this has been made explicit, the connection between the dialogical and the truth-functional notion of validity becomes transparent.

Let us see first the dialogical tableaux for normal logic as presented in Rahman/Rückert 1999 and improved in Blackburn 2001, though the notation there diverges slightly from the present one:

O-cases	**P**-cases
$(\mathbf{O})\Box A\ m$	$(\mathbf{P})\Box A\ m$
———	———
$<(\mathbf{P})?_{\Box/n\sharp}>(\mathbf{O})A\ n$ the context n need not to be new	$<(\mathbf{O})?_{\Box/n}>(\mathbf{P})A\ n$ the context n is new
$(\mathbf{O})\Diamond A\ m$	$(\mathbf{P})\Diamond A\ m$
———	———
$<(\mathbf{P})?_{\Diamond}>(\mathbf{O})A\ n$ the context n is new	$<(\mathbf{O})?_{\Diamond}>(\mathbf{P})A\ n\sharp$ the context n need not to be new

"m" and "n" stand for contexts; "$\sharp$" restricts the choices of **P** according to the properties of the accessibility relation defining the corresponding normal modal logics. Dialogical contexts always constitute a set of moves. These contexts may have a finite number, or a countable infinity of elements, semi-ordered by a relation of succession, obeying the very well known rules which define a tree. The thesis is assumed to have been stated at a dialogical context which constitutes the origin of the tree. The initial dialogical context is numbered 1. Its n immediate successors are numbered $1.i$ (for $i = 1 \ldots n$) and so on. An immediate successor of a context $m.n$ is said to be *of rank* $+1$, the immediate predecessor m of $m.n$ is said to be *of rank* -1, and so on for arbitrarily higher (lower) degree ranks.

I will leave the discussion of how to specify '$\sharp$' for the next section to present first the tableaux for non-normal dialogics:

O-cases	**P**-cases
$(\mathbf{O})\Box A\, m$	$(\mathbf{P})\Box A\, m$
$<(\mathbf{P})?_{\Box/n\sharp_{L_s}}>(\mathbf{O})A_{L_s}\, n$ the context n need not to be new; the logic at m is the standard one L_s	$<(\mathbf{O})?_{\Box/n_{L_i}}>(\mathbf{P})A_{L_i}\, n$ the context n is new; the logic L_i is different from the standard one L_s iff m is non-normal
$(\mathrm{O})\Diamond A\, m$	$(\mathrm{P})\Diamond A\, m$
$<(\mathbf{P})?_{\Diamond}>(\mathrm{O})A_{L_i}\, n$ the context n is new; the logic L_i is different from the standard L_s iff m is non-normal	$<(\mathbf{O})?_{\Diamond}>(\mathbf{P})A_{L_s}\, n\sharp$ the context n need not to be new; the logic at m is the standard logic L_s

We need the following rule concerning closure:
• *Closing branches*: No branch can be closed with the moves $(\mathbf{P})a$ and $(\mathbf{O})a$ if these moves correspond to games with different logics.

To produce **S0.5** add to the adequate implementation of the accessibility relations the following:

S0.5 normality conditions:

1 The dialogue's initial context has been assumed to be normal. No other context than the initial one will be considered as being normal.

2 The standard logic chosen by **P** is classical logic L_c.

3 The Proponent may not:

- choose a context where the logic is different from the standard one;
- change the logic of a given context m if m has been generated from a non-normal context.

To produce **S2**, add to the **S0.5**-rule the following:

(S2–normality conditions):

- If **O** has stated in a context m a formula of the form $\Box A$ (or if **P** has stated in m a formula of the form $\Diamond A$), then the context m can be assumed to be normal.

- A normal context can only be generated from a(nother) normal context.

The construction of the other tableaux is straightforward.

20.2.2 On how not to implement the accessibility relations

In dialogics, the properties of the accessibility relation could be implemented in the following way:

(SR-ST9.2K) (K): **P** may choose a (given) dialogical context of rank +1 relative to the context he is playing in.

(SR-ST9.2T) (T): **P** may choose either the same dialogical context where he is playing or he may choose a (given) dialogical context of rank +1 relative to the context he is playing in.

(SR-ST9.2B) (B): **P** may choose a (given) dialogical context of rank −1 (+1) relative to the context he is playing in, or stay in the same context.

(SR-ST9.2S4) (S4): **P** may choose a (given) dialogical context of rank >+1 relative to the context he is playing in, or stay in the same context.

(SR-ST9.2S5) (S5): **P** may choose any (given) dialogical context.

Moreover we could for example build the transitivity part of the rule for **S4** in the tableau rule in the following way:

$$\frac{\begin{array}{c}(\mathbf{O})\Box A\; m \\ n = m+1\end{array}}{<(\mathbf{P})?_{\Box/n}>(\mathbf{O})A_{L_s}\; n}$$

Actually, there is another technique to implement this and which is connected with the idea of finding in the object language formulae which express frame conditions: the idea has been used by Hintikka for the construction of tableaux and is thus known today as *Hintikka's strategy*. The idea is a bold one and captures the spirit of the axiomatic approaches. Let us formulate the rule in Hintikka's style leaving aside for the moment the choice of the logic:

$$\frac{\begin{array}{c}(\mathbf{O})\Box A\; m \\ n = m > +1\end{array}}{<(\mathbf{P})?_{\Box/n}>(\mathbf{O})\,\Box A\; n}$$

That is, if $\Box A$ holds at m then it should also hold at the context n provided n is accessible from m. The rule stems from the idea that transitivity is associated with the validity of the formula: $\Box A \rightarrow \Box\Box A$.

The "up-wards" transitivity of **S5** can be formulated similarly. Actually, the only device one needs is the one concerning **K**. Then, as soon as context has been "generated" the rules defining the other modal logics tells what formulae can be used to fill the opened context - Hintikka speaks of "filling rules. The simplicity and conceptual elegance of this strategy had made it very popular[9] and it is connected with a more radical formalisation strategy such as that of *hybrid languages*.[10] In the latter, the point is to fully translate the properties of accessibility relations into the object language of propositional modal logic, which has been extended with a device to "name contexts" such as "$@m$". The idea behind the @ operator is to distinguish the assertion that a given formula A can be defended in the dialogical context m from the dialogical context n where the assertion has been uttered – which could be different from m. Properties of the accessibility relation can in this case be formulated as propositions. One problem for the general application of Hintikka's strategy is that there are some frame conditions like irreflexivity, asymmetry, anti-symmetry, intransitivity and trichotomy which are not definable in orthodox modal languages. The aim of hybrid languages is to close this gap by enriching the modal language and apply then Hintikka's strategy.

The hybrid strategy seems at first sight, very appealing to our interpretation of non-normal modal logic where the concession of normality actually amounts to the concession of a rule defining the corresponding standard logic. If the standard logic is a modal one, then the concession, when formulated in the style of hybrid languages, amounts to add a premise. Now, if it is indeed a premise (stating frame conditions) then it seems a good idea to have this premise expressed in the same language as the other premises. For example in the following way:

$$\frac{\begin{array}{c}(\mathbf{O})\Box A \;\; @m \\ \Diamond\Diamond n \rightarrow \Diamond n \;\; @m\end{array}}{<(\mathbf{P})?_{\Box/n}>(\mathbf{O})A \;\; @n}$$

However, the application of both the Hintikka and the hybrid strategy in the context of non-normal logic should be done very carefully. If not we might, say in the **S3**, convert a non-normal context into a normal one by the assumption that the accessibility relation is transitive.[11] Moreover, we would come to the result that every non-normal logic with transitivity collapses into normality. But normality is a condition qualifying worlds and not about accessibility. In fact the point of logic as **S3** is that we could have transitivity without having necessitation. Certainly, defenders of Hintikka's and hybrid strategies might fight back introducing the proviso that their rules apply under the condition that

[9]See for example Fitting 1983, 37; Fitting/Mendelsohn 1998, 52; Girle 2000, 32–34.
[10]Cf. Blackburn 2001 and Blackburn/de RijkeVenema 2002.
[11]Cf. Girle 2000, 187 where the exercise 3.3.1. 2(a) shows how such a mistake slipped into his system.

the contexts in question are normal. In fact, Fitting uses such a strategy in his book of 1983 (274).

Anyway, this loss of generality awakes, at least to the author of the paper, a strange feeling. A feeling of being cheated: Transitivity talks about accessibility between contexts and not about necessitation in normal contexts. Hybrid languages seem to be the consequent and thorough development of a notion akin to Hintikka's strategy and perhaps pay the same price. Indeed, in the language of dialogics we would say that the propositionalisation of frame conditions amounts to producing a new (extension of a) logic without really changing either the local or the global semantics. It is analogue to the idea of producing classical from intuitionistic dialogic just by adding *tertium non datur* as a concession (or axiom) determined by the particular circumstances of a given context. Indeed, with this technique we can produce classical theorems within the intuitionistic local and global (or structural) semantics. Assume now that we are in the modal dialogic **K** and that in a given (dialogical) context the Opponent has attacked a necessary formula $a \vee b$ of the Proponent. Assume further that the Proponent has at his disposal a filling rule which allows him to "fill" this very context with a necessary formula of the Opponent, say, b.[12] Then obviously, **P** will win and strictly speaking, from the dialogical point of view, he always remains in **K**. One other way to see this is to realize that, what the "filling rules" do, is to allow appropriate "axioms" to be added to some contexts specified by these rules in order to extend the set of theorems of **K** without changing its semantics. As already acknowledged, the idea is elegant and perspicuous but it simply does not work so straightforwardly if non-normal contexts are to be included. Perhaps we should even learn from all this exercise that converting frame conditions into propositions drives us to a notion of the relation of accessibility which does not yet seem to have been fully understood.[13]

Acknowledgements
I would like to thank André Laks, director of the Maison des Sciences de l'Homme du Nord-Pas de Calais (MSH). In fact this work is part of the research projects Preuve and La science et ses contextes, attached to the MSH-Nord-Pas de Calais. Also thanks to Laurent Keiff (Lille 3) for interesting and lively discussions.

[12]Moreover, if the thesis were $\Box b \rightarrow \Box\Box(a \vee b)$ it would be valid.

[13]It could be even be fruitful to relate this problematic with *tonk*. From the dialogical point of view, *tonk* produces an extension into triviality because it has been introduced without semantic support (see Rahman/Keiff 2003). Here, if the semantics concerning the accessibility relation is not changed according to the classification of worlds into two disjoint, logic will collapse into another normal modal logic.

Appendix

A.1 A brief survey of dialogic

The aim here is to introduce very briefly the conceptual kernel of dialogic in the context of the dialogical reconstruction of first-order propositional calculus, in its classical and intuitionist versions.[14]

Let our language **L** be composed of the standard components of first-order logic (with four connectives $\vee$, $\wedge$, $\rightarrow$, $\neg$, and two quantifiers $\forall$, $\exists$), with small letters (a, b, c, ...) for prime formulæ, capital italic letters (*A, B, C,...*) for formulæ that may be complex, capital italic bold letters (***A*, *B*, *C*,...**) for predicators. Constants are noted τ_i, where $i \in \mathbb{N}$, and variables with the usual notation (x, y, z, ...). We will also need some special force symbols: ?..., and !..., where the dots stand for indices, filled with some adequate information that will be specified by appropriate rules. An *expression* of **L** is either a term, a formula or a special force symbol. **P** and **O** are two other special symbols of **L**, standing for the players of the games. Every expression e of our language can be augmented with labels **P** or **O** (written **P**$-e$ or **O**$-e$, called (*dialogically*) *signed expressions*), meaning in a game that the expression has been played by **P** or **O** (respectively). We use X and Y as variables for **P**, **O**, always assuming $X \neq Y$. Other more specific labels will be introduced where needed.

An *argumentation form or particle rule* is an abstract description of the way a formula, according to its principal logical constant, can be criticised, and how to answer the criticisms. It is abstract in the sense that this description can be carried out without reference to a determined context. In dialogic we say that these rules state the *local semantics*, for they show how the game runs locally, in the sense that what is at stake is only the critic and the answer to a given formula with one logical constant rather than the whole (logical) context where this formula is embedded. Hence, the particle rules fix the dialogical semantics of the logical constants of **L** in the following way:

	$\wedge$	$\vee$	$\rightarrow$
assertion	X$-A \wedge B$	X$-A \vee B$	X$-A \rightarrow B$
attack	Y$-?_L$, or Y$-?_R$	Y$-?_\vee$	Y$-A$
defense	X$-A$ or X$-B$ (resp.)	X$-A$, or X$-B$	X$-B$

	$\forall$	$\exists$	$\neg$
assertion	X$-\forall x A$	X$-\exists x A$	X$-\neg A$
attack	Y$-?_{\forall/\tau}$ for any τ chosen by Y.	Y$-?_\exists$ for any τ X may choose,	Y$-A$
defense	X$-A(x/\tau)$	X$-A(x/\tau)$	– (i.e. no defense)

(Where A and B are formulæ, and $A(x/\tau)$ is the result of the substitution of τ for every occurrence of the variable x in A.)

One more formal way to stress the locality of the semantics fixed by the particle rules is to see these rules as defining a state of a (structurally not yet determined) game. Namely:

Definition (*state of the game*): A *state of the game* is an ordered triple $< \rho,\ \sigma,\ A >$ where:

[14] Cf. Lorenzen 1958 and Lorenzen/Lorenz 1978. The present more modern version stems from Rahman/Keiff 2003.

- ρ stands for a role assignment either R, from players X, Y to only one element of the set ?(attack), !(defense) determining which player happens to occupy the challenger and which the defender role, or R', inverting the role assignment R of both players (e.g. if $R(X) =?$ and $R(Y) =!$, then $R'(X) =!$ and $R'(Y) =?$). The players perform their assigned role as challengers (defenders) by stating an attack (or asserting a defense) fixed by the corresponding rule.
- σ stands for an assignment function, substituting as usual individuals to variables.
- A stands for a dialogically labelled subformula A with respect to which the game will proceed.

Particle rules are seen here as determining which state of the game **S'** follows from a given state **S** without yet laying down the (structural) rules which describe the passage from **S** to **S'**. What state follows of $S =< R, \sigma, F >$ for the X-labelled formula F?

- *Negation particle rule*: If F is of the form $\neg A$ then
$S' =< R', \sigma, A >$, i.e. Y will have the role of defending A and X the role of (counter)-attacking A.
- *Conjunction particle rule*: If F is of the form $A \wedge B$ then
$S' =< R, \sigma, A >$ or $S'' =< R, \sigma, B >$, according to the choice of challenger $R(Y) =?$ between the attacks $?_L$ and $?_R$.
- *Disjunction particle rule*: If F is of the form $A \vee B$ then
$S' =< R, \sigma, A >$ or $S' =< R, \sigma, B >$, according to the choice of defender $R(X) =!$, reacting to the attack $?_\vee$ of the challenger $R(Y) =?$.
- *Subjunction particle rule*: If F is of the form $A \rightarrow B$, then
$S' =< R, \sigma, A >$ and the game might proceed to the state $S' =< R'', \sigma, B >$, or even the other way round according to the choice of the defender and reacting to the attack A of the challenger $R(X) =?$.
- *universal quantifier particle rule*: If F is of the form $\forall x Ax$ then $S' =< R, \sigma(x/\tau), A >$ for any constant τ chosen by the challenger $R(Y) =?$ while stating the attack $?_{\forall/\tau}$.
- *existential quantifier particle rule*: If F is of the form $\exists x Ax$ then $S' =< R, \sigma(x/\tau), A >$ for any constant chosen by the defender $R(X) =!$ reacting to the attack $?_\exists$ of the challenger $R(Y) =?$.

A dialogue can be seen as a sequence of labelled expressions, the labels carrying information on the game significance of these expressions. Dialogues are processes, so they are dynamically defined by the evolution of a game, which binds together all the labels mentioned. In other words, the set of expressions which is a complete dialogue can be dynamically determined by the rules of a game, specifying how the set can be extended from the original thesis formula. Particle rules are part of the definition of such a game, but we need to set the general organisation of the game, and this is the task of the *structural rules*. Actually structural rules can, while implementing the local semantics of the logical particles, determine a kind of game for a context where e.g. the aim is persuasion rather than logical validity. In these cases dialogic extends to a study of argumentation in a broader sense than the logical one. But *when the issue at stake is indeed testing validity*, i.e. when **P** can succeed with the use of the appropriate rules in defending the thesis against all possible allowed criticism by **O**, games should be thought of as furnishing the branches of a tree which displays the games relevant for testing the validity of the thesis. As a consequence of this definition of validity, each split of such a tree into two branches (dialogue games) should be considered as the outcome of a propositional choice of **O**. In other words when **O** defends a disjunction, he reacts to the attack against a conditional, and when he attacks a conjunction, he chooses to generate a new branch (dialogue). Dually **P** will not choose to change the dialogue (branch). In fact, from the point of view of games as actual (subjective) procedures (acts), it could happen that the subject playing as **O** (**P**) is not clever enough to see that his best strategy is to open (not to open) a new dialogue game (branch) anytime he can, but in this context where the issue is an inter-subjective concept of validity, which should lead to a straightforward construction of a system of tableaux, we simply assume that **O** makes the best possible move.

(SR-ST0) (***starting rule***): Expressions are numbered and alternately uttered by **P** and **O**. The thesis is uttered by **P**. All even-numbered expressions including the thesis are **P**-labelled, all odd-numbered expressions are **O** moves. Every move below the thesis is a reaction to an earlier move with another player label and performed according to the particle and the other structural rules.

(SR-ST1) (***winning rule***): A dialogue is closed iff it contains two copies of the same prime formula, one stated by **X** and the other one by **Y**, and neither of these copies occur within the brackets "<" and ">" (where any expression which has been bracketed between these signs in a dialogue either cannot be counterattacked in this dialogue, or it has been chosen in this dialogue not to be counterattacked). Otherwise it is open. The player who stated the thesis wins the dialogue iff the dialogue is closed. A dialogue is finished if it is closed or if no other move is allowed by the (other) structural and particle rules of the game. The player who started the dialogue as a challenger wins if the dialogue is finished and open.

(SR-ST2I) (***intuitionist ROUND closing rule***): In any move, each player may attack a (complex) formula asserted by his partner or he may defend himself against the last not already defended attack. Defences may be postponed as long as attacks can be performed. Only the latest open attack may be answered: if it is X's turn at position n and there are two open attacks m, l such that $m < l < n$, then X may not at position n defend himself against m.

(SR-ST2C) (***classical ROUND closing rule***): In any move, each player may attack a (complex) formula asserted by his partner or he may defend himself against *any attack* (including those which have already been defended).

(SR-ST3/SY) (***strategy branching rule***): At every propositional choice (i.e. when X defends a disjunction, reacts to the attack against a conditional or attacks a conjunction), X may motivate the generation of two dialogues differentiated only by the expressions produced by this choice. X might move into a second dialogue iff he loses the first chosen one. No other move will generate new dialogues.

(SR-ST4) (***formal use of prime formulæ***): **P** cannot introduce prime formulæ: any prime formula must be stated by **O** first. Prime formulæ can not be attacked.

(SR-ST5) (***no delaying tactics rule***):

- While playing with the *classical structural rule* **P** may perform once a new defense (attack) of an existential (universal) quantifier using a different constant (but not new) iff the first defense (attack) compelled **P** to introduce a new constant. No other repetitions are allowed.

- While playing with the *intuitionistic structural rule*, **P** may perform a repetition of an attack if and only if **O** has introduced a new prime formula which can now be used by **P**.

Definition (*Validity*): A tableau for (**P**)A (i.e. starting with (**P**)A) proves the validity of A iff the corresponding tableau is closed. That is, iff every dialogue generated by (**P**)A is closed.

Examples. In Fig. 1 the outer columns indicate the numerical label of the move, the inner columns state the number of a move targeted by an attack. Expressions are not listed following the order of the moves, but writing the defense on the same line as the corresponding attack, thus showing when a round is closed. Recall, from the particle rules, that the sign "–" signalises that there is no defense against the attack on a negation. In this example **P** wins because, after the **O**'s last attack in move 3, **P**, according to the (classical) rule **ST2C**, is allowed to defend himself (once more) from the attack in move 1 (in the same dialogue. **P** states his defense in move 4 though, actually, **O** did not repeat his attack - this fact has been signalised by inscribing the unrepeated attack between square brackets.

O				**P**	
				$a \vee \neg a$	0
1	$?_{\vee}$	0		$\neg a$	2
3	a	2		-	
[1]	[$?_{\vee}$]	[0]		$\neg a$	4

Fig. 1. **SDC** rules. **P** wins.

O				**P**	
				$a \vee \neg a$	0
1	$?_{\vee}$	0		$\neg a$	2
3	a	2		-	

Fig. 2. **SDI** rules. **O** wins.

In the game of Fig. 2, **O** wins because, after the challenger's last attack in move 3, **P**, according to the intuitionistic rule **SR-I**, is not allowed to defend himself (once more) from the attack in move 1.

Philosophical remarks: game as propositions. Particle rules determine dynamically how to extend a set of expressions from an initial assertion. In the game perspective, one of the more important features of these rules is that they determine, whenever there is a choice to be made, who will choose. This is what can be called the pragmatic dimension of the dialogical semantics for the logical constants. Indeed, the particle rules can be seen as a proto-semantics, i.e. a game scheme for a not yet determined game which when completed with the appropriate structural rules will render the game semantics, which in turn will build the notion of validity.

Actually by means of the particle rules games have been assigned to sentences (that is, to formulæ). But sentences are not games, so what is the nature of that assignment? The games associated to sentences are meant to be propositions (i.e. the constructions grasped by the (logical) language speakers). What is connected by logical connectives are not sentences but propositions. Moreover, in the dialogic, logical operators do not form sentences from simpler sentences, but games from simpler games. To explain a complex game, given the explanation of the simpler games (out) of which it is formed, is to add a rule which tells how to form new games from games already known: if we have the games A and B, the conjunction rule shows how we can form the game A B in order to assert this conjunction.

Now, particle rules have another important function: they not only set the basis of the semantics, and signalise how it could be related to the world of games - which is an outdoor world if the games are assigned to prime formulæ, but they also show how to perform the relation between sentences and propositions. Sentences are related to propositions by means of assertions, the content of which are propositions. Assertions are propositions endowed with a theory of force, which places logic in the realm of linguistic actions. The forces performing this connection between sentences and propositions are precisely the attack (?) and the defense (!). An attack is a demand for an assertion to be uttered. A defense is a response (to an attack) by acting so that you may utter the assertion (e.g. that A). Actually the assertion force is also assumed: utter the assertion that A only if you know how to win the game A.

Certainly the "know" introduces an epistemic moment, typical of assertions made by means of judgements. But it does not presuppose in principle the quality of knowledge required. The constructivist moment is only required if the epistemic notion is connected to a tight conception of what means that the player X *knows that there exists* a winning game or strategy for A.

A.2 Soundness and completeness of the tableaux systems

The tableau systems for non-normal logics presented above are essentially those of Fitting 1983, Girle 2000 and Priest 2001 without the use of Hintikka's strategy for the accessibility relation of the first two authors. I will not rewrite the proofs here and rely on the proofs of Fitting 1983 and Priest 2001. What I will do is to show how to transform the dialogical tableaux into the ones of the authors mentioned above. To see this notice that if the Opponent (=**T** in the signed non dialogical version of the tableau) is clever enough, on any occasion where he may choose logic he will choose one, where he assumes that the Proponent (=**F** in the signed non dialogical version of the tableau) will lose. In fact, if the tableau systems are thought as reconstructing the usual notion of validity of non-normal modal logic we must assume that it will be always the case that if **O** chooses a logic then **P** will lose however, notice that dialogically we must not assume this: **O** might lack some information and choose the wrong logic. One way to implement the assumption of the cleverness of the Opponent slightly more directly is to forbid **P** to answer to an attack on a necessary formula (or to attack a possible formula of the Opponent) stated at a context m unless this context is normal. Moreover, if we are interested in freeing ourselves from the interpretation of the contexts as representing situations where logic could be different, or more generally from any interpretation concerning the "structural inside" of non-normal contexts, the rules will amount to the following simplified formulation:

(O = **T**)-cases	(**P** = **F**)-cases
(O = **T**)$\Diamond A\ m$	(P = **F**)$\Box A\ m$
―――――――	―――――――
<(P)?$_\Diamond$>(O = **T**)$A\ n$ the context n is new the rule is activated iff m is normal	<(O)?$_{\Box/n}$>(P = **F**)$A\ n$ the context n is new the rule is activated iff m is normal

Furthermore, deleting from the tableau the expressions <(P)?$_\Diamond$> and <(O)?$_{\Box/n}$>, which have only a dialogical motivation, yields the usual tableau systems mentioned above.

References

Blackburn, P., 2001. "Modal logic as dialogical logic" in *Synthese* 127, *Special issue*, S. Rahman and H. Rückert (eds.), 57–93.

Blackburn P., de Rijke M., and Venema Y., 2002. *Modal Logic*, Cambridge, Cambridge University Press, 2002.

Cresswell, M. J., 1972. "Intensional logics and truth". *Journal of Philosophical Logic*, vol. 1, 2–15, 1972.

Fitting, M., 1983. *Proof Methods for Modal and Intuitionistic Logic*, D. Reidel, Dordrecht, 1983.

Fitting, M. and Mendelsohn R. L., 1998. *First-Order Modal Logic*, Dordrecht, Kluwer, 1998.

Girle, R., 1973. "Epistemic logic; language and concepts", *Logique et Analyse*, vol. 63–64, 359–373, 1973.

Girle R., 2000. *Modal Logics and Philosophy*, Montreal, McGill-Queen's University Press, 2000.

Grattan-Guinness, I., 1998. "Are other logics possible? MacColl's logic and some English reactions, 1905-1912". *Nordic Journal of Philosophical Logic*, vol. 3, 1, 1–16, 1998.

Hintikka J., 1975. "Impossible Possible Worlds Vindicated". *Journal of Symbolic Logic*, 4, 1975, 475–484; modified and reedited in Hintikka J. and M.B., The Logic of Epistemology and the Epistemology of Logic, Dordrecht, Kluwer, 63–72, 1989.

Kripke, S., 1965. "Semantical Analysis of Modal Logic II; non-normal modal propositional calculi." In J. W. Addison et alia (eds), *The Theory of Models*, Amsterdam, N. Holland, 202-220, 1965.

Lorenzen, P., 1958. "Logik und Agon". *Arti del XII Congresso Internationale de Filosofia*, Venezia. 187-194, 1958. (Reprinted in Lorenzen and Lorenz, 1–8 1978.)

Lorenzen, P. and Lorenz, K., 1978. *Dialogische Logik*. WBG, Darmstadt, 1978.

McCall, S., 1963. *Aristotle's Modal Syllogisms*. Amsterdam: North-Holland, 1963.

McCall, S., 1967. K "MacColl". In: P. Edwards (Ed.): *The Encyclopedia of Philosophy*, London: Macmillan, Vol. 4, 545–546, 1967.

MacColl, H., 1906. *Symbolic Logic and its applications*, London, 1906.

Priest, G., 1992. "What is a Non-Normal World?" *Logique et Analyse*, vol. 139–140, 291-302, 1992.

Priest G. "Editor's introduction". Special issue on "Impossible Worlds" of the *Notre Dame Journal of Formal Logic*, vol. 3/1, 481–487, 1998.

Priest, G., 2001. *An Introduction to Non-Classical Logic*. Cambridge, Cambridge University Press, 2001.

Rahman, S., 1997. "Hugh MacColl eine bibliographische Erschliessung seiner Hauptwerke und Notizen zu ihrer Rezeptionsgeschichte". *History and Philosophy of Logic*, vol. 18, 165–183, 1997.

Rahman, S., 1998. "Ways of understanding Hugh MacColl's concept of symbolic existence". *Nordic Journal of Philosophical Logic*, vol. 3, 1, 1998, 35–58.

Rahman, S., 2000. MacColl and George Boole on Hypotheticals". In J. Gasser (ed.), *A Boole Anthology*, Dordrecht, Synthese-Library Kluwer, 287–310, 2000.

Rahman, S. and Keiff, L., 2003. "On how to be a dialogician", to appear in D. Vandervecken (ed.), *Logic and Action*, Dordrecht, Kluwer, 2003.

Rahman, S. and Rückert, H., 2001 (eds.) "New Perspectives in Dialogical Logic". Special issue of *Synthese, 127*, 2001.

Rahman, S. and Rückert, H., 2001a. "Dialogische Modallogik (für T, B, S4, und S5)". *Logique et Analyse*, vol. 167-168. 243–282, 2001a.

Rantala, V., 1975. "Urn Models: a new kind of non-standard model for first-order logic." *Journal of Philosophical Logic*, 4, 455–474, 1975.

Read, S., 1998. "Hugh MacColl and the algebra of implication". *Nordic Journal of Philosophical Logic*, vol. 3, 1, 59–84, 1998.

Read, S., 1994. *Thinking About Logic*. Oxford, Oxford University Press, 1994.

Restall, G., 1993. "Simplified Semantics for Relevant Logics (and Some of their Rivals)", *Journal of Philosophical Logic*, vol. 22, 481–511, 1993.

Routley, R., Pluwood, V., Meyer, R. K. and Brady, R., 1982. *Relevant Logics and their Rivals*, Atascadero, Ridgeview, 1982.

Wolenski, J., 1998. "MacColl on Modalities". *Nordic Journal of Philosophical Logic*, vol. 3, 1, 1998, 133–140.

VI

APPENDICES

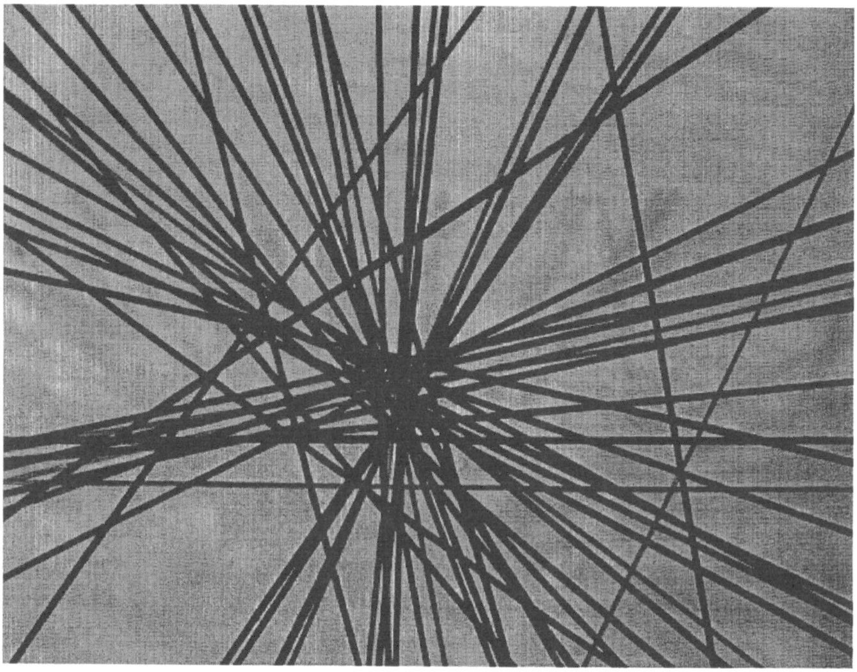

LOUIS JOLY AS A PLATONIST PAINTER

Roger Pouivet
University of Nancy 2/Archives Poincaré (CNRS)
Roger.pouivet@univ-nancy2.fr

Gerhard Heinzmann and the staff of Archives Poincaré thought that the participants of this conference would be interested in some of the works of Louis Joly: those that belong to his "mathematical" period. Drawing attention to this painter merely because his family lived in Toul, near Nancy, would be a bit misguided; for Joly himself did not live in Toul but in Paris, where he was an engineer at the National Cartographic Institute. He worked there professionally on all sorts of maps in search of techniques by which to relate real and topographic spaces. From the perspective of this conference, what is of special interest is the mathematical and pictorial project which preoccupied him during the fifties and the sixties.

About this project, he said: "I strive to achieve a *geometrical experimentation* related to an *axiomatic horizon*". In itself, this sentence seems rather obscure, or at least enigmatic. But by situating Joly within the recent history of art, we will perhaps be able to understand more clearly what he was up to.

Joly was fascinated by the works of a Parisian group which was active in the thirties, the so-called "Abstraction-Création" Group. Its leader was the Belgian painter, George Vantongerloo, who was himself a former participant in the Dutch De Stijl Group, of which Mondrian is the best-known artist. In the fifties, Joly belonged to a group of French artists who took the name "Groupe Mesure" and who had a common interest in geometry. The leading figure in this group was François Morellet. While it is important to consider Joly's historical situation in the twentieth century, we must not overlook the very old tradition to which he belongs: the tradition which comes from the ancient speculations on the "golden ratio" and continues through the development of geometrical perspective in the Renaissance.

J. van Benthem, G. Heinzmann, M. Rebuschi and H. Visser (eds.), The Age of Alternative Logics: Assessing Philosophy of Logic and Mathematics Today, 337–341.

The fundamental idea in all of these sources is that mathematics and painting are not opposed, as many of the Romantics seem to have thought, but are strongly interrelated. Joly thought of his work as an intellectual effort, controlled by mathematics, and not at all as a sentimental effusion. Some other painters in this "mathematical" movement seem to consider the relation between mathematics and painting as that of analogy: an external metaphorical relation. I think that François Morellet—far better known than Joly—adopted this metaphorical stance in his mathematical paintings. Not Joly. He thought that a painter could really create works of art out of mathematical constructions, for example, those associated with mathematical algorithms developed by Pascal, Cayley and Plücker.

Given this, are we to consider as Joly fundamentally (and perhaps naively) Platonist? Beauty for him seems have been not a property of sensible things in the material world, but of abstract mathematical objects in an ideal world. His efforts seem to be directed at providing phenomenal access to this formal beauty. But such a project seems contradictory. The harder one works on a sensible equivalent of the ideal model, the further away one is taken from real, abstract beauty. On the other hand, if one is too abstract, one simply ceases to be a painter, being in the end unable to create a sensible object with truly aesthetic properties. Surely, Joly faced this problem. He worked for days and weeks with very complicated mathematical calculi before taking up his brushes; but the act of painting was perhaps always too material to capture his ambition, even if we think that, from an aesthetic point of view, his painting is neither sufficiently careful nor particularly attractive.

I take no position here as to whether the idea of truly Platonist painting is or is not contradictory in itself. It is closer to my present purpose to ask whether Joly was really a Platonist. Perhaps is there another way—a non-Platonist way—to understand his project.

You may be familiar with the notion of *exemplification* proposed by Nelson Goodman in *Languages of Art*.[0] Roughly, A exemplifies B if and only if B denotes A and A instantiates B. Suppose that B is a mathematical algorithm and that A is one of Joly's "mathematical" paintings. Then, we could interpret his works as exemplifications of certain geometrical functions. Goodman distinguishes further between literal and metaphorical exemplification. If B metaphorically denotes A, A may also metaphorically exemplify B. In the spirit of this conference, we could say that metaphorical exemplification is a sub-relation of the converse of denotation! The advantage of such an interpretation is that Joly's project would then cease to appear contradictory, in the way that it did given the Platonist interpretation. Joly may be understood not as

[0]Nelson Goodman, Languages of Art, 2nd ed, Indianapolis : Hackett, 1976.

trying to instantiate abstract mathematical realities in the aesthetic world, but as attempting to exemplify the mathematical algorithms expressed in certain formulas.

But if this non-Platonist interpretation permits us to avoid the attribution of contradiction to Joly's project, it leads to another query. What is the difference between mathematical and aesthetic exemplification? More simply: are Joly's paintings *paintings* or rather mathematical figures, which present themselves as paintings but are finally something else? Is Joly, in the end, a painter, an artist? Or is he rather a geometer?

I think that the Goodmanian interpretation gives more reason than the Platonist one to see Joly as a painter and an artist. Why? Because, as Goodman says, metaphorical exemplification, even if it is not a criterion of the aesthetic, can be considered as a *symptom* of the aesthetic. Of course, someone might ask whether Joly's paintings are not in fact literal exemplifications of mathematical algorithms, and therefore geometrical figures, rather than artistic images. But I think that the answer is that they are absolutely non-literal. What supports this view? The answer is simple: Joly offers his paintings *as* paintings! He exhibits them *as* paintings. The stance that renders them metaphorical is simply the artistic one. Does this mean that Joly shares with Marcel Duchamp the now completely hackneyed idea of the ready-made? In a sense, yes. Marcel Duchamp offered us ready-made, material artefacts. Joly was far more subversive when he proposed to offer us ready-made, mathematical algorithms. With Joly's work, the metaphor lay in the transition from the mathematical realm to the artistic one, exactly as the Duchampian transfer consisted in moving from industrial objects to artistic ones.

It seems to me that we have here discovered the reason why Joly did not attempt to "aestheticize" his works. He did not try to be aesthetically impressive by adding visual effects to the simple figuration of mathematical functions. He did not try to give a pleasant, and simply allusive, representation of mathematical functions, but meant to present them as mathematical ready-mades, with the fewest possible artistic embellishments. But to what end?

In a paper on Virgina Woolf and our knowledge of the external world, Jaakko Hintikka point out a strong, but unexplored, relationship between the literary works of the Bloomsbury Group, especially Virginia Woolf, and the philosophy of Moore and Russell. Hintikka describes certain parts of Woolf's novels as "fictionalised epistemology".[1] It is not, he explains, that philosophical ideas are the topic of her novels but that the ideas are included in the texture of those works. This seems to me very close to what Joly had in mind. In speaking of the metaphorical *exemplification* of mathematical functions in Joly's painting,

[1] I used the French edition of this text : "Virginia Woolf et notre connaissance du monde extérieur", in Jaakko Hintikka, *La vérité est-elle ineffable ?*, Paris : Vrin, 1994.

I meant to emphasize that Joly was not trying to *illustrate* mathematics. In literature, one encounters the kind of novel in which each character illustrates a philosophical position. Even very great novelists—Thomas Mann, for example, in the *Magic Mountain*—give the impression that their characters are creatures of philosophy courses. In painting, there is the same risk. In facing it, Joly cleaves more closely to Woolf's method than to that of Thomas Mann. He is not making an *allusion* to mathematical functions, but is really trying to *manifest* their aesthetic character. By analogy with the expression "fictionalised epistemology", used by Hintikka about Woolf, one could speak, in the case of Joly, about "painted mathematics".

As I explained earlier, metaphorical exemplification is a kind of reference in which something instantiates a predicate that denotes it. Allusion involves another kind of reference. A alludes to B in both cases where it denotes something C that exemplifies B or exemplifies something C that denotes B. In both cases, the reference is indirect and can be completely external. For example, if you characterize a philosopher by saying that he is an Hegelian, meaning that his works exemplify the predicate "enormous in scope and depth", then the term "Hegelian" denotes those works of Hegel that exemplify this predicate; but this does not mean that the philosopher whose thought you characterize is himself a follower of Hegel. Indeed, we might say, in *this* sense, that Russell is Hegelian. An example of the second case is one in which a football team, say the French team, is called the *roosters*, on the grounds that roosters exemplify some predicate that denotes the French football team (I hesitate here between "proud" and "arrogant"). Joly's paintings do not *allude* to mathematical functions. Joly has the remarkable ambition of realizing a far more direct relation between his paintings and mathematics: the relation captured by Goodman's notion of metaphorical exemplification.

Even if my interpretation were to be accepted, it could still be objected that Joly's mathematical paintings are not *chefs-d'oeuvre*; and it might even be maintained that they are completely lacking in aesthetic merit. How might such a criticism be answered? I could try to side-step this issue by reminding you that I am not an art critic, but merely a modest philosopher, completely unable to evaluate the aesthetic merit Joly's works. But even if I think, with Goodman, that aestheticians tend to overvalue questions of aesthetic merit, I know also that these questions cannot be avoided in the end, for we humans are not only *rational* animals but also *axiological* animals. So let me make a stab at answering this charge.

I have here taken for granted that Joly's paintings lack the quality of being easy on the eye. His painting is perhaps too much intellectual for its aesthetic merit to be assessed in the usual ways. But compared to the paintings of Magritte, or even Francis Bacon, for example (and I apologize to those who like them)—that is, paintings saturated by what I take to be a crude philosophical symbolism—it seems to me that Joly's works are not lacking in aesthetic merit. Rather, their merit is of a rather special kind. Joly does not pretend to deliver any speculative, post-metaphysical message about the inner psyche, the status of images, or the condition of post-modern man. He simply takes seriously the requirement that he should know why the lines he traces on planks must lie precisely where he traces them. Joly is not preoccupied with the human condition, individual or collective, but is interested in something that he believed to be more real and robust.

Now, even if, using the nominalistic semiotics of Goodman, I defended a non-Platonist interpretation of Joly's paintings, Joly was perhaps a Platonist in another—and far more interesting—sense. His paintings are intended not as romantic expressions of subjectivity, but as the products of objective research. Finally, Joly's stance amounts to a kind of scientific asceticism. Painting does not aim at creating things which are easy on the eye, as in vulgar impressionism. Painting is a more serious thing. Joly chose what he conceived as the difficult path of mathematical painting, not the easy path of pleasant painting. If we respect this approach, aesthetic merit must be seen to lie in the creative effort as much as in its results. The strenuous task that Joly imposes on himself—to find an artistic way to exemplify mathematical functions—his asceticism, and his seriousness in this research project, all possess the kind of value Plato attributes to the intellectual formation of character. The formal requirements that Joly set for painting amount really to moral constraints. This is what is truly Platonist in Joly's work.

Like Vantongerloo especially, Joly was not interested in phenomenal appearances or, more generally, in *mimesis*. Nor was he interested, like some Cubists, in other, less mimetic, approaches to the representation of reality. He was not interested at all in imitation. The forms that appear in his paintings had to be justified by something which escapes all contingency. And it seems to me that this requirement was something like an austere and Platonic moral in painting. I leave it to you to judge whether or not this Platonic morality is, in the end, much too austere.

INDEX